15th Joint Conference on Chemistry (JCC 2020)

IOP Conference Series: Materials Science and Engineering
Volume 959

Salatiga, Indonesia
9 September 2020

Editors:

Ferdy S. Rondonuwu
Didit B. Nugroho

ISBN: 978-1-7138-3200-3
ISSN: 1757-8981

Printed from e-media with permission by:

Curran Associates, Inc.
57 Morehouse Lane
Red Hook, NY 12571

Some format issues inherent in the e-media version may also appear in this print version.

Institute of Physics
Dirac House, Temple Back
Bristol BS1 6BE UK

Phone: 44 1 17 929 7481
Fax: 44 1 17 920 0979

techtracking@iop.org

Additional copies of this publication are available from:

Curran Associates, Inc.
57 Morehouse Lane
Red Hook, NY 12571 USA
Phone: 845-758-0400
Fax: 845-758-2633
Email: curran@proceedings.com
Web: www.proceedings.com

TABLE OF CONTENTS

ANALYTICAL CHEMISTRY

THE STUDY OF COMPRESSION RELAXATION TIME OF SILICON RUBBER USING OPTICAL METHOD 1
W Hidayat, E Prasetyo, A Marzuki, H P Yanuar, A Setyawan

MOLECULAR MODELLING AS A DOWNSTREAM EFFORT IN THE DISCOVERY OF NOVEL ANIONIC SULPHATE SURFACTANTS 6
P Iswanto, E V Y Delsy

GLUCOSE-WATER INTERACTIONS AT INCREASING CONCENTRATIONS AND TEMPERATURES AS REVEALED BY NEAR-INFRARED SPECTROSCOPY 10
F S Rondonuwu, A Setiawan, J Muninggar, F F Karwur

COMPUTATIONAL CHEMISTRY

AB INITIO COMPUTATIONAL STUDY: ADSORPTION OF CO MOLECULE ON VARIOUS BASE METAL OXIDE SURFACES 17
I Kurnia, P Siahaan, A Suseno

CHEMISTRY EDUCATION

SELF-ASSESSMENT OF CHEMISTRY LABORATORY BASIC SKILLS USING PERFORMANCE SCORING RUBRICS AT THE CHEMISTRY TEACHER TRAINING 25
I B N Sudria, I W Redhana, I W Suja, I N Suardana

THE EFFECTIVENESS OF ONLINE LEARNING BY EDPUZZLE IN POLYMER MATERIALS ON STUDENTS' PROBLEM-SOLVING SKILLS 33
Giyanto, L Heliawaty, B Rubini

ENVIROMENTAL

CHEMICALLY INDUCED DEMULSIFICATION OF WATER IN CRUDE OIL EMULSION FROM EAST KALIMANTAN OIL FIELDS 42
A P Gustianthy, D Febriantini, B Purnomo, S U Muzayanha, R Noverdi

THE ADDITION EFFECT OF PSEUDOMONAS AERUGINOSA ON BIODEGRADATION OF METHYL ORANGE DYE BY BROWN-ROT FUNGUS FOMITOPSIS PINICOLA 47
A S Purnomo, D M Rahayu, R Nawfa, S R Putra

POLYELECTROLYTE COMPLEX (PEC) FILM BASED ON CHITOSAN AS POTENTIAL EDIBLE FILMS AND THEIR ANTIBACTERIAL ACTIVITY TEST 56
N Ismillayli, I G A S Andayani, R Honiar, B Mariana, R K Sanjaya, D Hermanto

RESPONSE SURFACE METHODOLOGY OPTIMIZATION OF CONGO RED DYE ADSORPTION ONTO MNFE-LDH ADSORBENT 62
L V Tan, N T H Tham, D T Sy

IDENTIFICATION AND SIMULATION MODEL OF MIXED SOLUTIONS OF DIETHANOLAMINE (DEA) AND METHYL DIETHANOLAMINE (MDEA) FOR REMOVAL OF ACID GASES 68
D Febriantini, S Pebriani, Usman, B Purnomo, A K Rianto, A S Nugraha, N Khoirunnisa

ENERGY

STUDY OF FINAL TEMPERATURE AND HEATING RATE VARIATION TO PYROLYSIS OF ACACIA (ACACIA MANGIUM W.) WOOD WASTE 74
Pranoto, K D Nugrahaningtyas, R N O Putri

MATERIAL

CARBON AND ZINC OXIDE SYNTHESIZED BY GELATIN TEMPLATE AS POTENTIAL MATERIAL FOR FIGHT VIRUSES COVID-19: FUTURE POTENTIAL MATERIAL........ 91
M Ulfa, D Prasetoko, M Fajar

SYNTHESIS AND CHARACTERIZATIONS OF C-3-NITROPHENYLCALIX[4]RESORCINARENE AS A POTENTIAL CHEMOSENSOR FOR LA(III) IONS........ 98
K T A Priyangga, Y S Kurniawan, L Yuliati

SYNTHESIS AND CHARACTERIZATION OF SILVER NANOPARTICLES SUPPORTED ON HIGHLY ORDERED BIPHENYLENE-BRIDGED PERIODIC MESOPOROUS ORGANOSILICA 104
A M Khikmah, Y K Krisnandi, I Abdullah

SYNTHESIS OF NANOCOMPOSITES CELLULOSE-FE3O4/ZNO AS NOVEL GREEN CATALYST FOR BIODIESEL PRODUCTION FROM COCONUT OIL 110
S N A R Azahra, Helmiyati

ADSORPTION CAPABILITY FOR CONGO RED ON EXFOLIATED GRAPHENE-DECORATED COFE2O4 NANOCOMPOSITE: KINETIC, ISOTHERM, THERMODYNAMIC AND RECYCLABILITY STUDIES 120
L V Tan, N T Hong-Tham, P V Thinh

SILVER NANOPARTICLES (AGNPS) AS PHOTOCATALYST IN THE PHOTODEGRADATION OF RHEMAZOL BRILLIANT BLUE 129
I E Suprihatin, G A D Lestari, R Mardhani, V Edoway

SYNTHESIS OF CARBON NANOTUBES (CNT) BY CHEMICAL VAPOR DEPOSITION (CVD) USING A BIOGAS-BASED CARBON PRECURSOR: A REVIEW........ 135
D D Saputri, A M Jan'ah, T E Saraswati

SOL-GEL SYNTHESIS AND PHOTOCATALYTIC ACTIVITY OF BIOMIMETIC CALCIUM MANGANESE OXIDE CATALYSTS........ 142
W Setyarini, Nurhayati, N Dewi, R Anggraini, S S Siregar, E Amiruddin, A Awaluddin

SYNTHESIS OF INDIUM SULFIDE PHOTOANODE, ITS CHARACTERIZATION AND APPLICATION FOR DEGRADATION OF METHYLENE BLUE AND METHYL ORANGE........ 148
Gunawan, A Haris, W Septina

NATURAL CHEMISTRY

THE POTENCY OF HIBISCUS TILIACEUS LEAVES AS ANTIOXIDANT AND ANTICANCER AGENTS VIA INDUCTION OF APOPTOSIS AGAINST MCF-7 CELLS 155
Y Andriani, M Sababathy, H Amir, PR Sarjono, DF Syamsumir, S Sugiwati, MNI Kassim

THE EFFECT OF LEACHING AGENT ON MOLECULARLY IMPRINTED MEMBRANE UREA TRANSPORT PROCESS BASED ON POLYEUGENOXY ACETIC ACID 163
M C Djunaidi, T Wahyuni, R A Lusiana, D S Widodo, Pardoyo

FORMULATION OF THE LEMONGRASS (CYMBOPOGON CITRATUS) ESSENTIAL OIL-BASED ECO-FRIENDLY DIFFUSE SOLUTION 172
T T K Ngan, T T Hien, P H Danh, L T H Nhan, L X Tien

APPLICATION OF COCONUT OIL FROM BEN TRE PROVINCE (VIETNAM) AS THE MAIN DETERGENT FOR BODY WASH PRODUCTS 183
T T K Ngan, T T Hien, N T C Quyen, P N Q Anh, L T H Nhan, M H Cang, D D Nhat, N D Phuc, L G Bach

DEVELOPMENT AND EVALUATION OF SHAMPOO PRODUCTS BASED ON COCONUT OIL SOURCE FROM BEN TRE PROVINCE (VIETNAM) 192
T T K Ngan, T T Hien, L T H Nhan, M H Cang, P H Danh, N D Phuc, L G Bach

STUDY ON THE FORMULA TO PRODUCE SHAMPOO DERIVED FROM LEMONGRASS (CYMBOPOGON CITRATUS (D.C.) STAPF.) ESSENTIAL OIL 201
T T Hien, P T Nhut, L G Bach, N P T Nhan, L T H Nhan

THE EFFECT OF CHEMICAL REFINING PROCESS ON THE PHYSICOCHEMICAL AND FATTY ACID COMPOSITION OF WATER MELON SEED OIL (CITRULUS LANATUS L.) 211
H Soetjipto, C P Septiyarini, S Hartini

OPTIMIZATION INFUSION CONDITION ON FLAVONOID AND ANTIOXIDANT ACTIVITY OF HERB BEVERAGES USING RESPONSE SURFACE METHODOLOGY 219
Y Martono, J Muninggar

HYDROLYSIS OF SORGHUM STARCH TO GLUCOSE USING ORGANIC ACID CATALYST FROM ROSELLA FLOWER EXTRACT (HIBISCUS SABDARIFFA L.) 226
Riniati, S H Abdulloh, R H Fauziyah, Istiqomaturohmah

INTERACTION OF MUTANT PBP2A AND BIOACTIVE COMPOUNDS FROM STREPTOMYCES WITH ANTI-MRSA ACTIVITIES 231
A Laksono, A Asnani, P Iswanto

THE AGRONOMIC PERFORMANCE AND ARTEMISININ CONTENT OF COLCHICINE-INDUCED POLYPLOID GENOTYPES ARTEMISIA CINA 238
M M Herawati, E Pudjihartati, T D Kurnia, A W Setiawan

ORGANIC SYNTHESIS

SYNTHESIS OF POLY PYRIDINE-2-YLMETHYL 2-(EUGENOXY) ACETATE (PMEOA) AS A METAL MIXTURE CARRIER 244
M C Djunaidi, H Cahyono, Ismiyarto, D Siswanta, Jumina

Author Index

IOP Conf. Series: Materials Science and Engineering **959** (2020) 011001 doi:10.1088/1757-899X/959/1/011001

Speech by the chairman of the committee

Ladies and gentlemen, welcome to the 15th Joint Conference on Chemistry (JCC) 2020.

1. The honorable, Rector of Universitas Kristen Satya Wacana.
2. The honorable, Dean of Faculty of Science and Mathematics, Universitas Kristen Satya Wacana.
3. The honorable, head of Department/Study Program, members of Chemistry Consortium (Univ. Diponegoro, Univ. Negeri Semarang, Univ. Sebelas Maret Surakarta, and Univ. Jend. Soedirman Purwokerto).
4. The honorable, Keynote Speakers.
5. The honorable, invited presenters, presenters, and all the conference participants.

First of all, we thank God for this extraordinary opportunity, so that we can hold the 15th JCC 2020, during pandemic COVID-19. At first, we were still confused and doubtful about the implementation of this conference. It also happened to be the first time we become the organizers of this prestigious Conference. But once again thanks to God's blessing and support from our colleagues, leaders, and support from the extraordinary Consortium, so that we can move forward with certainty.

At this moment, the challenge of a very bad situation due to the COVID-19 pandemic has spread to almost all countries, including Indonesia, so that we must face it together. Every component of this country must start moving according to its duties and portions. Likewise, with this consortium of five universities, the annual agenda in the form of an international conference must continue on time. Even though COVID-19 has not subsided yet, the 6 months that have just passed have educated us to be able to walk with Corona, which we don't know yet, when it will completely disappear. Although the plan has changed drastically due to the current condition, the 15th JCC will be held online and only for one day.

We invite four Keynote Speakers: Prof. Hadi Noor from Universitas Teknologi Malaysia, Prof Hidetoshi Sato from Kwansei gakuin University, Japan, Prof. Ming -Kang Tsai from National Taiwan Normal University, Taiwan, and Dr. Tatas HP Brotosudarmo, Dipl. Chem., Ph.D from Indonesia, which also the previous Chief of HKI period 2018. Several invited speakers from members of the consortium also attend this 15th JCC. We also want to report that there are 60 papers that come from various places, for ex. Taiwan, Japan, Vietnam, Malaysia.

IOP Conf. Series: Materials Science and Engineering **959** (2020) 011001 doi:10.1088/1757-899X/959/1/011001

The 15th JCC activity would not have been possible without the help of many parties, therefore allow me as the chairman of the committee to say many thanks to all parties that we cannot mention one by one. Especially for my internal committee members, I would like to express my highest appreciation for the hard work and extraordinary cooperation for the implementation of the 15th JCC. We also thank all the 15th JCC participants who have joined online. Finally, we apologize if in the implementation there is something that is not pleasing to the hearts of all the participants. Thank you and have a good conference.

IOP Conf. Series: Materials Science and Engineering **959** (2020) 011002 doi:10.1088/1757-899X/959/1/011002

Peer review declaration

All papers published in this volume of *IOP Conference Series: Materials Science and Engineering* have been peer reviewed through processes administered by the Editors. Reviews were conducted by expert referees to the professional and scientific standards expected of a proceedings journal published by IOP Publishing.

- **Type of peer review:** Single-blind (Authors do not know who the reviewers are. The reviewers know who the authors are. The members of Reviewer Team reviewed papers by the relevance to the topics, paper format requirements, similarity, and English proofread.)

- **Conference submission management system:** https://callforpapers.uksw.edu/index.php/jcc

- **Number of submissions received:** 39

- **Number of submissions sent for review:** 39

- **Number of submissions accepted:** 38

- **Acceptance Rate (Number of Submissions Accepted / Number of Submissions Received X 100):** 97.44%

- **Average number of reviews per paper:** 3

- **Total number of reviewers involved:** 16

- **Any additional info on review process:** -

- **Contact person for queries:** Didit Budi Nugroho, D.Sc. (didit.budinugroho@uksw.edu)

IOP Conf. Series: Materials Science and Engineering **959** (2020) 012001 doi:10.1088/1757-899X/959/1/012001

The study of compression relaxation time of silicon rubber using optical method

W Hidayat[1,*], E Prasetyo[2], A Marzuki[2], H P Yanuar[2] and A Setyawan[3]
[1] Duta Bangsa University, Surakarta, Indonesia
[2] Department of Physics, Sebelas Maret University, Surakarta, Indonesia
[3] Department of Civil Engineering, Sebelas Maret University, Surakarta, Indonesia

[*] E-mail: wahyu_hidayat@udb.ac.id

Abstract. When rubber is subjected to a load, deformation will occur. Microscopically, the rubber bond structure is transformed from one equilibrium corresponds to the unloaded rubber that corresponds to the loaded rubber. As with any other transformation process, this process will take time. This paper experimentally demonstrates the relaxation time dependence on the mass of load observed at a constant temperature. The experiment was carried out by inserting the fiber sensor was into a silicon rubber bar. As the rubber bar was loaded, the intensity of light coming out of the fiber decreases exponentially with time. It is shown that the relaxation time increases with the increase of mass loaded on the fiber. By knowing this, the results of the fiber sensor can be adjusted according to the time-based review of rubber deformation.

1. Introduction

Rubber is a class of material that everybody knows because of its excellent properties: elastic, stretchy, strong, waterproof, and durable. With these properties in mind, rubber has been used for a wide variety of applications such as the use of home appliances, laboratory, military, automotive, nuclear,aerospace, and sensor packaging.

The use of rubber for sensor packaging will provide not only protection to the main sensing element, but also to adjust the function of the main sensing element. The example of the rubber that used to adjust the function of the main sensing element is mechanical fiber sensors such as rail pad sensors [1], structural health monitoring [2], weight measuring system [3,4], and weigh in motion (WIM). In this study, we applied a rubber for weight in motion sensor. In weigh in motion (WIM) fiber sensor, an optical fiber sensor is embedded or sandwiched in a rubber. The measuring principle is then similar to that of spring based weight measuring systems that work based on Hooke's law. WIM technology is a technology for detecting the weight of a vehicle moving at a certain speed on the road by measuring the wheel axle load of the vehicle while it is running. When there is an emphasis on the WIM sensor, there will be a change in the intensity of the optical signal that is transmitted in the optical fiber. This change in optical signal corresponds to the weight of the load on it. This can be used to detect the weight of a load. Since this object is measured in motion, the relaxation time of the rubber should be as short as possible.This occurence cause a differences optical signal measurement between movement and un-movement conditions.

In this paper, we present how the relaxation time of the rubber (silicon rubber) depends on the mass of the load. Measurement was carried out by detecting light coming out of the fiber sensor inserting into the rubber bar.

IOP Conf. Series: Materials Science and Engineering **959** (2020) 012001 doi:10.1088/1757-899X/959/1/012001

2. Experimental method and working principle

This experiment was carried out using silicon rubber (RTV 588) inserted with the optical fiber sensor made of polymer optical fiber (POF). The optical fiber sensor was designed to work based on bending loss. Figure 1 is a schematic diagram explaining the basic principle of the sensor.POF was coiled onto a cylindrical tube with a diameter of 1.0 cm. The light was launched from one end and detected its intensity from the other end. As the coil is loaded (figure 1), part of the light at the far left and right sides of the fiber will be radiated out of the fiber core. Larger force (load) will result in a smaller radius of coil curvature at X position. Applying Snell's law to trace the incident angle of light reflected back and forth along with the fiber, it can be seen that more light will be radiated out of the core as the load increases. As shown, the incident angle at the far left and right of the fiber decreases as the load (Force *F*) is increased. Since the radius of the coil is already chosen less than a critical radius, this decrease of the incident angle results in part of the propagated light is transmitted across the fiber clad.

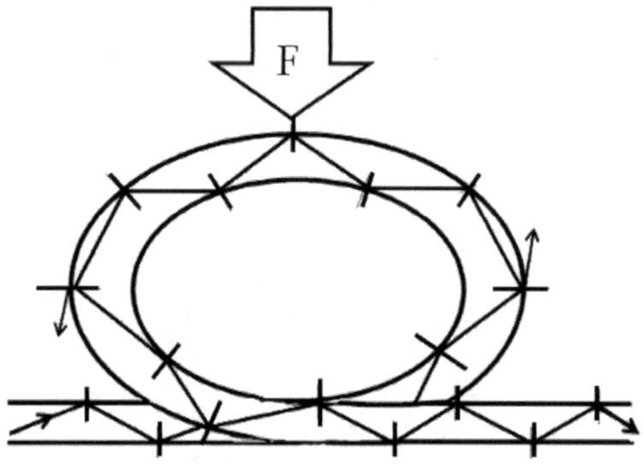

Figure 1. Working principle of the sensing element used in the experiment.

Figure 2 is a schematic diagram of the experiment. Light is launched from one end of the fiber. Light is then split by the optical splitter into two paths: reference fiber and modulated fiber. Reference fiber is connected to the fiber sensor which is inserted into silicon rubber and the modulated fiber is connected to the fiber attenuator. Before loading, fiber attenuator is adjusted so that light intensity detected by photo-detector 1 (I_m) equals to that detected by photo-detector 2 (I_r) or the transmittance T defined by $T = I_m/I_r$ is equal to one. As the rubber bar is loaded, I_m decreases. Since I_r is kept constant (taken as a reference), decreasing I_m means T will decrease. Once the decreasing light intensity (represented by T) over time is obtained, relaxation or decaying time relating to a process of rubber deformation can be constructed.

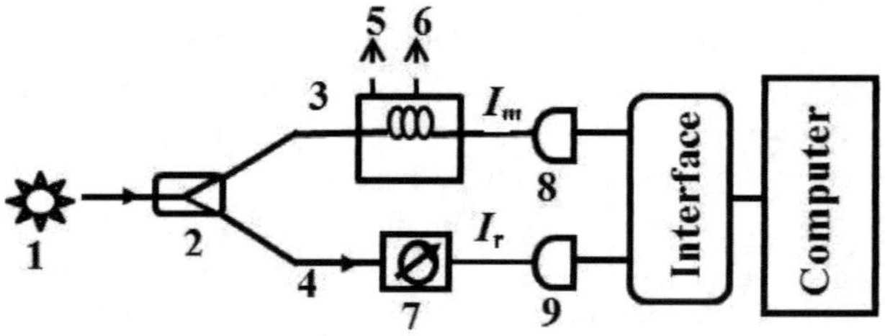

Figure 2. An experimental setup used to measure rubber relaxation time. (1) light source, (2) splitter, (3) modulated fiber, (4) reference fiber, (5) silicone rubber, (6) fiber sensor, (7) attenuator, (8) photo-detector 1 and (9) photo-detector 2.

3. Results and discussion

The sensing mechanism to record the history of rubber deformation due to normal force was made by sandwiching a fiber sensor into silicon rubber. The fiber sensor was designed to work based on bending loss. In an experiment reported in this paper (figure 1), polymer optical fiber (POF) was coiled such that its cross-section shape was circular. As the fiber coil is loaded with a weight F (figure 1), the fiber cross-section changes its shape from circular to elliptical. The vertical diameter of the circular coil is compressed into a minor axis(of the ellipse) by S_D. Figure 3 shows the transmittance change as the vertical diameter of the coil is compressed by S_D which corresponds to the weight or force $F = m_i.g$ where F is force, m_i is load mass, and g is the gravitational acceleration. We have derived the equation that relates the percentage of light radiated out of the fiberdue to bending loss (α) as a function of the amount of the diameter reduction S_D [5]:

$$\alpha = C_1 \exp\left(-C_2 \frac{r^2 - rS_D + \frac{S_D 2}{4}}{\sqrt{r^2 + rS_D - \frac{S_D{}^2}{4}}} \right) \quad (1)$$

where r is circular radius and C_1 and C_2 are constants. It is seen from equation (1) that bending loss (α) increases as S_D is increased.In another word, light transmitted through the fiber sensor decreases as S_D increases.

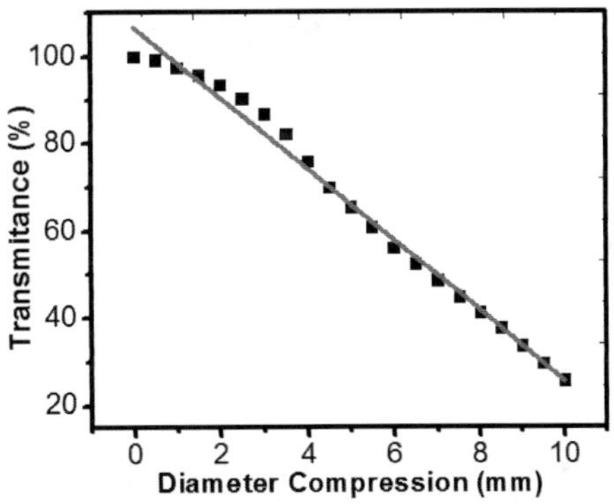

Figure 3. Reduction of light intensity coming out of the fiber sensor as the rubber bar is compressed by S_D which corresponds to the load mass m.

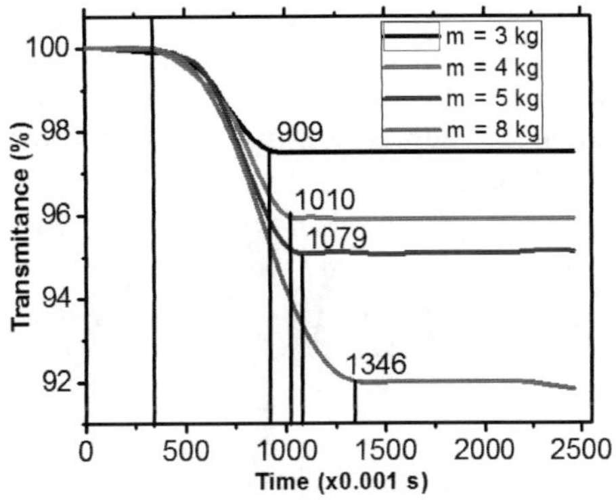

Figure 4. Light transmitted through fiber sensor inserted into silicon rubber loaded with different load mass.

IOP Conf. Series: Materials Science and Engineering **959** (2020) 012001 doi:10.1088/1757-899X/959/1/012001

Figure 4 shows the comparison of transmittance change as the different values of mass is loaded on the rubber bar. Before loading, light transmitted through a fiber sensor is in a constant maximum value over time. As load is given, light transmitted through the fiber drops exponentially until a constant transmittance is achieved. As schematically explained in figure 1, since a decreasing transmittance (increasing bending loss) means shortening the vertical diameter of the fiber coil, the way the transmittance changes over time can then describes the dynamics of structural relaxation in rubber [6,7].

As a normal external force is applied to the silicon rubber bar, an internal intermolecular force arises that oppose the external force. If the external force is not too great, the internal force may be sufficient to resist the external force and allow the rubber to assume a new equilibrium.Assuming the rubber as a collection of springs, this internal force can be regarded as astrong damping force,loading a mass on the rubber can lead to over-damped oscillation. This means that there is no periodic motion and the amplitude dies away as a modified exponential. The maximum amplitude at which the rubber does not move any more concludes that the rubber is in a new equilibrium. In transmittance curves, as shown in figure 4, this state is shown as a lower constant transmittance. As shown, rubber deformation as well as the time required for the rubber to go to its corresponding new equilibrium increases with the increase of the loaded mass (figure 5). This is easily explained using Hooke's law that heavier load will result in a larger value of deformation and therefore longer decay time. This finding suggests that the experimental method used to determine the relaxation time as described herein can be used as a tool to select an appropriate rubber used to a base for weight sensor as well as a tool for studying time relaxation in a rubber.

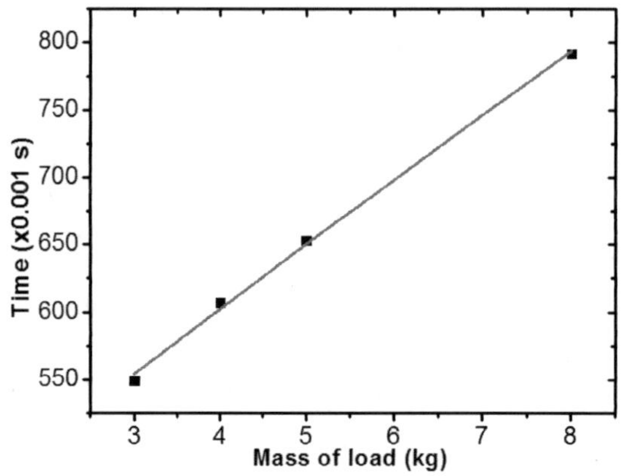

Figure 5. Time required by silicon rubber subjected to a load to relax to a new equilibrium position.

4. Conclusion

We have demonstrated a method to experimentally monitor time relaxation in rubber subjected to different normal loads. In the proposed method, sensing was carried out by inserting a fiber sensor into a silicon rubber. We have shown the history of light transmitted through this fiber sensor from a state at which rubber is not loaded to that is loaded. In this method, light transmitted through a fiber sensor can be associated with the rubber deformation resulted from loading. From this finding, we have suggested that fiber sensor inserted rubber made in this method might be used for weight sensing as well as the history of rubber deformation.

References

[1] Yoon H J, Song K Y, Kim J S and Kim D S 2011 *NDT E Int.* **44** 637
[2] Schotzko T, Reuter M and Lang W 2015 *Polym. Test* **48** 31
[3] Vieira J C, Morais O M F, Vasques C M A and de Oliveira R 2015 *Meas.* **61** 58

IOP Conf. Series: Materials Science and Engineering **959** (2020) 012001 doi:10.1088/1757-899X/959/1/012001

[4] Purwanto H, Fitriani U R, Dwijosutomo A and Marzuki A 2016 *J. Pyhs. Conf. Ser.* **776** 012106
[5] Marzuki A, Heritanto M, Setiyadi I D and Koesuma S 2015 *J. Pyhs. Conf. Ser.* **622** 012059
[6] Yang C and Oyadiji S O 2016 *Sensor Actuat A-Phys.* **244** 1
[7] Roveri N, Carcaterra A and Sestieri A 2015 *Mech. Syst. Signal. Process.* **60–61** 14

Molecular modelling as a downstream effort in the discovery of novel anionic sulphate surfactants

P Iswanto[*] and E V Y Delsy
Chemistry Department, Faculty of Mathematics and Natural Sciences,
Jenderal Soedirman University, Purwokerto, Indonesia
Jl. Dr. Soeparno, Karangwangkal, Purwokerto 53122

[*] E-mail: ponco.iswanto@unsoed.ac.id

Abstract. Sulphate anionic surfactant is a type of surfactant that is widely used as a cleaning agent. The discovery and development of anionic sulphate surfactant molecule is necessary because it has a huge impact, both in science and economics. Molecular modelling to produce new anionic sulphate surfactant molecules has been carried out. The mathematical equation of the Quantitative Structure-Property Relationship (QSPR) based on Ab Initio has been obtained. Molecules that were candidates for structural modelling and modification were those that have the smallest critical micelle concentration (CMC). The smallest experimental data was $C_{16}H_{33}SO_4Na$ with a CMC value = 0.000579 M. The result of this study was $C_{17}H_{35}SO_4Na$ molecules with a theoretical CMC value of 0.000515 M.

1. Introduction

The population in the world is increasing every year, indirectly it will encourage the use of cleaning materials, namely detergents used for washing clothes, household utensils, transportation and industrial fields for the past 40 years [1]. Detergent contains ingredients that have surface active agent (surfactants). Detergents that are sold freely in the market usually contain 20-40% surfactants, while the rest are chemicals called additives or detergent builders which function to increase the clean power of detergents. Generally, after the use of surfactants, the waste is simply disposed of into the environment. The surfactant content contained in detergents is generally a type of anionic surfactant. Anionic surfactants are a type of surfactant that is non-degradable, so their high use causes environmental pollution [2]. One of the parameters used to determine the performance of the surfactant is the Critical Micelle Concentration (CMC). CMC is the concentration of saturated surfactant in an emulsion or it can also be interpreted as the concentration of surfactant when there is a physical change of two phases that do not mix with each other [3].

There are two ways, classic and modern, to obtain new molecules. The classical method is done in the laboratory using laboratory materials and equipment. The weakness of the classical method is causing environmental pollution, while the modern method is done by using computer technology and software in it. The modern method is known as the computational chemistry approach. The advantage of the computational chemistry approach is that it is environmentally friendly [4].

The computation method used is the ab initio method with a large basis set (6-31G**). The ab initio method was chosen because the results of the calculations obtained have a high degree of accuracy in providing information about the physical properties of a molecule [5]. The computational chemistry approach taken to design new molecules is the analysis Quantitative Structure-Property

Relationship (QSPR). In producing a novel molecular design, it cannot be separated from the experimental data that has been done by previous researchers. Experimental data in the form of physicochemical properties are used to create mathematical equations [6]. Mathematical equations are used to predict the physicochemical properties of new molecules. The aim of this research was to obtain the best QSPR mathematical equation which can be used to find the theoretical CMC value for the needs of the new sulphate group anionic surfactant compound.

2. Methods

2.1. Material and equipment

The tools and materials needed in this research were: One computer with a Core i5 Processor, 2 GB RAM; Hyperchem 8.0 Software; 3-dimensional structural model for anionic sulphate surfactant molecules; Experimental data of CMC value of 26 anionic surfactant molecules, see table 1.

Table 1. The experimental data on the critical micelle concentration (CMC) value of the 26 molecules of anionic sulphate surfactant. Atom H is not included in the structure [7].

Number	Name of Molecule	Log CMC Experiment	Number	Name of Molecule	Log CMC Experiment
1	$C_{16}SO_4Na$	–3.237	14	$C_7C(C_6)SO_4Na$	–2.013
2	$C_{14}SO_4Na$	–2.658	15	$C_9C(C_4)SO_4Na$	–2.171
3	$C_{12}C(C_2)SO_4Na$	–2.658	16	$C_8C(C_8)SO_4Na$	–2.629
4	$C_{10}C(C_4)SO_4Na$	–2.469	17	$C_7C(C_7)SO_4Na$	–2.177
5	$C_{11}SO_4Na$	–1.783	18	$C_{15}SO_4Na$	–2.921
6	C_9SO_4Na	–1.209	19	$C_7C(C_7)CSO_4Na$	–2.523
7	$C_6C(C_6)SO_4Na$	–1.714	20	$C_{10}C(C_5)SO_4Na$	–2.269
8	$C_{10}SO_4Na$	–1.481	21	$C_{12}C(C)SO_4Na$	–1.081
9	$C_8C(C)SO_4Na$	–1.328	22	$C_{13}SO_4Na$	–2.367
10	$C_5C(C_5)SO_4Na$	–1.081	23	$C_9C(C_9)SO_4Na$	–3.027
11	C_8SO_4Na	–0.854	24	$C_{10}EO_2SO_4Na$	–1.925
12	$C_{12}SO_4Na$	–2.066	25	$C_{12}C(C_3)SO_4Na$	–2.764
13	$C_{11}C(C)SO_4Na$	–2.187	26	$C_{12}EOSO_4Na$	–2.396

2.2. Molecular modelling

Molecular modelling begins with depictions on the Hyperchem 8.0 Program canvas. This study also used the molecular structure data of anionic surfactants along with the CMC values obtained from the experimental results. The chemical structure of the anionic sulphate surfactant molecule used in this study was made a two-dimensional (2D) model using the Hyperchem 8.0 device. The model is then equipped with a hydrogen atom on each atom, and is formed into a three-dimensional (3D) structure using the Build menu (Add H and Model Build). The picture of the anionic sulphate surfactant molecule is then saved in .HIN format [8]. The surfactant molecules described were optimized to form the most stable structure.

2.3. CMC calculation

In order to calculate CMC the sulphate anionic surfactant molecule was calculated based on equation (1). The mathematical equations to predict [9] resulting from Ab Initio's Quantitative Structure-Properties Relationship were:

$$\lg \text{CMC} = -5.665 + 0.042(\text{BM}) - 0.095(\text{Pol}) - 0.012(\text{Vvdw}) - 0.027(\text{MD}) + 1.418(\text{QC1}) - 14.562(\text{QC2}) \quad (1)$$

The lg CMC is the logarithmic of the CMC, after which it is constant. BM and Pol were molecular weight and polarizability. Then Vvdw and MD were van der Waals volumes and dipole moments. Then QC1 and QC2 were the net charges of atoms C1 (C polar) and C2 (C non polar), see equation (1).

3. Result and discussion

3.1. Molecular modelling of surfactant sulphate group

Anionic surfactants are surfactants whose alkyl portion is attached to an anion. Its hydrophilic nature comes from the ionic head which is usually a sulphate group or a sulfonate group [4]. One of the important properties that can indicate the quality of the surfactant is the value of CMC. CMC is an important property of a surfactant which indicates the critical concentration limit of the surfactant in a solution. Surfactants have good quality if the CMC value is getting smaller.

This study used the sulphate group anionic surfactant in a linear chain. Molecular modelling of sulphate group surfactants was carried out on the Hyperchem 8.0 canvas on the build menu. The modelling of the structure of the sulphate group anionic surfactant on the Hyperchem 8.0 canvas is shown in figure 1. One molecule of the sulphate anionic surfactant was a molecule of Sodium Lauryl Sulphate, $C_{17}H_{35}SO_4Na$.

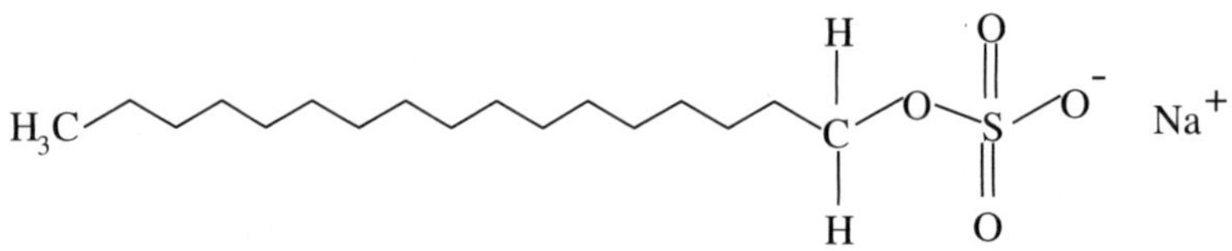

Figure 1. Molecular modelling of anionic surfactants for the sulphate group, Sodium Lauryl Sulphate, $C_{17}H_{35}SO_4Na$.

3.2. Geometry optimization and CMC calculation

Geometry optimization is a process of changing the conformation of molecular structures to obtain the lowest energy. Energy optimization can also be said as an energy minimization stage. Optimization of geometry aims to obtain the structure of compounds that are in stable conditions, namely compounds with the lowest potential energy. Optimization of the geometric structure of the sulphate group anionic surfactants was carried out using the Ab Initio method with the basis set is 6-31G** [5].

After the optimization was complete, the $C_{17}H_{35}SO_4Na$ molecule was continued by calculating the theoretical CMC value or calculated CMC [10].

$$\begin{aligned}
\lg \text{CMC} &= -5.665 + 0.042(\text{BM}) - 0.095(\text{Pol}) - 0.012(\text{Vvdw}) - 0.027(\text{MD}) \\
&\quad + 1.418(\text{QC1}) - 14.562(\text{QC2}) \\
&= -5.665 + 0.042.\ (358.51) - 0.095.\ (34.18) - 0.012(1180.60) - 0.027(9.485) \\
&\quad + 1.418(0.10534) - 14.562(-0.332393) \\
&= -5.665 + 15.05742 - 3.2471 - 14.1672 - 0.256095 + 0.14937212 + 4.84031 \\
&= -3.288 \\
\text{CMC} &= 10^{-3.288}\ \text{M} = 0.000515\ \text{M}
\end{aligned}$$

The theoretical CMC value or the calculation result of the $C_{17}H_{35}SO_4Na$ molecule obtained was 0.000515 M and this was actually lower than the $C_{16}H_{33}SO_4Na$ molecule which was 0.000579 M.

4. Conclusion

The results of the study concluded that the mathematical equations resulting from the quantitative structure-properties relationship of Ab Initio could be used to predict sulphate anionic surfactants. The theoretical CMC value was $C_{17}H_{35}SO_4Na$ of 0.000515 M.

Acknowledgment

We, *Soedirman Group for Computational Chemistry*, Physical Chemistry Laboratory, Department of Chemistry, Faculty of Mathematics and Natural Sciences, Unsoed, would like to express our gratitude for the financial support. This support is provided through Competency Improvement Research Grant, LPPM Unsoed, with a Contract Number: T/584/UN23.18/PT.01.03/2020.

References

[1] Azarmi R and Ali A 2015 *J. Chem. Pharm. Res.* **7** 632
[2] Babajanzadeh B, Sherizadeh S and Ranji H 2019 *J. Sci.* **3** 94
[3] Reningtyas R 2015 *Eksergi* **12** 12
[4] Iswanto P, Delsy E V Y, Setiawan E and Putra F A 2019 *Molekul* **14** 78
[5] Young D C 2001 *Computational Chemistry: A Practical Guide for Applying Techniques to Real-World Problems* (New York: John Willey & Sons, Inc.)
[6] Iswanto P, Chasani M, Delsy E V Y, Harjono, Tahir I and Hanafi M 2011 *Indo. J. Chem.* **11** 31
[7] Huibers P D T 1999 *Langmuir* **15** 7546
[8] Hypercube 2007 *Hyperchem Release 8.0 for Windows* (Florida: USA)
[9] Ananto A D 2017 *Al-Kimia* **5** 100
[10] Lamon L, Asturiol D, Vilchez A, Ruperez-Illescas R, Cabellos J, Richarz A and Worth A 2019 *Comput. Toxicol.* **9** 143

IOP Conf. Series: Materials Science and Engineering **959** (2020) 012003 doi:10.1088/1757-899X/959/1/012003

Glucose-water interactions at increasing concentrations and temperatures as revealed by Near-Infrared Spectroscopy

F S Rondonuwu[1,3,*], A Setiawan[2,3], J Muninggar[3] and F F Karwur[4]
[1] NIR Center, Faculty of Science and Mathematics, Universitas Kristen Satya Wacana, Diponegoro 52–60, Salatiga 50711, Indonesia
[2] Study Center for Multidisciplinary Applied Research and Technology, Universitas Kristen Satya Wacana, Salatiga, Indonesia
[3] Physics Department, Faculty of Science and Mathematics, Universitas Kristen Satya Wacana, Diponegoro 52–60, Salatiga 50711, Indonesia
[4] Faculty of Health Sciences, Universitas Kristen Satya Wacana, Diponegoro 52–60, Salatiga 50711, Indonesia.

* E-mail: ferdy.rondonuwu@uksw.edu

Abstract. Water is an essential molecule and one of the most intensive research subjects, yet it is a peculiar molecule. The non-invasive measurement of blood glucose levels by near-infrared spectroscopy requires detailed information about the glucose-water interactions. The glucose in water induces strong hydrogen bonds and may influence the tetrahedral structure of water molecules. This knowledge is essential in constructing a non-invasive calibration of the blood glucose prediction model. How they are related to the absorption spectra of water and glucose is particularly valuable to comprehend. The evaluations were carried out using near-infrared spectroscopy, tracing changes in absorption intensity and shifts in corresponding peaks with glucose varying at an increasing concentration from a 0-0.9 molar fraction. The near-infrared spectra analysed were around 6900 cm^{-1} and 4716 cm^{-1} corresponding to the overtone and the combination band of water and glucose absorptions, respectively. The analysis suggests that glucose prefers to enter a water cluster, rather than bind to free molecules, and induce bond breaking at low glucose concentrations at which the molecular fraction is less than 0.05.

1. Introduction

Water is available in abundance on Earth and is essential to sustain life. The adult human body is made up of approximately 60% water, while a newborn baby is about 78% [1]. Whole blood consists of about 50% water, while plasma contains up to approximately 90% water. Water plays a critical function, including as nutrition for all living beings as it is the building material for cells, a medium for transporting carbohydrates and proteins to all parts of the body. Together with glucose, it functions as a cryoprotective agent [2]. Hence, in various phases, this molecule has been the subject of exciting research for decades by many researchers [3,4]. The conformations in both solid and liquid forms of water were studied through the atoms' parameters of energy, bond length, structure, and orientation. How water molecules interact with other water molecules to form water clusters were also investigated [5,6]. Water clusters are influenced by additives such as glucose, where this additive can induce a structure breaker or a structure maker as well as limit the translational and rotational mobility of water [7,8]. Even then, water remains a mystery [9,10].

Near-infrared is the electromagnetic absorption of water overtones and the combination band of symmetry and antisymmetric stretching and bending vibration modes [11]. Therefore, near-infrared spectroscopy is a great tool that has become a hallmark of rapidly advancing analytical techniques in the last few decades, including the nature of anharmonic molecular vibrations. Advances in NIR spectroscopy have also provided valuable information in quantum mechanical models and calculations regarding molecular vibrations [12].

This paper aims to report the near-infrared study of the interactions between water and glucose as an additive with the absorption in the spectral region of 5552–9000 cm^{-1}, corresponding to the absorption of overtone and combination band of the vibration mode. The dynamics of a glucose solution at various concentrations and temperatures and the effects of glucose on the water clusters are discussed. This information is particularly useful in designing a non-invasive measurement of glucose concentrations in blood using near-infrared spectroscopy.

2. Experimental method

2.1 Sample preparation

The D(+)-glucose monohydrate sample used in this study was purchased from Merch without further treatment. A total of 14 samples of glucose were dissolved in distilled water prepared with a mole fraction (c) of 2, 4, 6, 11, 17, 24, 32, 41, 51, 62, 76, and 91 x 10^{-3}. The D(+)-glucose in each solution sample was dissolved using a magnetic stirrer. Since the glucose solution can easily suspend, each sample was then stirred for a second time just before being loaded to the measurement cell ensuring even molecule distribution.

2.2 Data acquisition

The near-infrared absorption spectra were obtained by scanning samples loaded into a cuvette with a geometrical length of 1 mm using the BUCHI NIRFLEX N500 spectrometer (with liquid accessory). The targeted spectral regions were 5552–9000 cm^{-1} with an interval of 4 cm^{-1}. Thus, the total data points in the specified spectral regions are 862. Each spectrum was averaged over 32 scans to achieve an acceptable s/n ratio. Sample temperatures were set at 25 °C and 60 °C using a built-in temperature controller (a repeatability of 0.5 °C and stability less than 0.2 °C).

2.3 Data analysis

Near-infrared spectra in almost all measurement modes often undergo baseline fluctuations. However, in this case, the baseline fluctuations were corrected by pushing each spectrum to zero at 9340 cm^{-1}. No additional corrections, including multiplicative scattering and smoothing, were applied. The absorption intensity in the spectral gap of 4820-5552 cm^{-1} was intentionally set to zero due to the saturated water absorption around 5500 cm^{-1}. Since water absorption has approximately double the magnitude of absorption compared to glucose, the spectral evolution due to the increase of glucose mole fraction was not quite profound in the glucose-solution spectra. Therefore, the absorption spectrum of pure water was then subtracted from each spectrum measured at the same temperature. Hence, only the changes in the absorption spectrum due to the addition of glucose are shown. The same method was applied to groups of spectra measured at various temperatures with the same concentration. In this case, the background subtraction was carried out using a spectrum from the lowest concentration and temperature.

3. Results and discussion

Strong absorption from a combination band of symmetric overtone stretching and bending modes vibrations originated from water OH bond appears in the spectral region around 5200 cm^{-1}. The combination band of bending, overtone symmetric and antisymmetric stretching modes appears in an area of around 6900 cm^{-1}. On the other hand, the glucose absorption dissolved in water at various mole fractions appears in the spectral region 5700–6600 cm^{-1}, as shown in figure 1(a). There are two groups of spectra in this figure, those measured at 25 °C (black) and 60 °C (red). Absorption spectra measured at 25 °C have a maximum absorption peak at 6880 cm^{-1}, while at 60 °C, the maximum absorption peak appears at 7000 cm^{-1}. There is a peak shift in absorption when the sample measured from lower to higher temperatures or vice versa. These results are consistent with previous reports observed in the range of 5400–4600 cm^{-1} [13] and around 7000 cm^{-1} [14]. When the number of glucose molecules increases, the water absorption intensity at 6880 cm^{-1} and 7000 cm^{-1} decreases. The decreasing absorption intensity is due to the reduced number of OH bond water molecules.

Along with increasing glucose concentration, the absorption band in the spectral region of 5700–6600 cm^{-1} increases along with the increased glucose concentration. To carefully examine how the absorption spectra change with the addition of glucose, we subtracted the absorption spectra measured at 25 °C and 60 °C with the corresponding one at zero concentration for a baseline and depicted in figure 1(b). In this particular figure, the ordinate axis represents the change in absorption (ΔA), which has a negative value for a decrease in absorption intensity and positive value for increases in absorption intensity while glucose increases. Such spectral behavior implies absorption around 5552–6540 cm^{-1} is originated from glucose molecules. Increasing glucose mole fraction, enabling decreasing absorption of water in the spectral region 6540–9000 cm^{-1}.

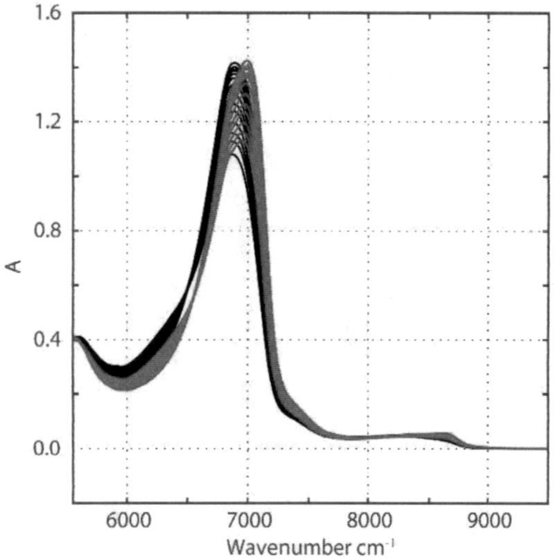

(a)

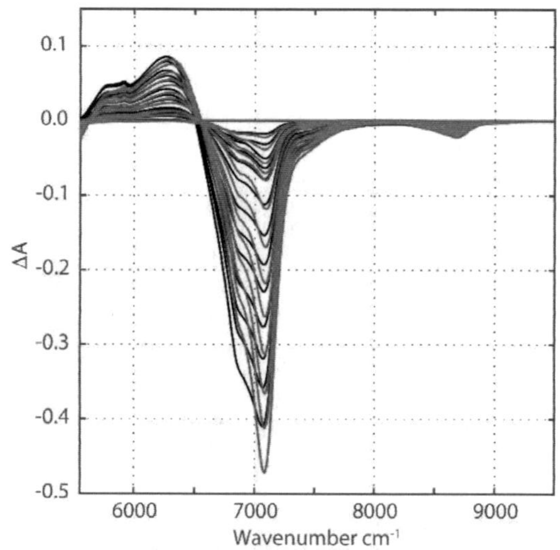

(b)

Figure 1. (a) Near-infrared absorption spectra (A) of a glucose solution in concentrations of 0, 2, 4, 6, 11, 17, 24, 32, 41, 51, 62, 76, and 91 x 10^{-3} mole fraction measured at 25 °C (black) and 60 °C (red). (b) Difference spectra (ΔA), absorption spectra minus the spectrum at 0 mole fraction for the set of spectra measured at 25 °C (black) and 60 °C (red).

To examine how the temperature modifies the spectrum of pure water and a glucose solution with a concentration of 0.091 mole fraction, a spectral acquisition of those samples was carried out at a temperature of 25–60 °C, as represented in Figure 2(a). The spectrum of pure water at various temperatures is represented with a black line, while the glucose solution at a molar fraction of 0.091 is represented with a red line. The heating effect of pure water and glucose solutions causes the water peak position to shift to the higher energy side. The spectra difference, i.e., the spectrum that has been subtracted from the spectrum of pure water at 25 °C, is shown in Figure 2(b). This shift eventually originated from the competition of the two water absorption peaks at 6880 cm^{-1} (suppressing) and 7000 cm^{-1} (growing) when the temperature increases. However, both peak positions remain invariant in the course of the heating process. Similar behavior is also valid for a glucose solution at a 0.091-mole fraction. These results indicate that, unlike water absorption, glucose absorption is insusceptible to temperature variations.

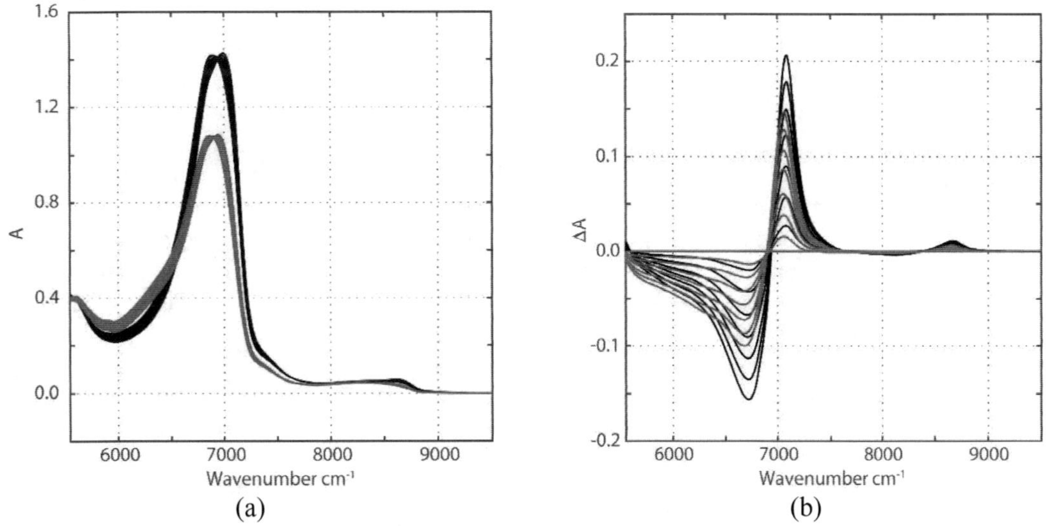

Figure 2. Near-infrared absorption spectra (A) measured at temperature from 25 °C to 60 °C (interval 5 °C) for (a) of pure water (black) and a glucose solution at concentrations of 91 x 10^{-3} mole fraction (red). (b) Difference spectra (ΔA), absorption spectra minus spectrum at 25 °C for a set of pure water spectra (black), and a glucose solution (red).

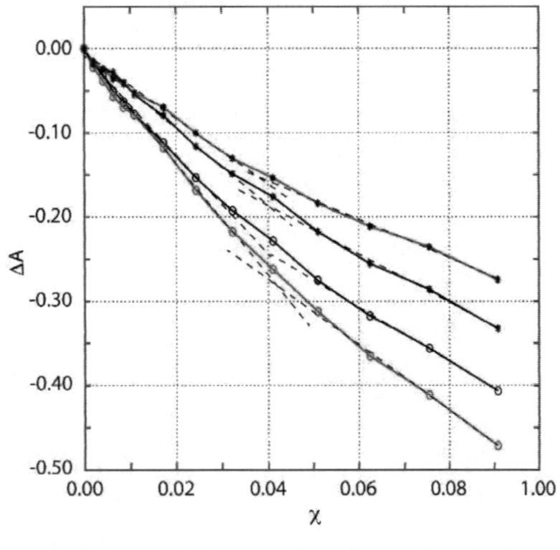

Figure 3. Plot of ΔA as a function of mole fractions (χ) measured at 25 °C (black) and 60 °C (red) for water peak positions of 6868 cm^{-1} (*) and 7084^{-1} (o).

Figure 3 shows a plot of the difference absorbance (ΔA) for the increase of glucose mole fractions at 25 °C (black) and 60 °C (red) and two absorbance positions, namely 6868 cm^{-1} (*) and 7084 cm^{-1} (o). All peaks decrease when glucose is increasing. At 60 °C, the decrease in absorption intensity at 7084 cm^{-1} appears faster than that of the same peak at 25 °C. In contrast, the reduction rate of the absorption peak intensity at 6868 cm^{-1} was faster at 25 °C, which suggests that the water molecules at high temperatures more rapidly reduce the vibrational mode from antisymmetric to symmetric as the glucose mole fraction increases. Probably the distance between water molecules is elongated when glucose is added and is further enhanced as the temperature gets higher.

The decreases in absorbance intensities do not linearly diminish with the increasing glucose mole fraction. There are two decreasing phases; the first phase occurs below the mole fraction of 0.4 and the second phase is at or above the mole fraction of 0.04. The two decreasing phases appeared for the absorption peaks of 6868 cm^{-1} and 7084 cm^{-1} and at temperatures of 25 °C and 60 °C. The phase changes at a mole fraction of 0.4 can probably be related to water's nature, which can be in a cluster or as individual molecules (monomeric). Such a water molecules change their form continuously between monomeric and cluster forms in such a way that every time a water molecule changes from a monomeric to a part of the cluster, one of the cluster's water molecules changes back into a monomer and vice versa.

At low concentrations, glucose tends to enter the water cluster rather than bind to a free molecule in the monomeric form. In this way, the glucose molecule establishes a strong H-bonding resulting in a stable solution. However, the presence of glucose also induces the breaking of hydrogen bonds both peripherally and within clusters. Consequently, the cluster size was reduced, and so glucose functions as a structure breaker. Such a break mechanism is indicated by the increase in water absorption as the overall number of H-bonds increases. Meanwhile, free molecules tend to be close to each other (probably for sterically reasons) to increase their overall stability. As a result, water absorption will increase. When the glucose concentration increases, the water absorption resumes to its initial state. To a certain extent, the interactions between glucose will significantly dominant, so that water forms its clusters again. Thus, it functions as a structure maker. The threshold between the structure breaker and the structure maker for glucose is a mole fraction of 0.04. In addition, a shift in the absorption peak near 7000 cm^{-1} was observed. The shift became prominent when the water absorption spectra at the increasing concentration of glucose were represented by their 2nd derivative spectra, as shown in figure 4. The position of the water absorption peaks shifts to the lower energy side when the glucose concentration increases, both at 25 °C and 60 °C. This shift occurs monotonically below the 0.04 mole fraction and then changes a bit beyond the 0.04 mole fraction, as depicted in figure 5. This threshold border at a 0.04 mole fraction is consistent with the above results. So, both the change in the intensity of water absorption and the analysis of the shift in the peak position near 7000 cm^{-1} support the earlier mentioned mechanism of glucose as a structure breaker and structure maker [13, 14] as well as the result of computational calculations using molecular dynamics simulations [15].

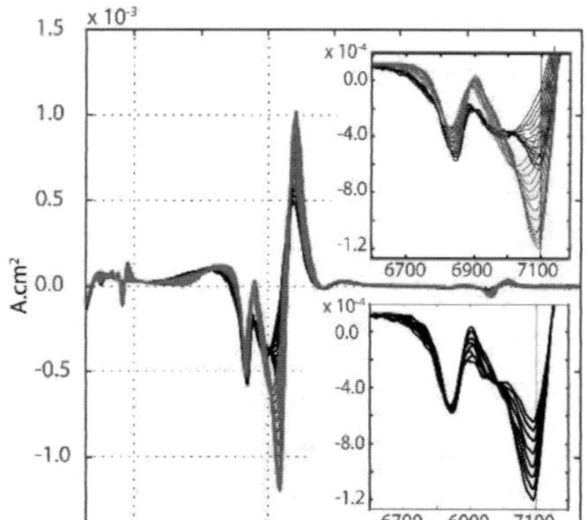

Figure 4. Calculated second derivative spectra of a glucose solution at 0, 2, 4, 6, 11, 17, 24, 32, 41, 51, 62, 76 and 91 x 10^{-3} mole fraction measured at 25 °C (black) and 60 °C (red). The inset in the upper-right corner shows a portion of the spectra for clarity, while the inset in the lower-right corner shows a similar calculation for pure water for comparison.

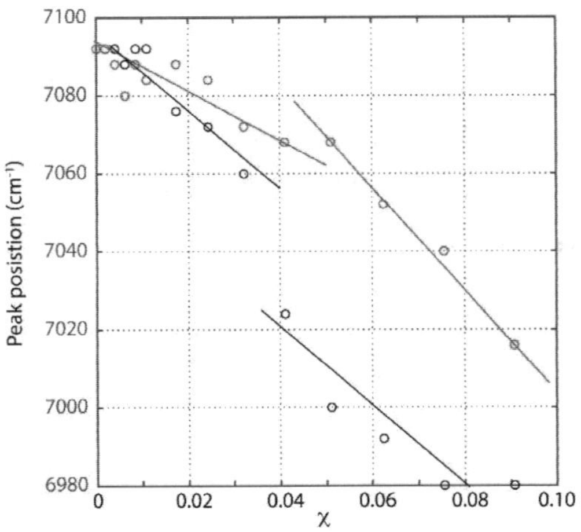

Figure 5. The shifts of water peak positions as a function of the molar fraction (χ) measured at 25 °C (black) and 60 °C (red).

4. Conclusion

The near-infrared spectrum has been measured for increasing a glucose concentration of 0-0.09 mole fractions. In addition, the near-infrared spectra for a glucose solution at temperatures ranging from 25 °C to 60 °C have been measured. Compared to water absorption, glucose absorption is relatively low, which is about two magnitudes lower than that of water, but glucose absorption can also be observed from an apparent decrease in the water absorption intensity at 6868 nm^{-1} and 7084 nm^{-1}. When glucose with a concentration of less than a 0.04-mole fraction is added to water, it tends to insert itself into the water cluster and form a stable hydrogen bond with water. At concentrations above a 0.04-mole fraction, a glucose solution tends to interact with other glucose. Meanwhile, water tends to revert to form clusters as before. Increasing the sample temperature does not affect the water absorption peak. However, increasing the sample temperature pushes the absorption intensity of the anti-symmetric stretching and symmetric stretching modes lower. In contrast to the temperature, increasing the glucose concentration induces a shift in the absorption peak of the anti-symmetric stretching vibration mode towards the lower energy side, supporting the statement that glucose functions as a cluster breaker at mole fraction less than a 0.04 and a cluster maker at mole fraction higher than 0.04.

Acknowledgments
The author is thankful to the Directorate of Research and Community Service, the Minister of Education and Culture of the Republic of Indonesia for funding this research in the 2020 funding year with the Higher Education Leading Basic Research scheme (PDUPT) through Satya Wacana Christian University with proposal ID: 05f61621-2266-407c -9a14-0751f1ecc41a. FSR would also like to thank Tafip Haryanto for helping prepare the NIR equipment and Sheila and Dina Samalukang for the sample preparations.

References

[1] Mitchell H H, Hamilton T S, Streggerda F R and Bean H W 1945 *J. Biol. Chem.* **158** 625
[2] Elliott G D, Wang S and Fuller B J 2017 *Cryobiology* **76** 74
[3] Philip B 2008 *Chem. Rev.* **108** 74
[4] Tian X, Jiang L, Yuan Y, Wang M Q, Guo Y Z, Zeng X J, Li M L and Pu X M 2013 *J. Mol. Model.* **19** 2525
[5] Ohmine I and Saito S 1999 *Acc. Chem. Res.* **32** 74
[6] Dixit S, Crain J, Poon W C, Finney J L and Soper A K 2002 *Nature* **416** 829
[7] S. Perticaroli, P. Sassi, A. Morresi and M. Paolantoni 2008 *J. Raman Spectrosc.* **39** 227
[8] Moran G R and Jeffrey K R 1999 *J. Chem. Phys.* **110** 3472
[9] Clarke C, Woods R J, Gluska J, Cooper A, Nutley M A and Boons G J 2001 *J. Am. Chem. Soc.* **123** 12238

[10] Bellissent-Funel M C, Hassanali A, Havenith M, Henchman R, Pohl P, Sterpone F, Van Der Spoel D and Garcia A E 2016 *Chem. Rev.* **116** 7673

[11] Tzenkova R, Kovacs Z and Kubota Y 2015 Aquaphotomics: near infrared spectroscopy and water states in biological systems *Membrane Hydration: The Role of Water in the Structure and Function of Biological Membrane* vol 71, ed E A Dislalvo (Springer) p 189

[12] Bec K B and Huck C W 2017 *Front. Chem.* **7** 48

[13] Arfin A, Mafy N N, Rahman M M, Mollah M Y A and Susan Md A B H 2014 *RSC Adv.* **4** 50906

[14] Cui X, Cai W and Shao X 2016 *RSC Adv.* **6** 105729

[15] Chen C, Li W Z, Song Y C, Weng L D and Zhang N 2012 *Comput. Theor. Chem.* **984** 85

IOP Conf. Series: Materials Science and Engineering **959** (2020) 012004 doi:10.1088/1757-899X/959/1/012004

Ab initio computational study: Adsorption of CO molecule on various base metal oxide surfaces

I Kurnia, P Siahaan* and A Suseno
Department of Chemistry, Faculty of Science and Mathematics, Diponegoro University

* E-mail: siahaan.parsaoran@live.undip.ac.id

Abstract. The use of some base metal oxide already widely studied for having beneficial properties in CO oxidation catalytic system. The properties of these base metals such as lower working temperature, high CO affinity, and economically cheaper would be beneficial for this solid system catalyst. The study of various base metal oxides shown these kinds of characteristics, but there still a few analyses of comparison among the various base oxide metals used as the support system for the oxidation reaction catalytic system. This study would give a comparison in the difference of the adsorption energy of various base oxide metals. This adsorption energy is one factor that promoting the activity of the catalytic system due to its affinity to the reactant on their surfaces. The DFT ab initio calculation for this comparison shows that NiO is the base metal oxide that having the highest CO adsorption energy. This oxide could be a potential support system combined with potential solid oxidative catalytic system.

1. Introduction

Carbon monoxide (CO) is a kind of gas that is formed as a product of incomplete carbon combustion reaction, which is a common process found in the environment [1]. Despite the common happening of this gas, the presence of this compound is extremely increasing since the age of industrial revolution, contributing to the high level of this gas in the atmosphere beyond the tolerated limit in some regions. By the knowledge of the hazard of CO gas, efforts have been conducted to minimalize the CO emission from some artificial combustion processes. The incomplete combustion, which is actually an oxidation process of carbon, giving a formula to this problem. The common simple and widespread idea is to complete those oxidation process of unfinished CO, by oxidizing this CO gas to become less hazardous CO_2 gas [2]. This conversion scheme is commonly done by the use of catalytic system, and some of those best known so far are the use of noble metals such as Ag and Pt which giving a good conversion yield [3]. Although these catalysts are considered as good enough, the exploration for catalyst of this process is still widely done, to meet the better conversion and efficiency. Another improvement aspects searched are the activity in lower temperature, selectivity, stability, and also the economic aspect on making those catalyst [1].

Several oxides known for having support properties for this catalytic process. Some of them are Al_2O_3, SiO_2, and MgO known as inert support, and some are considerable active support such as CeO_2, Fe_2O_3, MnO_2, and TiO_2, especially with the ability to facilitate the oxidation-reduction reaction and facilitating the oxygen mobility which are important parameters in the redox reaction [4]. This classification is based on the observation in which Au nanoparticles are exhibiting lower intrinsic activities when supported on inert oxides, but exhibiting higher intrinsic activities when supported on active oxides that could contribute to the activation and supply of oxygen for the reaction. Some cheaper base transition metals such as Mn, Fe, Co, Ni, and Cu have been studied as well, known for potential activity as redox catalyst by the combination of those metals with CeO_2 [3,5,6]. The variation of the metals with CeO_2 giving different results, depends on the interaction of the metals particle with CO molecule. Stronger metal-CO interaction would give the increase of adsorption of the CO molecule on the surface particle, which directly influence the higher oxidation process [3].

The previous studies show that various base metal oxides are having potential activity as a support for this CO oxidation catalyst systems. One of the parameters supporting this activity is that the interaction of the bases metals and the CO molecules in the surface of the solid system. However, there has not been any comparative study to show the difference of the activity of these potential base metals by using the exactly same method. Thus, in this study it will compare those base metals activity, by using the parameters of their interaction with CO molecule on the surface of their metal oxide.

The ab-initio term means "from first principle" which means the total wave function, orbital, and energy are calculated respectively without any empirical parameters. This research used DFT ab-initio computational method, to determine the stable structure and interaction energy of each modelled surface systems. The model used are basic metals oxides lattice, respectively used are TiO_2, Cr_2O_3, MnO_2, Fe_2O_3, CO_3O_4, NiO, Cu_2O, and ZnO which are common base metal oxides, with the interaction of CO molecules on each of their surfaces. All those base metal oxides are modelled to compare the interaction energy with CO molecules on their surfaces, in which this adsorption is one of the parameters directly influencing the activity of the support system for CO oxidation catalytic process.

2. Computational methods

All calculations are performed using Linux-based Quantum ESPRESSO 6.5 machine [7,8]. The input files are created by using notepad++ 7.7 and all modelled interactions are visualized using BURAI 1.3. The calculations of all the models are using first principle Density Functional Theory. The model was created as a slab in 2x2x2 supercell, with a vacuum space gap between lattice. This vacuum was purposed as the mimic of the interphase over the surface of solid system. The vacuum are set to be 15Å to ensure there is no interaction between each of slab surface [9]. The second process is that to model the interaction of CO molecules adsorbed in the surfaces of each oxide systems, with C atom bond to the surface metal atom. All the cell parameters and models are set as the same with the first calculation process [5].

The metals oxides modelled in this study were the commonly found structure of those base metals. TiO_2 crystal was known to be able to use as a support system for catalytic CO oxidation system [4]. This structure modelled as rutile structured and crystallizes in the tetragonal P4_2/mnm space group which then turn to be slab model. While Cr_2O_3 crystal is also known to have ability as an active support for oxidative catalytic system, with the modelled common Cr_2O_3 structure of orthorhombic, Pbcn, 60 space group [10]. The oxide chosen for Mn was MnO_2 that exhibiting catalytic activity in oxidation reaction better than other oxides of Mn [11]. The structure of modelled MnO_2 was rutile structured and crystallizes in the tetragonal P4_2/mnm space group. Then for Fe the structure is Fe_2O_3 which also generating activity in CO oxidation [5]. The next is CO_3O_4 which already wide studied as a support in CO oxidation [12,13]. The modelled structure was cubic lattice with Fd-3m space group. For Ni oxide, it is chosen NiO for calculation with cubic structure with Fm-3m space group [14]. Cu is also widely studied for CO oxidation support system, and in this study the modelled structure was cubic structure of Cu_2O with Pn-3m space group [14,15]. Then the last is ZnO modelled in Wurtzite structure in the hexagonal P6_3mc space group to be calculated [15].

IOP Conf. Series: Materials Science and Engineering **959** (2020) 012004 doi:10.1088/1757-899X/959/1/012004

3. Results and discussion

3.1. Optimization of the metal oxide slabs

Before studying the interaction, computational calculation was done to the metal oxide slabs structure first to determine the energy of each of the oxide structure so that it can be used to calculate the interaction energy with CO in the following part. The calculated structure of metal oxides are TiO_2, Cr_2O_3, MnO_2, Fe_2O_3, CO_3O_4, NiO, Cu_2O, and ZnO, while there is no oxide of Sc and V since there is no relevant reference saying their application for CO oxidation catalyst for both of the elements.

All these structures were then modelled as 2x2x2 supercell and turned become slab structure, to mimic the surface condition of each of the oxide systems [9]. This kind of slab surface is shown in figure 1.

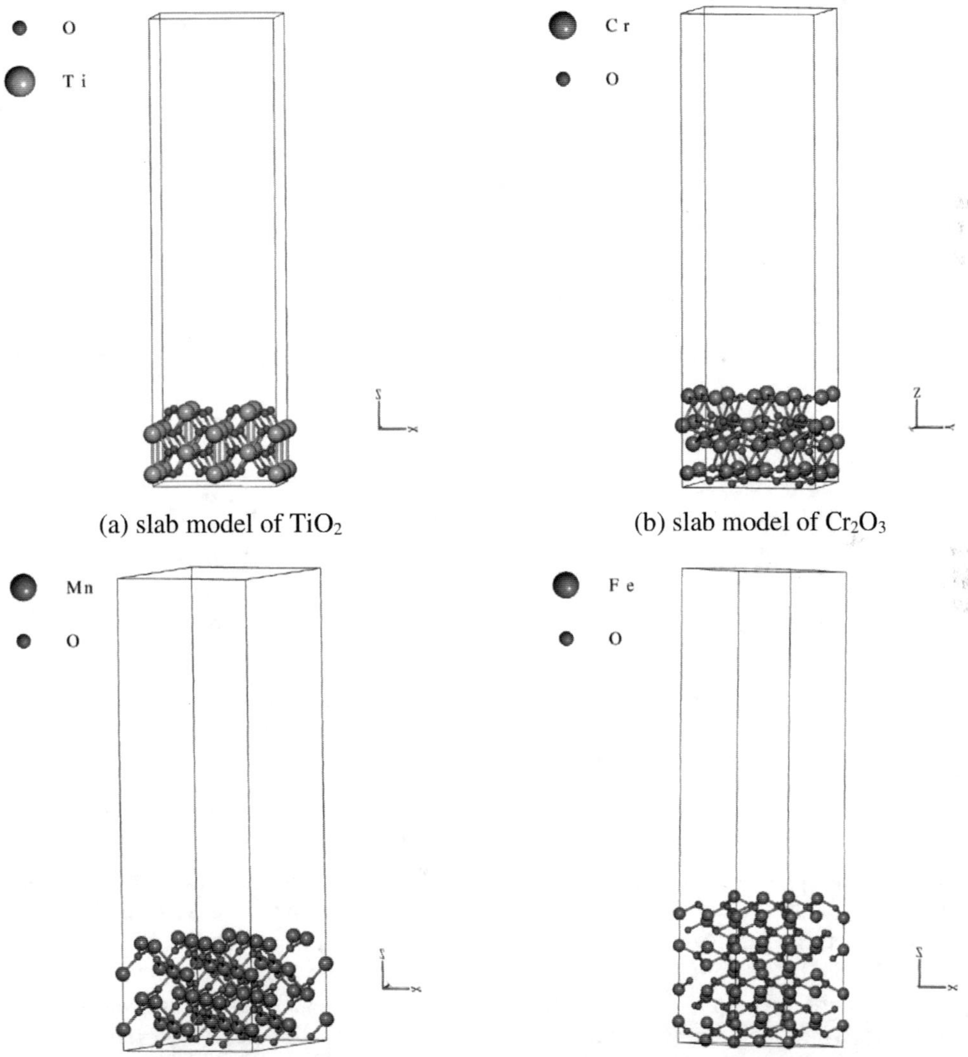

(a) slab model of TiO_2 (b) slab model of Cr_2O_3

(c) slab model of MnO_2 (d) slab model of Fe_2O_3

Figure 1. The slab modelled structure of each base metal oxides.

IOP Conf. Series: Materials Science and Engineering **959** (2020) 012004 doi:10.1088/1757-899X/959/1/012004

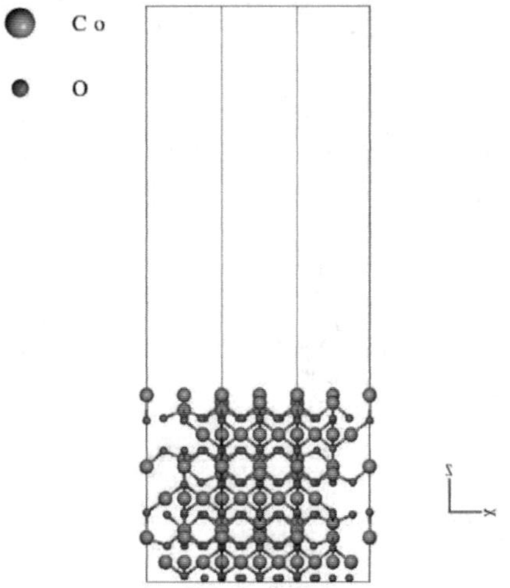

(e) slab model of CO_3O_4

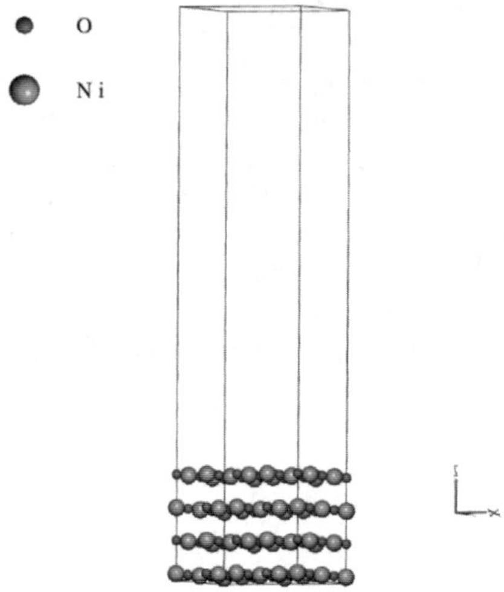

(f) slab model of NiO

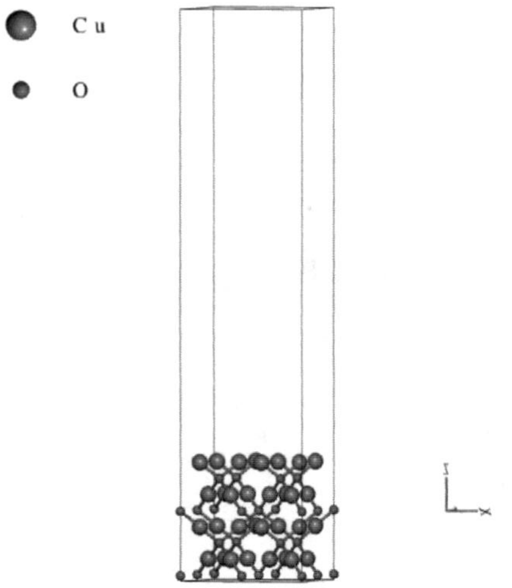

(g) slab model of Cu_2O

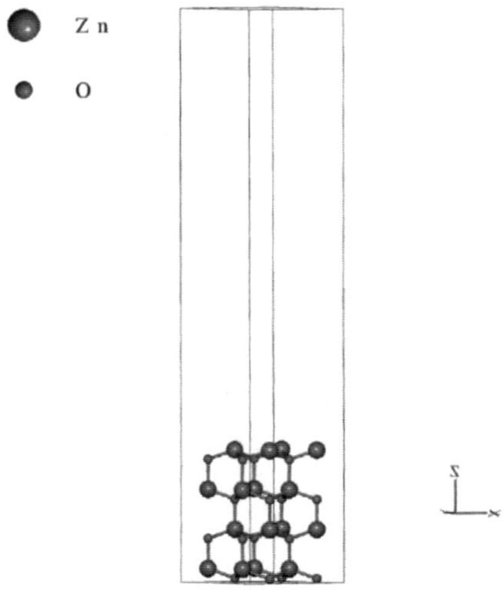

(h) slab model of ZnO

Figure 1. Continued.

Those structure are created to model and mimic the surface part of a solid system. These models are computationally calculated in pw relax scf with DFT basis to optimize the slab structure and to gained the final self-consistent lattice energy for each of the structure. The calculation translates the three bottom atom of each structure in fixed position, while the above atoms are mobile during the optimization to get the relaxed optimized structure and the lowest energy of the oxides. The calculation giving the final result of optimized energy for each oxide slabs as in table 1.

These energies are the result of the convergence structure for each system, as the most stabilized relaxed structure. The result shows unpattern number of energies for these base metal oxides. This wide gap of crystal energy reaches 2000 Ry for some oxides indicate that the energy doesn't simply depend on the periodicity of the element oxide. There is no sign of linear correlation of metals atomic number of each oxide to the resulted energies. This result can be as a result of further variables such as the difference in the structure and also the electronic properties of each oxide as well.

This optimized structure is the final calculated structure of the slab with of the lowest energies of each oxide slab. These slab structures and energies are studied for further analysis in the interaction of these structures with the adsorbed CO molecules on their surfaces.

Table 1. The optimized energy of each slab metal oxide

Oxide slab	Optimized energy (Ry)
TiO_2	–2891.369625
Cr_2O_3	–1775.37291
MnO_2	–4367.846306
Fe_2O_3	–828.575096
CO_3O_4	–2095.995745
NiO	–1883.14872
Cu_2O	–2197.515964
ZnO	–1266.583221

3.2. Calculation of adsorbed CO models

The adsorption of CO molecules in the surface of each oxide slabs are modelled so that it will generate the adsorption energy of CO in each of the surfaces. The position of CO in all of the surface was set to be directly above the outermost metal atom, as it known to be the active site of the adsorption [3]. The distance of the CO molecules to the outermost metal atoms was set to be 2Å, regarding that it will be calculated to its optimized structure. Here shown the visualization of adsorbed CO in every oxide surfaces.

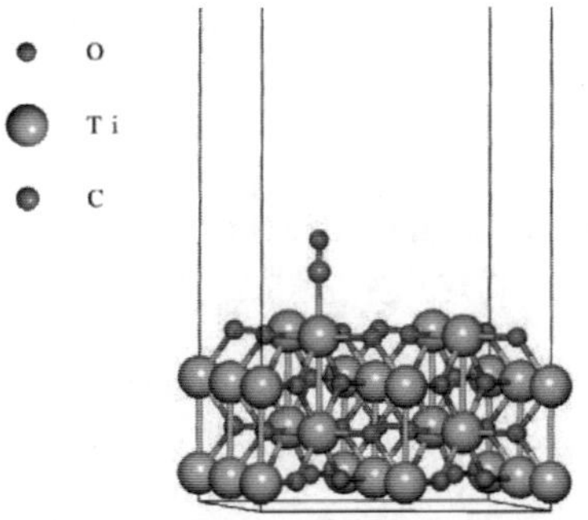

(a) adsorbed CO on TiO_2 surface

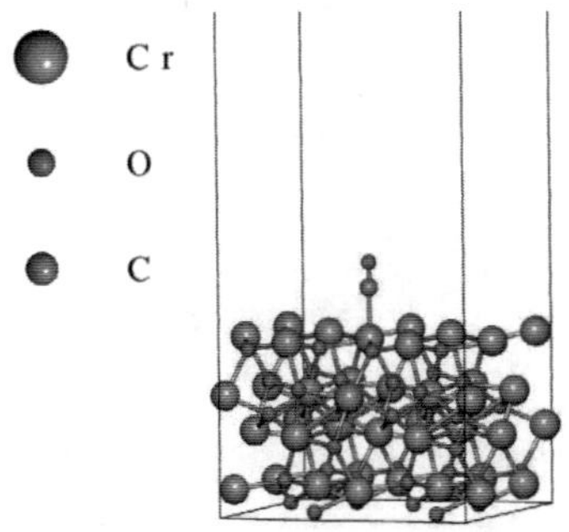

(b) adsorbed CO on Cr_2O_3 surface

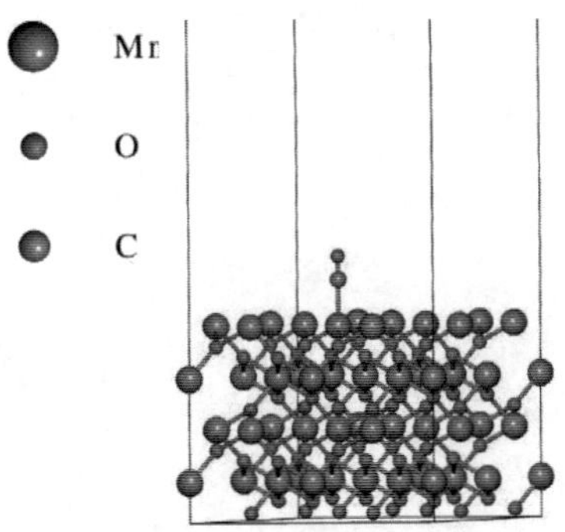

(c) adsorbed CO on MnO_2 surface

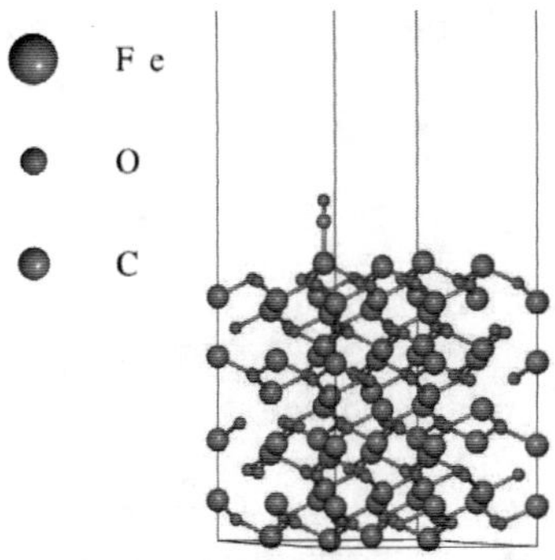

(d) adsorbed CO on Fe_2O_3 surface

Figure 2. Visualization of adsorbed CO on base metal oxide slab surfaces.

IOP Conf. Series: Materials Science and Engineering **959** (2020) 012004 doi:10.1088/1757-899X/959/1/012004

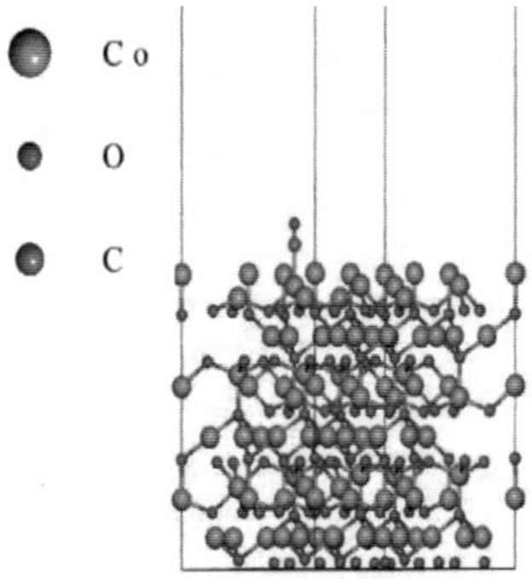

(e) adsorbed CO on CO_3O_4 surface

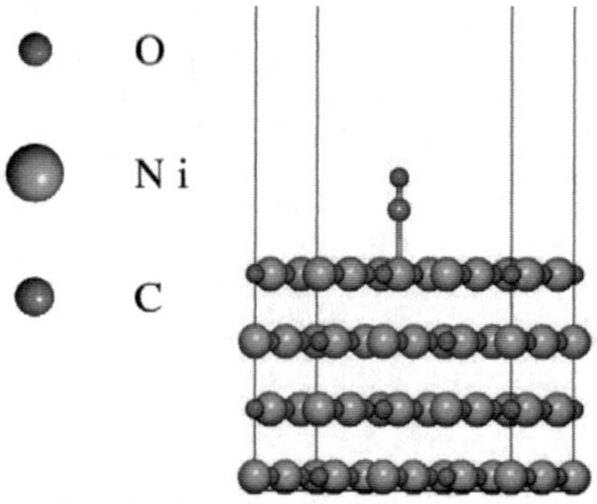

(f) adsorbed CO on NiO surface

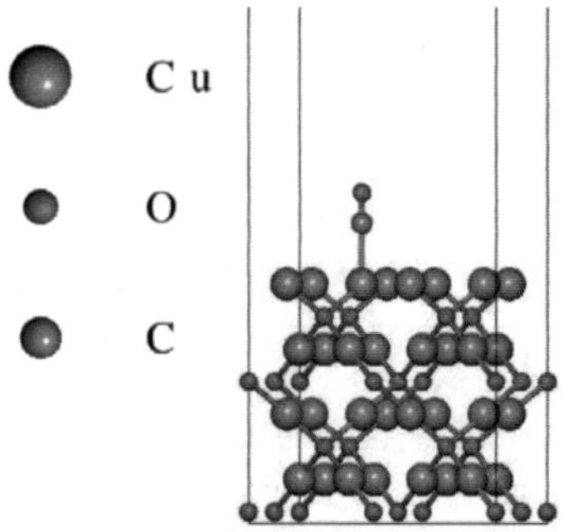

(g) adsorbed CO on Cu_2O surface

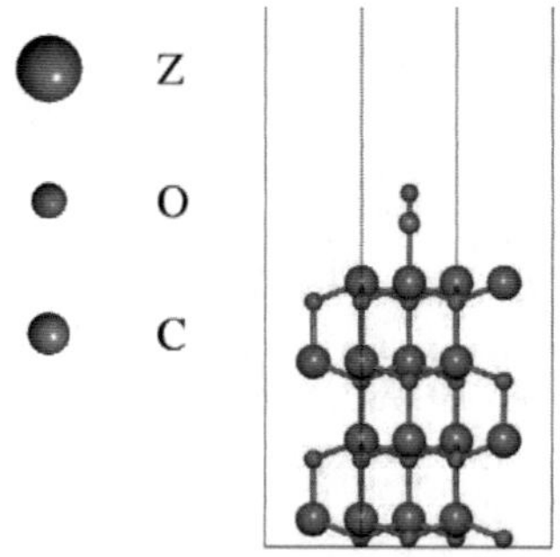

(h) adsorbed CO on ZnO surface

Figure 2. Continued.

Table 2. Adsorption energy of CO on each base metal oxide surfaces

Oxide slab	- CO Adsorption energy (Ry)	- CO Adsorption energy (kcal/mol)
TiO_2	-7,51508464	-2357,05483
Cr_2O_3	-7,61239045	-2387,57413
MnO_2	-7,55402713	-2369,26887
Fe_2O_3	-7,62020367	-2390,02469
CO_3O_4	-7,63996909	-2396,22398
NiO	-7,65798318	-2401,87398
Cu_2O	-7,61570149	-2388,61261
ZnO	-7,58351789	-2378,51845

The adsorbed CO was then calculated and the generated optimum energy used further for the calculation of the adsorption energy. The calculated parameters including the distance interaction and the lattice system energies which will be used further calculation of the adsorption energies. The adsorption energy is correlated to the interaction energy between the two components of each system, which are the metal oxide slab and the CO molecules.

The adsorption energy is calculated based on the difference of energies of the slab from the first calculation with the energies of the slab while adsorbing CO in the surface as follow [9], and the resulted adsorption energy of CO in each of the surfaces shown down below.

$$-E_{ads} = E(adCO\ slab) - E(oxide\ slab) - E\ (CO)$$

The data shows the differences of CO adsorption energies on each metal oxide surfaces. It doesn't show any linear trends, and can be said that the adsorption energy of CO doesn't seems to change periodically. The interaction of CO with metals doesn't simply follow the periodicity of the periodic system; thus, it cannot be predicted based only the periodicity of the element. The difference of the adsorption energy might be influenced by the specific structure of the surface. It can be seen that for TiO_2 and MnO_2 both are having rutile crystal structure and both are considered among the lowest adsorption energy for CO molecule. In contrast, crystals in cubic structure such as CO_3O_4, NiO, and Cu_2O are among the highest adsorption energy for CO molecule on their surfaces. Thus, the probability of CO adsorption is varied on different structure, in which the kind of the structure is one of determining factors affecting to the CO affinity. The other factor could be the specific electronical properties for each base metal, which are not studied further in this work.

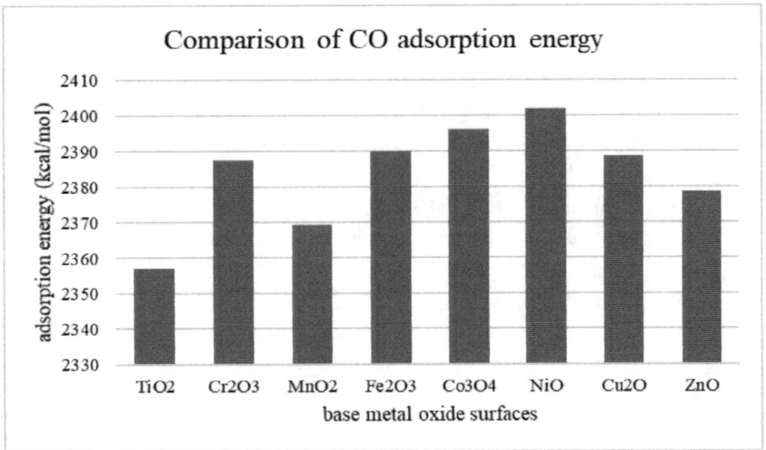

Figure 3. The comparison of CO adsorption energies among the base metal oxide surfaces.

The focus is that from the data we get that NiO surface is having the highest adsorption energy for CO molecules, while TiO_2 have the least CO adsorption energy. Here we can say that some base metal oxides could have better affinity to CO rather than the others, which could lead to higher CO oxidation activity. Some base metals having higher CO adsorption energy such as Cr_2O_3, Fe_2O_3, CO_3O_4, NiO, and Cu_2O would become a better active support for CO oxidation catalytic system rather than that the other base metal oxides, regarding that they are easier to adsorb CO on their surfaces. From this adsorption step it can be drawn that for system with better adsorption capability, it will lead to a better catalytic performance due to the efficiency of interphase interaction on the surface [3]. This trend could be brought to the consideration for NiO rather than other oxides with its strong adsorption energies that could led to a better catalytic performance. This oxide can be an optional support system for CO oxidation catalytic system combined with other conventional active system.

4. Conclusion

The conducted calculation giving result that some base metal oxides are having relatively higher adsorption energy for CO adsorption process on their solid surfaces system, while the rest base metal oxide tends to have lower adsorption energy. This result shows that NiO surface having the higher affinity to CO molecules, which could led to a better efficiency for the catalytic system. Then it could probably become potential support metal oxide for CO oxidation catalytic system which would bring the beneficial properties of this oxide, which are relatively cheaper and also lower working temperature for solid CO catalytic system.

Acknowledgement
Thanks to EFISON Lisantek who has providing the High Performance Computing (HPC) service for all computational processes in this research, and also thanks to the Directorate General of Higher Education, and also thanks to the Faculty of Science and Mathematics of Diponegoro University, who has funded this research through the 2020 research and technology funding scheme.

References

[1] Valechha D, Lokhande S, Klementova M, Subrt J, Rayalu S and Labhsetwar N 2011 *J. Mater. Chem.* **21** 3718
[2] Bruk L *et al.* 2018 *Nanomaterials* **8**
[3] Park Y, Kim S K, Pradhan D and Sohn Y 2014 *Chem. Eng. J.* **250** 25
[4] Qian K *et al.* 2009 *J. Mol. Catal. A Chem.* **306** 40
[5] Chen H T and Chang J G 2011 *J. Phys. Chem. C* **115** 14745
[6] Hsu L C, Tsai M K, Lu Y H and Chen H T 2013 *J. Phys. Chem. C* **117** 433
[7] Giannozzi P *et al.* 2009 *J. Phys. Condens. Matter* **21**
[8] Enkovaara J, Rostgaard C and Mortensen J J 2017 *J. Phys. Condensed Matter* **29** 465901
[9] Yang Z, He B, Lu Z and Hermansson K 2010 *J. Phys. Chem. C* **114** 4486
[10] Prasad R and Singh P 2012 *Catal. Rev.-Sci. Eng.* **54** 224
[11] Ramesh K, Chen L, Chen F, Liu Y, Wang Z and Han Y F 2008 *Catal. Today* **131** 477
[12] Xie X, Li Y, Liu Z Q, Haruta M and Shen W 2009 *Nature* **458** 746
[13] Yu Y, Takei T, Ohashi H, He H, Zhang X and Haruta M 2009 *J. Catal.* **267** 121
[14] Han S W, Kim D H, Jeong M G, Park K J and Kim Y D 2016 *Chem. Eng. J.* **283** 992
[15] Pillai U R and Deevi S 2006 *Appl. Catal. B Environ.* **64** 146

IOP Conf. Series: Materials Science and Engineering **959** (2020) 012005 doi:10.1088/1757-899X/959/1/012005

Self-assessment of chemistry laboratory basic skills using performance scoring rubrics at the chemistry teacher training

I B N Sudria*, I W Redhana, I W Suja and I N Suardana
Universitas Pendidikan Ganesha, Jalan Udayana 11 Singaraja Bali 81116 Indonesia

* E-mail: ibnsudria.undiksha@gmail.com

Abstract. This student self-assessment of chemistry laboratory basic skill performances is intended to overcome the heavy time-consuming and very intensive involvement of lecturer as the only performance assessor, promote and accelerate the skill improvement based on learner self-performance assessment. The study involved 26 students undertaking Basic Chemistry subject at Chemistry Education Department of Universitas Pendidikan Ganesha in Bali. The self-assessment applied the scoring rubric developed by Sudria and Sya'ban. Each item of the 22 assessment items applies five-scaled grading performance indicators as very good, good, sufficient, poor, and very poor. Every skill item was assessed at least three time along the semester and analyzed twice namely at the middle of the semester as data S1 and near the end of the semester as data S2. The same performance assessment was also conducted by a lecturer team as the laboratory work examination at the end of the course as reference data of S3. This study showed significant correlation of S1 and S2 toward S3 with score correlation amounts of 0.520 for r1-3 and 0.588 for r2-3 which were above the critical value of 0.374 at 95% confidence. Guidance and monitoring during laboratory work supported to the success of the self-assessment study.

1. Introduction

Stiggins [1,2] recommends the use of performance scoring rubrics to assess skills and products. Sudria [3,4] developed an analytic performance scoring rubric to assess basic Chemistry laboratory skills in teacher training education and also can be used for secondary school. The availability of the instrument to conduct skill performance assessment facilitates to figure progress along an exercise program of which the assessment results for certain time periods such at the middle and end of the training project can be compiled. But the use of the assessing rubrics of laboratory work skills only by lecturer/teacher in the assessment caused intensive activities and heavy time-consuming for the educator [3]. Meanwhile, prospective teachers must have sufficient performance in basic Chemistry laboratory skills and require the best skills training as needed for professional teaching. A reasonable strategy is needed to increase the feasibility of the use of scoring rubric in delivering performance assessment. Concerning the effectiveness of performance improvement based on self-performance assessment of students and reduction of lecturer workload in using performance assessing rubrics, self-assessment of performance is a prospective alternative solution in the assessment of basic chemistry laboratory skills for effective progress. Self-assessment with rubric assessment for basic laboratory works skills is needed either as group works or independent activities on chemistry in limited social situations such as the Covid-19 pandemic. Careless of chemical handling and apparatus setting was often observed for teacher training students in several laboratory work courses, even when they conduct field practice of chemistry teaching at secondary school, such as horizontal position of chemical filled dropping pipette, use of a same pipette for taking different liquids or solutions even without cleaning first of the pipette, untightened clamp/s on a stand such on filtering apparatus fitting, and placement of glass wares in a risky position. The careless of hazardous chemical handling and apparatus fitting due the poor laboratory work skills could raise endanger in laboratory works.

Chemistry laboratory work optimally apply basic skills which strongly supports science inquiry learning as one recommended for the scientific learning approach. Scientific method is the reference for creative and critical learning through investigation in the scientific learning approach. Well understanding of basic chemistry requires data supports collected with the use of laboratory basic skills in preparing quality of prerequisite conceptions for further chemical lessons. Mastering the basic skills will empower and accelerates further construction of science conceptions. Chemistry laboratory basic skills are also life skills in processing daily goods for daily demands and also a tool to develop thinking skills and creativity.

Performance assessment rubric facilitates authentic assessment, the appropriateness of the goal, target, method and instrument of skill and product assessment [1]. Learners who are assessed through performance scoring rubric are assigned a task, and then they perform the desired skills which are observed and recorded by assessors or via self-assessment [5] using the performance scoring rubric. "*A rubric is a rating scale that consists of ordered categories, together with descriptions of criteria that may include exemplars, which are used to sort student-produced responses into levels of achievement*" [6]. They stated further that "*rubrics as vehicles that can provide a useful mechanism to translate achievements into assessment terms*". Rubric has four aspects: criteria, examples, scale, and standards [7]. A*nalytical rubric* is used to assess specific learning aspects. Effective performance assessment requires same perception of scoring criteria of grading of performance quality indicators. The scoring criteria should be clearly socialized at the beginning of the assessment program.

Self-assessment is "*the involvement of students in identifying standards and/or criteria to apply to their work and making judgments about the extent to which they met these criteria and standards*" [8]. Student self-assessment involved the students in monitoring and making judgments about aspects of their own learning. Students should have opportunities to make adjustment in learning based on their self-performance assessment [9]. Self-assessment contributed to preferred feedback, more open and supportive learning environment, responsibility assignments expectations, proactive actions, and to develop a more interactive role [10]. Self-assessment is a key strategy to involve students in taking more responsibility for their own learning.

The same perception and clear understanding of quality gradation scoring criteria of performance quality indicators among learners and assessors before the administration of self-assessment program will lead to effective assessment conduct. This will effectively focus the learners and assessors on practicing the performance quality of the desired skills, makes it is easier to assign the related assessment tasks and feed back to the learners during practices, so administration of the performance assessment would be feasible. This paper showed the feasibility administration supported by significant score correlation between Chemistry laboratory basic skill performance assessment results recorded via student self-assessment and that assessed by the lecturer team through final examination of the basic skill performances at the end of the course.

2. Method

This correlation research on learning aims at supporting the validity of self-assessment of chemistry laboratory basic skill performance. The research involved one class of 26 students undertaking Basic Chemistry II at Chemistry Education Department of *Universitas Pendidikan Ganesha*. Situation of self-assessment were created due to it had not normally observed. Guidance of conducting self-assessments were given for every laboratory session. Assignment to conduct administrated individual student self-assessment for the whole laboratory skills were conducted in two sessions. First session was conducted at the middle semester involving topics of chemical kinetics (P1) and chemical equilibrium (P2) which resulted average score data of S1. Second session was conducted near the end of semester involving topics of solution (P3) and redox (P4) which resulted average score data S2. As a reference of correlation

IOP Conf. Series: Materials Science and Engineering **959** (2020) 012005 doi:10.1088/1757-899X/959/1/012005

method, the same performances of the whole specific skills were assessed by a lecturer team at the end of the course as final exam of the accommodated laboratory basic skills.

Table 1. The organization of specific skills accommodated in the units

No.	Basic skills	Events			
		1st Half smt.		2nd Half smt.	
		P1	P2	P3	P4
1	Taking pure solid substances from a storing container	√	√	√	√
2	Taking pure liquid solution from its container	√	√	√	√
3	Taking a liquid or a solution from a container provided for whole learners in a class	√	√	√	√
4	Demonstrating laboratory equipment and use of chemical safely	√	√	√	√
5	Using fume hood	√	√	√	√
6	Lighting Bunsen burner		√		√
7	Handling and disposing chemical waste safely	√	√	√	√
8	Weigh chemicals	√	√	√	√
9	Using volumetric equipment	√	√	√	
10	Decanting a liquid/solution from a mixture		√	√	√
11	Using a dropping pipette to transfer a liquid/solution	√	√	√	√
12	Preparing a certain concentration of a solution from a pure substance	√	√	√	√
13	Preparing a certain concentration of a solution by dilution	√	√	√	
14	Using a stand to support the equipment / glassware		√	√	√
15	Setting filtering tools	√	√	√	√
16	Filtering liquid/solution phase from its containing mixture		√	√	
17	Flowing gas which is assisted by heating to a liquid solvent to make the its solution		√	√	√
18	Cleaning glassware	√	√	√	√
19	Observing a cloud/colloidal dispersion as a chemical reaction product in a solution	√		√	√
20	Observing color change especially for color intensity due to chemical reaction		√	√	√
21	Identifying the chemical reaction on heating which characterized by gas formation		√	√	√
22	Identifying a chemical product by smelling (wafting) safely		√	√	√

2.1. Materials

The laboratory basic skills performance instrument includes 22 items. Each item assesses a basic skill. The whole sessions of self-assessments of the basic skills involved two sessions. Each session contained the whole items. The first session involved two topics revealing the whole basic skills for self-assessment session I at the middle of the semester and the other two topics revealing the whole basic skills for self-assessment session II at about the end of the semester as presented in table 1. Several items did not reveal for each topic with respect to the topic characteristics.

2.2. Instruments

This research used instrument of the analytical performance scoring rubric developed by Sudria [3]. The laboratory basic skills performance instrument originally included 19 items with the Cronbach's alpha reliability coefficient of 0.880 [4], then it extended up to consisted of 22 items in this study. The reliability of the used instrument in this study is also as a result to support the feasibility of this conducted self-assessment. The reliability is assigned by Pearson correlation coefficients between the scores on the laboratory basic skills achieved by students via respectively self-assessment collected at the middle and end of the semester toward the scores achieved at final exam of the laboratory basic skill performances scored by the lecturers.

Each item measured a specific aspect of chemistry laboratory basic skill. The item involved 5 (five) grading quality categories of the skill performance to be chosen which matched to the observed performance indicator. The quality grading scores were 5 for very good, 4 for good, 3 for satisfactory, 2 for poor, and 1 for very poor observed performance. The learners who do-self-assessment or assessors just put check in the cell of the observed performance indicator category for the observed student. One example of the self-assessment rubric item for filtering skill is shown as table 2.

Table 2. Format of performance scoring rubrics with an example item for filtering skill

Score	Quality grading indicators of skill/product performances	Observed performance			
Practice		1st	2nd	3rd	…
Filtering					
5	Use stand/support safely, fold filter paper in right way without leaking, avoid the increase of pressure inside the filtrate container (there is a way air to come out).				
4	Use a stand and fold/fix filter paper not so properly but without leaking, avoid the increase of pressure the filtrate container				
3	Without stand/support, no leaking on the filter paper, avoid increase of pressure inside the filtrate container				
2	Do not avoid the increase of pressure inside the filtrate container (there is no way air to come out).				
1	Use leaking filter paper **or** loosely sets the stand				

Each specific skill was observed several times through self-assessment along the course revealing it. Mapping the assessed skills along the semester is showed at table 1. All specific aspects were also observed once as final exam at the end of the course.

2.3. Procedure

The procedure used in this study involved several steps. At the beginning of the performance assessment program, the lecturer team socialized and discussed the scoring rubric criteria to subjects of this study, students undertook the Foundation Chemistry II subject and also explained the assessment tasks included in the worksheet for the whole of the course in which self-assessment units were organized following the topics/sub-topics stated in the table 1. Then, student self-assessments were assigned to the students by applying rubric items of the scoring rubric units for topics/sub-topics of chemical kinetics (unit P1), chemical equilibrium (unit P2), solutions (unit P3), and redox (unit P4). Lecturer team also assessed each specific skill accommodated in each unit for several samples of the students throughout program conduct with special focus on well and bad skill performances observed. The data were used by lecturer for giving reflection of the session and for improvement of the skills in further exercises. At the end of the course, the lecturer team conducted assessment for the whole specific skills once as final exam of the practices of the basic laboratory skills as scores S3 (as a reference). Average self-assessment of the basic skill performance scores of each student for first half semester (P1 and P2) as S2 scores and for second half semester (P3 and P4) as S2 scores were calculated. Finally, the correlation coefficient among student self-assessment performance scores (S1 and S2) and S3 scores which were assigned by lecture team were determined using product moment method as $r_{1\text{-}3}$ and $r_{2\text{-}4}$ to conclude the validity of the student self-assessment. At the end of the assessment program, the responses of subjects to this performance self-assessment study also recorded using a questioner. Feasibility of the student performance self-assessment administration confirmed by the magnitude of correlation coefficient found (empirical validity), recorded student responses to the self-assessment program, and improvement of situation of the self-assessment administration along the course.

3. Results

3.1. Correlation of performance scores assessed by student himself and by lecturers
Average scores recorded through student self-assessment in the first half of semester (S1), second half of semester (S2) and the performance final exam scores assigned by lecturer (S3) shown in table 3. Correlation coefficients were 0.520 for $r_{1\text{-}3}$ and 0.588 for $r_{2\text{-}3}$ which were above the critical value (>0.374) at 95% confidence.

3.2. Learner comments on performance self-assessment
The subjects of this student chemistry laboratory basic skill performance self-assessment program gave responses as shown in the Table 3. The subject responses in five categories namely strongly agreed (SA), agreed (A), moderate (M), disagreed (D), and strongly disagreed (SD) were shown in table 4.

Majority learners (about 90%) gave positive responses to involve in self performance assessment of the basic skills. This positive response showed that the benefits of self-assessment on student learning mentioned by many authors were recognized by students.

4. Discussion
The self-assessment was difficult and slow implementation at the beginning. In the region of this conducted research, the students are rarely involved in laboratory work in secondary schools. It takes time to understand and remember the adequate five options of descriptive qualitative grading indicators of skill performance such as for filtering skill showed in table 2. After that, the steady improvement on the laboratory basic skills took place along the semester. The significant correlations of student performance scores by self-assessment either collected at the middle and near the end of the semester toward student performance scores on final performance exam assessed by the lecture team. The average performance scores were steadily in good category since the middle of the semester. The students appreciated the self-assessment program. Those indicated the feasibility of this conduction of laboratory basic skill self-assessment.

The performance scoring rubric item states quality levels of performance indicator descriptions for one dimension of the laboratory basic skill. The options of the graded quality indicator descriptions range from the best to the worst to be selected according to the observed/recognized skill of the assessed learner. Using this scoring rubric with qualitative description options is more objective than giving quantitative score. The results of the assessment in the form of a quality description of the performance provide direct input if there is an inappropriate performance of the skill, then for improvement. The selected option automatically and consistently gave a quantitative score of the observed skill as long as it is chosen objectively as the skill quality performed. On the other hand, skills assessment without the support of an assessment rubric tends to be inconsistent, especially by learners, so it is less objective

IOP Conf. Series: Materials Science and Engineering **959** (2020) 012005 doi:10.1088/1757-899X/959/1/012005

and does not provide direct information about inappropriate skills that need to be improved. Each skill dimension was assessed at least three times throughout the scheduled semester according to the topic / sub-topic of the laboratory work that accommodates it. Involvement of students in co-scheduling self-assessment basic skills practices makes self-assessment more feasible.

Table 3. Average scores of chemistry laboratory basic skill performance.

Subjects (No.)	Average scores		
	Student self-assessment		Lecturer conducted assessment at end semester
	1st half Smt.	2nd half Smt.	
1	4.2	4.3	4.7
2	4.0	4.3	4.7
3	3.8	4.3	4.5
4	4.4	4.4	4.7
5	4.5	4.9	4.6
6	4.2	4.8	4.6
7	4.4	4.3	4.1
8	4.2	4.3	4.4
9	4.6	4.6	4.4
10	4.0	4.5	4.3
11	4.0	4.2	4.0
12	3.6	4.0	4.0
13	3.9	4.4	4.3
14	4.6	4.5	4.5
15	3.7	4.3	4.5
16	3.5	4.3	4.2
17	3.9	4.6	4.3
18	4.1	4.6	4.3
19	4.5	4.6	4.6
20	4.2	4.6	4.4
21	3.9	4.0	4.0
22	3.7	4.0	3.9
23	4.1	4.0	4.5
24	3.7	4.2	3.8
25	4.4	4.8	4.5
26	4.2	4.7	4.4
Average	4.09	4.40	4.35

Table 4. Several learner responses to self-performance assessment administration of Chemistry laboratory basic skills

IOP Conf. Series: Materials Science and Engineering **959** (2020) 012005 doi:10.1088/1757-899X/959/1/012005

No	Response descriptions	Number of responses				
		SA	A	M	D	SD
1	The rubric stated clear quality performance indicator grading for learners and assessors	18	8	0	0	0
2	Assigning descriptive observation of performance quality grading indicators ware more objective then giving a quantitative score	12	10	4	0	0
3	Confuse and very slow to apply scoring rubric items at the beginning self-performance assessment program. It was not a problems for the second session and further	5	20	1	0	0
4	Criteria of scoring rubric must be properly socialized at the beginning of the performance assessment program	15	9	2	0	0
5	Practiced tasks must be socialized properly at the beginning of the performance assessment program	20	6	0	0	0
6	Student self-performance assessment stimulated optimal improvement on student learning	13	11	2	0	0
7	Performance scoring rubric helped learners to recognize benchmark of the competence	11	12	3	0	0
8	Students agreed to assess Chemistry laboratory skills through self-assessment of performance	6	12	7	1	0
9	Mastering Chemistry laboratory basic skills improved self-confident of the students to properly conduct laboratory works	8	16	2	0	0
Average responses for the whole aspects		12.0	11.6	2.3	0.1	0.0
%		46.2	44.4	9.0	0.4	0.0

The socialization of the scoring rubric at the beginning of the semester and the clear self-assessment procedure had many impacts contributing to the success. The rubric provides a reference for the dimensions of the expected skill aspects and could challenge the best performances. The involvement of students in scheduling of assessed skills training throughout the semester contribute to practice with their own initiative which tends to promote commitment to improving skills independently. Guidance at the beginning and accompaniment facilities especially with demonstration of basic laboratory skills by lecturer team when students ask for help along exercising periods accelerated the development of the basic skills as well as the capability to conduct self-assessment. The understanding of the rubric assessment criteria improved along with self-assessment of their performance. Those potentials synergistically empower self-performance assessments to accelerate the skill improvement. Significant positive correlation between the average score of self-assessment and the score given by the lecturer team in accordance with student appreciation.

This study agreed to many reported studies [11,12]. Benefits of self-assessment [5] were reemphasized such as contribution to improve the students' understanding of assessment, feelings of control over one's own learning, self-worth, self-motivated to independent learners, honest with themselves when setting targeted skills, a greater sense of ownership of their work, demanding enthusiastic to learning, the potential to help students to develop important professional skills using higher-order thinking, improve ability to focus on key elements of an assignment, increase effectiveness in identifying strengths and weaknesses in their work, provide students with an opportunity to participate in the assessment process which increases their motivation, force student assessors to have a more conscious understanding of the processes involved in the learning activity, and to be used as a means of producing formative learning oriented feedback to complete the learning cycle and encourage the ongoing development of skills. Self-assessment of chemistry basic skills would be useful in limited social distancing such on the covid-19 pandemic situation.

5. Conclusion

Learners self-assessment on their performance of the Chemistry laboratory basic skills were feasible to be conducted for Basic Chemistry subject at teacher training education. The learners recognized the benefits of self-performance assessment of Chemistry laboratory basic skills in chemistry learning.

Acknowledgments

Our thanks go to research institute of Ganesha University of Education for funding support, and also to teaching staffs, laboratory technicians, as well learners at Chemistry Teacher Training Education at Ganesha University of Education for their involvement and supports.

References

[1] Stiggins R J 1994 *Student-Centered Clasroom assessment* (New York: Maxwell Macmillan International)
[2] van Helvoot A A 2010 *J. Inf. Lit.* **4** 22
[3] Sudria I B N and Sya'ban S 2008 *J. Penelitian Pengembangan Pendidikan* **2** 30
[4] Sudria I B N and Siregar M 2009 *J. Pendidikan Pengajaran* **42** 222
[5] Andrade H 2007 *Educ. Leadersh.* **65** 60
[6] Shafer W D, Swanson G, Bene N and Newberry G *Appl. Meas. Educ.* **14** 151
[7] Kuo S A 2007 *Educ. Research. J.* **22** 179
[8] McDonald B and Boud D 2003 *Assess. Educ.* **10**
[9] Ndoye A 2017 *Int. J. Teach. Learn. High. Educ.* **29** 255
[10] McMillan J H and Jessica H 2008 *Educ. Horiz.* **87** 40
[11] Papanthymou A and Darra D 2019 *J. Educ. Learn.* **8** 48

IOP Conf. Series: Materials Science and Engineering **959** (2020) 012006 doi:10.1088/1757-899X/959/1/012006

The effectiveness of online learning by EdPuzzle in polymer materials on students' problem-solving skills

Giyanto[1,2,*], L Heliawaty[1] and B Rubini[1]
[1] Department of Science Education, Graduate School of Pakuan University Bogor, Indonesia
[2] SMK Negeri 1 Gegerbitung Sukabumi

* E-mail: giyanto081578663380@gmail.com

Abstract. This study aims to produce EdPuzzle content in polymer materials and examine its effectiveness on students' problem-solving skills. The method used in this research is experimental research, employing a pre-test post-test one-group design through online learning. The subject of this study is 70 tenth grade students. Data were collected using students' problem-solving skills tests and analysis of the learning process to determine the effectiveness of online learning. The data obtained were then analyzed using quantitative descriptive. The results showed that learning using EdPuzzle is statistically very significant in improving students' problem-solving skills and is quite effective in online learning. The study concludes that EdPuzzle is quite effective to improve students' problem-solving skills which are characterized by an increase in the score of pre-test to post-test, with an N-gain of 63%. The implementation of online learning EdPuzzle is quite effective with an effectiveness score reaching 91%.

1. Introduction

In this digital era, there are learning trends that are increasingly developing. One such trend is that learning can be taken place anytime anywhere. Previously, learning was identical to face-to-face in class and was limited to certain hours, now learning can be done anywhere and anytime. Various applications are available that allow learning to be carried out not only in schools. Students can learn anywhere and anytime using the applications found on their mobile phones. Some applications provide two-way learning features, which allows direct interaction between teachers and students. Learning trends like this continue to develop as the digital world grows rapidly. This learning trend is characterized by e-learning and other applications related to learning [1, 2, 3].

Flipped Learning is one of the learning trends that makes students learn independently at home before entering class in the form of presentation slides, videos, or e-modules provided by the teacher. Nowadays, many countries have implemented the flipped learning method as a method to make the learning process more enjoyable. By student-centered learning, the activities in learning by flipped learning involve active student participation and make students enjoy the learning process. Flipped Learning requires students to watch videos or read the material before entering class. Students can access videos from the internet, so this concept can also be used for distance learning processes that do not require certain space and time. By watching videos related to the lesson, students gain prior knowledge before starting lessons in the class, both conventional and online classes.

One of the flipped learning types is the use of EdPuzzle applications. EdPuzzle is a platform that can help teachers carry out interactive video learning between teachers and students. One of EdPuzzle's strengths is that it helps students learn through video following their respective learning speed and ability [4, 5, 6]. By using EdPuzzle, the teacher can collect data related to student interactions with the video, for example, the number of times that students watch, answer questions, or provide responses. The video can be edited directly with the video clipping feature to take just the part that is needed. Besides, the teacher can also insert their voice into the video and also slip the questions between the videos that students will watch to test their understanding of the video they watch [7]. This is what makes learning with video more interactive. Thus, EdPuzzle can be used as alternative learning in the digital age.

In the digital age where problems are increasingly complex now, problem-solving skills for students become very important. This is because every human being is inseparable from problems. These problems range from personality problems, learning problems, and popular problems to the attention of the world. Plastic waste is one of the problems that are currently a serious problem in the world and especially in Indonesia. In Chemistry subject, material about plastics is studied in the topic of polymer chemistry. Learning chemistry in polymer material is expected to be able to develop students' problem-solving skills so that the impact of plastic waste problems can be reduced.

The students' problem-solving skills in various schools are still low. It is found that students have difficulty in determining the form of the solution of a problem. This was revealed by the low number of students in waste management both in schools and the surrounding environment. Based on the Focus Group Discussion (FGD) conducted by MGMP chemistry teachers in Sukabumi Regency, students are still ignorant of waste segregation and care in disposing of garbage, especially plastic waste. This condition is alarming, considering that students have received material about polymers in which there is material about plastic polymers and their impact on the environment. Learning about polymers should lead students to have an awareness of the environment and be able to develop their problem-solving skills.

Current learning must be able to encourage students to develop their problem-solving skills as research conducted by Wang [8], Gwo-Jen & Pei-Ying [9], and D'Agostino [10] and follow the trends of 21st-century learning. Problem-solving skills are needed so that students can face all problems when engaging in society. The 21st-century learning trends in the form of learning based on flipped learning are needed today with various advantages such as increased creativity and critical thinking [11, 12] as well as independence and achievement [5, 13, 14]. Learning can be done online based on existing applications, such as gamification of learning with Kahoot, e-module, quizzes, EdPuzzle, and others [15, 16].

The teacher must quickly adapt to various digital learning media through various applications. In this COVID-19 pandemic condition, there is a considerable gap between learning in various schools with the ideal conditions of learning following the demands of 21st-century learning. There are still relatively few teachers who use flipped learning in learning. If this condition is not immediately resolved by implementing and testing the effectiveness of learning with a digital platform (such as EdPuzzle), it will have a wide impact on the low quality of education. So, it is very important to immediately carry out digital-based innovative learning following the conditions and needs of students.

The purpose of this research is to create EdPuzzle content and examine EdPuzzle online learning on students' problem-solving skills. The topic of learning is limited to polymer chemistry. Analysis of students' problem-solving skills is limited to a video file about the problem of plastic waste in Indonesia published by the Indonesian Institute of Sciences (LIPI). Also, this study measures the effectiveness of online learning by using EdPuzzle that has been produced as an impact of the learning process due to the COVID-19 pandemic.

2. Methods

This study was designed using an experimental research method that refers to Sugiyono [17]. The experiment aims to examine the impact of using EdPuzzle on students' problem-solving skills. Student selection is done by purposive sampling referring to Arikunto [18]. Data is collected through tests to measure students' problem-solving skills. The test of problem-solving skills is in the form of 10 essay questions, by giving score 10 for each question, thus, the maximum score is 100. The test questions are tested with a one-group pretest-posttest design technique that refers to Fraenkel *et al.* [19]. The schematic is shown in table 1.

Table 1. The scheme of the group pretest-postest design.

Pretest	Treatment	Posttest
T_1	X	T_2

T1 as a pretest is done before students are treated (Treatment); X as a treatment is given through EdPuzzle learning, and T2 as a posttest is done after the sample is given treatment. Through this pretest-posttest, students' problem-solving skills are measured before and after treatment.

The subject of the study was 70 students in grade 10. Tests are carried out online using Google Form by accommodating 5 aspects of problem-solving skills indicators based on categories from Rahman [20] and Polya [21], namely aspects of defining a problem, examining a problem, planning a solution, implementing a solution made and evaluating. Problem-solving skills data in the form of grades are then analyzed by determining the N-Gain based on the Hake [22] category, as shown in table 2.

Table 2. The N-Gain category.

N-Gain (%)	Interpretation
$(g) \geq 70$	high
$70 > (g) \geq 30$	medium
$(g) < 30$	low

The effectiveness of EdPuzzle online learning is calculated based on feedback from the number of students who watched the video and completed the assignment compared to the total number of students. The data obtained were then analyzed using quantitative descriptive.

The research procedure includes the planning stage, the implementation stage, and the final stage. The planning stage is curriculum analysis (analysis of core competencies, basic competencies, learning indicators, and learning objectives), preparing the completeness of learning instruments and research instruments. The implementation stage includes the pretest, the implementation of learning using EdPuzzle, then ends with a post-test to measure students' problem-solving skills. While the final stage is by conducting data analysis, discussion, and drawing research conclusions.

Before testing the significant difference between the pretest and posttest, it is necessary to do a data normality test. In statistics, this is important because it is the main requirement in parametric statistics. The data normality was tested through the Kolmogorov Smirnov test with SPSS version 20. The normality test with the Kolmogorov Smirnov test was chosen because the number of samples was more than 50. The normality test aims to find out whether the residual score (not on each research variable) is normally distributed or not. The interpretation of the Kolmogorov Smirnov normality test is as follows. If the significance score is in the column of Asymp. Sig (2-tailed) or probability > 0.05, so the data is said to be normally distributed. If the residual score is normally distributed then it is followed by a paired sample t-test, but if the residual score is not normally distributed, the significant difference test is performed on a non-parametric statistical test, namely Wilcoxon test [23].

To find out whether there is a significant difference in the average between the results of the pretest and posttest of problem-solving skills, it will be tested through paired sample t-test. This test is used for two sample/paired data. A paired sample t-test was processed using SPSS version 20. The basis of the research decision is if the score of Sig. (2-tailed) $p < 0.05$, then there is a significant difference between the score of problem-solving skills in the pretest and posttest data. If the Sig. (2-tailed) $p > 0.05$, there is no significant difference between the scores in the pretest and posttest data. A paired sample t-test is part of parametric statistics. Therefore, as a rule in parametric statistics, the research data must be normally distributed.

3. Result and discussion

EdPuzzle content that has been created contains learning videos and questions that have been inserted in it to ensure that students have watched and understood the contents of the video. The front page of EdPuzzle is shown in figure 1. The EdPuzzle link can be accessed at https://EdPuzzle.com/media/5eb647ea124f433ee8f6e619.

Figure 1. EdPuzzle homepage.

The results of the study in the form of data analysis show an increase in the pretest to posttest with an N-Gain of 63%, and the mean increasing from 31 to 74 after learning using EdPuzzle, as shown in table 3.

Before the t-test is performed, the data normality test is first tested, based on the SPSS data output. It is known that the significance score of Asymp. Sig. (2-tailed) of 0.923 is greater than 0.05, then according to the basis of decision making in the Kolmogorov-Smirnov normality test, it can be concluded that the data are normally distributed. The significance test of the difference between pre-test and post-test with paired test t-test using SPSS version 20 software shows the number of 0,000, which means it is very significant, as shown in table 4.

Table 3. Result of the pre-postest analysis of problem-solving skills.

EdPuzzle data	*Pretest*	*Posttest*
Number of students	70	70
Lowest score	10	50
Highest score	70	100
Mean (average score)	31	74
% N-Gain	63 % (medium)	

Table 4. The result of the normality test and t-test

Data	Sig. score	Interpretation
Pre-Posttest	Normality test = 0.923	Data distributed normally
	Paired sample t-test = 0.00	Very significant

Table 5. The result of pre-posttest and posttest analysis on problem solving skills.

Indicators of problem-solving skills	% N-Gain	Criteria
Defining problems	70	High
Checking problems	72	High
Planning solution	53	Medium
Implementing solution	60	Medium
Evaluating	66	Medium

As for the results of the pre-test–post-test analysis of each indicator, it was found that the problem-solving skills in the aspects of defining and checking problems with an N-Gain of 70% and 72% (high category), while the aspect of planning a solution got the lowest score with an N-gain of 53% (medium category), as presented in table 5.

From the research conducted, it was found that an increase in the pre-posttest score with an N-gain of 63% (medium category). This proves that learning with flipped learning with EdPuzzle can improve students' problem-solving skills. This finding is strengthened by t-test = 0.000 that this learning is very significant in improving students' problem-solving skills in polymer materials. Increased problem-solving skills are caused by the learning video content that is displayed, so it can stimulate students to think at a higher level. In this polymer chemistry learning, a video about the problem of waste in Indonesia which is so severe. The video presents facts about the number and constraints in waste management. This condition causes students to feel challenged to participate in solving these problems starting from the lowest level and themselves. Some research conducted by Lewis [24] and Sarmantayev *et al.* [25] stated that stimulation in the form of videos, narratives, and presentation slides plays an important role in generating students' metacognition skills.

The increase in the score of problem-solving skills that is very significant is also caused by this learning carried out personally. In face-to-face learning that is carried out in class, the teacher faces many students, so in this digital age, students can carry out their learning process personally through gadgets by utilizing online learning. This personal learning has the advantage of providing opportunities for students to be able to focus on learning without any distractions from others or in other words this learning can increase students' learning independence [16, 26, 27]. For this reason, learning with EdPuzzle can be used as alternative learning in the Covid-19 pandemic. Although learning is personal, it does not close the space for discussion among students online.

The high N-gain in the aspect of defining and checking problems (high category) shows that students master problems. This is due to the interactive activities of watching videos and answering questions on the sidelines of the video that make students truly understand the concepts discussed. The video also presented facts and problems that are following the conditions of their surroundings/contextual. So, EdPuzzle can improve students' concept mastery. This finding is supported by many researchers [5, 9, 13, 14] that flipped learning can improve students' concept mastery and achievements. This ability is the main asset for the skills to plan problem-solving.

This study also found the fact that among the five indicators of problem-solving skills, the aspect of planning problem-solving got the smallest N-Gain score (53%). This is supported by data during the learning process. When students answer questions on the sidelines of the video, students are not quite right in ranking the priority scale for handling plastic waste. Students answer only with short answers, without breaking down into systematic solution plans. An example of students' answers that are not systematic in explaining is shown in figure 2.

However, the following processes make students more familiar with the context of the problem. This is indicated by the relatively higher N-Gain score in the aspect of implementing the solution (60%) and evaluating the solution (66%). The high score in both aspects is caused during learning, students are led to think critically, so they can implement and evaluate the solutions offered [8, 10, 28].

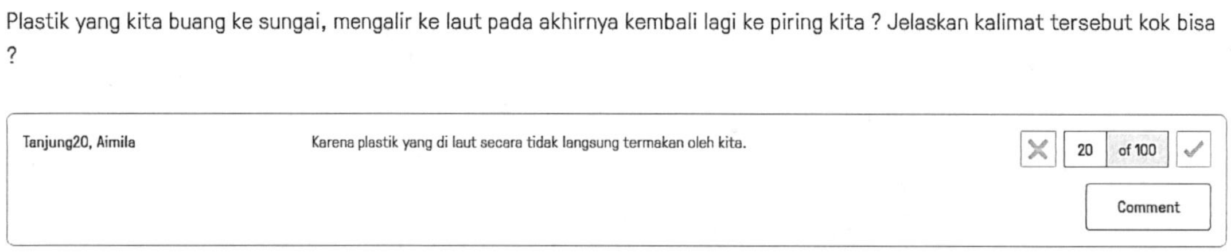

Figure 2. An example of students' responses (not systematic).

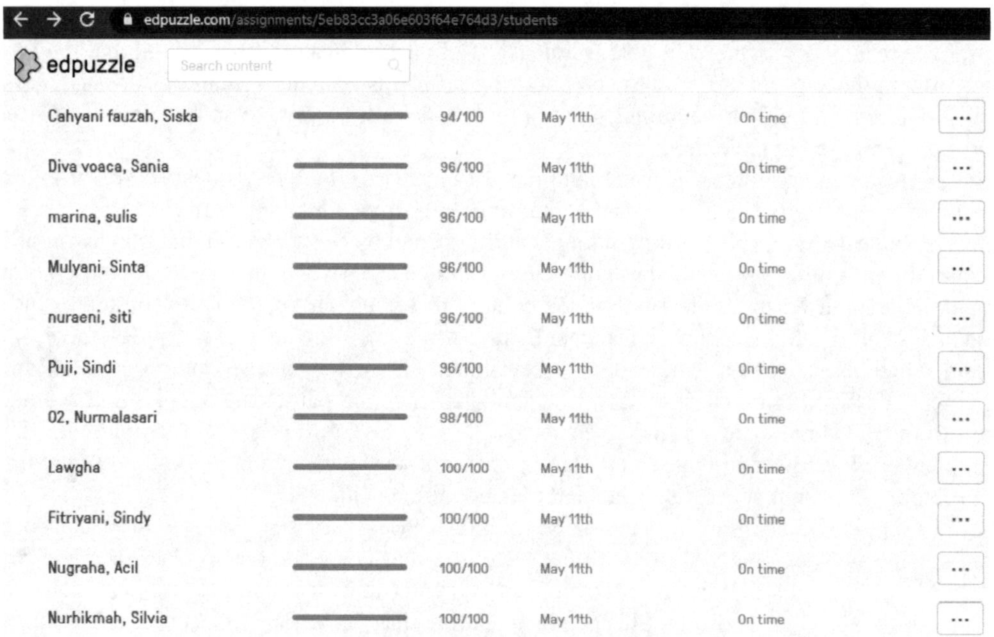

Figure 3. Students' activities monitored through grade book.

Students' activities are monitored through the grade book, EdPuzzle feature that contains scores, videos that have been watched, and questions that have been answered, as shown in figure 3. Analysis of the data in the grade book found that many students answered not according to what was wanted in the video show. Students answer according to their perceptions, relating to the solutions they offer. This shows that learning with EdPuzzle can develop students' critical and creative thinking [11, 12].

From the grade book data, the effectiveness of EdPuzzle can be determined in online learning by 91% which shows that students are enthusiastic about implementing the learning process, conducting feedback, and evaluating themselves in the form of tests. The high effectiveness cannot be separated from the advantages of choosing EdPuzzle in learning for students, namely user friendly, which is proven at the time of installation that the teacher only shares video tutorials through the WhatsApp group and students can carry out this process themselves. Another advantage is that this application is relatively lightweight which does not require a lot of android memory. So that even students with low android specifications do not experience problems.

For the teacher, as an administrator, EdPuzzle has some advantages: 1) this application is free; 2) this application can change any video to flip according to learning needs; 3) the video used can be from resources from others who have posted to YouTube, etc.; 4) we can upload our video. Besides, the teacher can see the progress of the percentage of completion, so it is known who has finished watching the video and embed questions throughout the video. The video will stop automatically and move forward after someone responds. The teacher can provide feedback to be displayed after the response is sent so that the sender receives immediate feedback. This makes learning more interactive [15, 29, 30].

Strengthening the last aspect of problem-solving, namely evaluating solutions, is carried out with online face-to-face activities with the zoom meeting application, as shown in figure 4.

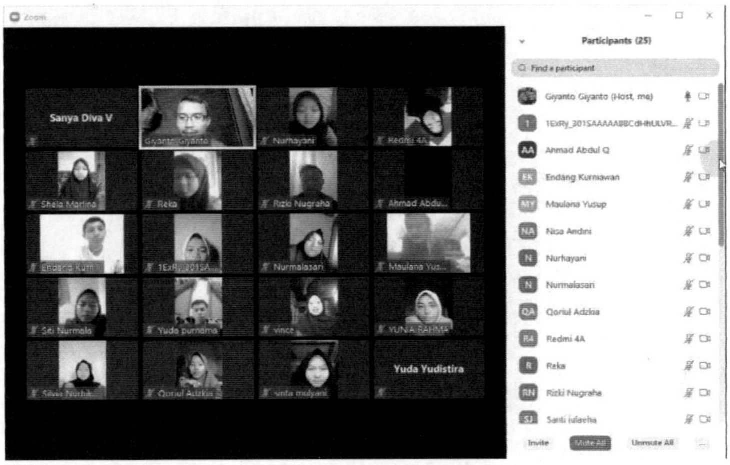

Figure 4. Strengthening activities; aspect of evaluating solutions.

At this stage, students brainstorm among other students and the teacher roles as the facilitator. Some students convey ideas related to problems, plan solutions, and implement them. The obstacles encountered were revealed, then other students responded trying to solve various obstacles that were conveyed by other students. This results in a discussion among students. This is done to reduce the disadvantages of using EdPuzzle, namely the lack of discussion during learning so that some studies show the ineffectiveness of flipped learning in learning as research by Geist [31].

From the explanation given, it is clear that the use of EdPuzzle can significantly improve students' problem-solving skills in polymer chemistry. This is confirmed by the findings of Wang [8] and Gwo-Jen and Pei-Ying [9]. EdPuzzle is also quite effective in online learning with various advantages and disadvantages. In the condition of the COVID-19 pandemic and the rapid development of information technology, EdPuzzle can be chosen as an alternative learning media that is effective and efficient.

4. Conclusions

Flipped learning using EdPuzzle media effectively improves students' problem-solving skills. The five aspects of problem-solving skills have an N-Gain percentage in the medium-high category. Overall, the N-Gain pretest-posttest achievement was 63% and in the medium category. The level of significance statistically with the Sig. (2-tailed) = 0.00 less than 0.05 which means that learning using EdPuzzle can increase the ability of the problem, is very significant. Online learning with EdPuzzle media is also quite effective, based on students' feedback in doing assignments by 91%.

References

[1] Latip A 2019 *Pembelajaran di era digital* URL: https://www.kompasiana.com/altip/5d11d195097f3668d07f3b82/tren-pembelajaran-di-era-digital?page=2. Accessed on 14 May 2020
[2] Costa B, Seok S, and Kinsell C 2019 Mobile game-based learning *Advanced Methodologies and Technologies in Modern Education Delivery* pp 809–824
[3] Xu D, Zhang M, Brett E and Hung S J 2019 Learning anytime, anywhere: a spatio-temporal analysis for online learning *Interactive Learning Environments* pp 1–15
[4] Kay R and Dermott K 2019 Flipped vs. Traditional Classrooms in High School Chemistry: A Case for Emphasizing Quality of Implementation *Soc. for Information Tech. & Teacher Edu. Int. Conf.* pp 2299–2305
[5] Lin L and Hung I K 2019 *Educ. Tech. Research Dev* **67** 1573
[6] Andujar A, Ramiro S and Martínez M S C 2020 *Sustainability* **12** 1
[7] Song Y, Jong M S, Chang M and Chen W 2017 *Educ. Tech. & Soc.* **20** 180
[8] Wang F H 2017 *Comp. & Educ.* **114** 79
[9] Gwo-Jen H and Pei-Ying C 2019 *J. Interact. Learn. Env.* 1
[10] Agostino D and Alfred T 2019 *SoTL Commons Conf.* **67**
[11] Al-Zahrani A M 2015 *British J. of Educ. Technol.* **46** 1133
[12] Rodríguez, Chueca J, Molina-García A, García-Aranda C, Pérez J and Rodríguez E 2020 *Env. Educ. Res.* **26** 238
[13] Bhagat K K, Chang C N and Chang C Y 2016 *J. of Educ. Tech. & Soc.* **19** 134
[14] Hwang G J and Chiu-Lin L 2017 *J. of Educ. Technol. & Soc.* **20** 184
[15] Mischel L J 2019 *Manag. Teach. Rev.* **4** 283
[16] Silverajah V G and Govindaraj A 2018 *Proc. the 10th Int. Conf. Educ. Technol. Comp.* **259**
[17] Sugiyono 2017 *Metode Penelitian Kualitatif, Kuantitaif dan R & D* (Bandung: Alfabeta)
[18] Arikunto S 2017 *Pengembangan instrumen penelitian dan penilaian program* (Yogyakarta: Pustaka Pelajar)
[19] Fraenkel J R and Helen H H 2012 *How To Design And Evaluate Research In Education* (New York: McGraw Hill)
[20] Rahman M 2019 *Asian J. Interdiscip. Res.* **2** 64
[21] Polya G 1973 *How To Solve It* (New Jersey: Princeton University Press)
[22] Hake R R 1999 *Analyzing change/gain scores* Unpublished [online] URL: http://www. physics. indiana. edu/~ sdi/AnalyzingChange-Gain.pdf
[23] Singh N U, Roy A, Tripathi A K and Kendall's W 2013 Non Parametric Tests: Hands on SPSS *ICAR Research Complex for NEH Region Umiam Meghalaya*
[24] Lewis H 2019 Supporting the development of young children's metacognition through the use of video-stimulated reflective dialogue *Early Child Development and Care* **189** 1842-1858
[25] Sarmantayev A S, Ishanov P, Sadykov K, Abdrakhmanov Z, Imanbetov A and Iskakov M 2020 *Space Cul. India* **7** 255
[26] Jensen J L, Kummer T A and Godoy P D 2015 *CBE-Life Sci. Educ.* **14**
[27] Tsai Y R 2019 *Comp. Assisted Lang. Learn.* DOI: 10.1080/09588221.2019.1650779
[28] Cáceres M, Nussbaum M, González F and Gardulski V 2019 *Interact. Lear. Env.* DOI: 10.1080/10494820.2019.1619595

IOP Conf. Series: Materials Science and Engineering **959** (2020) 012006 doi:10.1088/1757-899X/959/1/012006

[29] Jimenez J M, Penichet-Tomás A, Pueo B and Carbonell-Martinez J A 2017 *Proc. 9th Int. Conf. Educ. New Learn. Technol.* pp 3927–3933

[30] Palma L O, Tobías P J B, Prieto M C, León F J M and Ruiz Á A M 2018 Use of Kahoot and EdPuzzle by smartphone in the classroom: The design of a methodological proposal *Learning Technology for Education Challenges* eds L Uden, G Sanchez, D Liberona and S R Gonzalez (Springer International Publishing) pp 37–47

[31] Geist M J, Larimore D, Rawiszer H and Al Sager A W 2015 *Nurs. Educ. Perspect.* **36** 114

IOP Conf. Series: Materials Science and Engineering **959** (2020) 012007 doi:10.1088/1757-899X/959/1/012007

Chemically induced demulsification of water in crude oil emulsion from East Kalimantan oil fields

A P Gustianthy, D Febriantini, B Purnomo*, S U Muzayanha and R Noverdi
Downstream Research & Technology Innovation, Innovation & New Ventures
PT Pertamina (Persero)
Jl. Raya Bekasi KM 20, Pulo Gadung, Jakarta 13920, Indonesia

* E-mail: bambangpurnomo@pertamina.com

Abstract. The application of demulsifiers on dewatering of crude oil emulsions from East Kalimantan oil fields is reported in this work. Two types of experiment were conducted: the bottle test in which emulsion resolution was carried out after demulsification process within the time intervals of 0, 5, 10, 15, 20, 25, 30, 40, 50, 60, 120 and 180 min at temperature of 40 °C followed by the appearance of separated aqueous layer, and identification of variety of fractions or compounds in crude oils that promote and stabilize these emulsions. Demulsification was successfully obtained by adding PT Pertamina (Persero)'s commercial demulsifier series of Pertadem B-02 and Pertadem B-06. Pertadem B-02 is the best candidate for destabilizing water in oil emulsions. The optimal percentage demulsification efficiency (DE) achieved by Pertadem B-02 was 100% with just concentration of 18 ppm for field A and 97% for field B. Moreover, demulsification activities in the water in oil emulsions by Pertadem B-02 were faster compared to Pertadem B-06. Fractions identifications showed emulsifying agents in crude oil of Field A that promote very unstable emulsions compared to crude oil of Field B from East Kalimantan.

1. Introduction

Water in crude oil emulsions have significant practical importance during uspstream and downstream operations and pose serious challenges on the economic values of petroleum industry [1]. Emulsified water in crude oils may contain chloride that generate corrosion problems to downstream refining equipments [2]. Thus, effective separations of emulsified water from feedstock oils of refineries is necessary to enhance the commercial value of the crude [3]. Certain fractions of crude oil have been reported their contribution on the formation of water in oil emulsions, including waxes, asphaltenes and resins that might be found in both the dissolved and particulate form [4]. Studies reported that crude oil emulsions are stabilized by naturally occuring surfactants or emulsifier which partially soluble in the oil and water phases and have a strong tendency to move and stabilize the interfacial layer of water droplets [5]. High concentrations of praticles and surfactants in crude oils promoted prevention of water droplet coalescence; producing stable emulsions [4].

A process aimed to separate the emulsion into two phases is called demulsification [6]. There are three primary methods for demulsification of highly stable crude oils: physical, chemical and biological in which the effectiveness of these methods can be measured by their ability to break the emulsion stability until the two phases separate [6,7]. Chemical demulsification is conducted by using chemical compounds in a small concentration (usually 10–1000 ppm) and is one of the very crucial techniques to minimize the formation of crude oil emulsion and to improve phase separation process [3,8]. Study on the effect of the demulsifier's chemical structure is needed to break these emulsions in fast and cost-effective separation.

The fundamental mean of chemical demulsification is that the adsorption kinetics and surface affinity of the demulsifiers gradually replace the naturally occuring emulsifiers within the water-oil interfacial film, and that might eventually destablize the interfacial layer by some proposed mechanism: competitively adsorbing at the interfacial region, replacing natural emulsifiers of asphaltenic aggregates, breaking up the asphaltenic layer, along with lowering interfacial tension between the hydrocarbon and aqueous phase, leading to increasing the kinetics of water droplets coalescence [5,8].

The aim of the present study is to investigate the influence of two types of PT Pertamina (Persero)'s commercial demulsifier series of Pertadem B-02 and Pertadem B-06 on breaking the emulsion of crude oils from East Kalimantan oil fields. Two experimente were conducted: the bottle test in which emulsion resolution was carried out after demulsification process within the time intervals followed by the observance of separated aqueous layer, and identification of variety of fractions or compounds in crude oils that promote and stabilize these emulsions. Moreover, specific demulsifier formulations can be modified to the type of crude oil and the nature and fractions of emulsifying agents of asphaltenes and resins in crude oils.

2. Materials & methods

2.1. Materials

The crude oil sample was obtained from two oil fields located in East Kalimantan, Indonesia: field A and field B. Two homologous series of demulsifier of PT Pertamina (Persero)'s Pertadem B-02 and Pertadem B-06 were employed to perform demulsification test.

2.2. Demulsification experiments

To evaluate the abililty of these chemicals to break the water in crude oil emulsions, certain amounts of Pertadem B-02 and Pertadem B-06 were added to the prepared water in crude oil emulsion and bottle test was conducted [9]. A 100-mL batch of each emulsion of crude oil from different fields was prepared by mixing the formation water (that comes with crude oil) (60 vol %) and crude oil (40 vol %) in a standard 100-mL uniform-shaped tube with scale. A small amount of Pertadem B-02 and Pertadem B-06 was added to the mixture of each batch of crude oils with final concentration of 18 ppm. Homogenization was performed by mixing the emulsion 150 times by hand after and before adding the demulsifier. The mixture of Pertadem series with emulsion was heated at 40 °C for 180 min in a thermostatic water bath. The samples were examined periodically in 0, 5, 10, 15, 20, 25, 30, 40, 50, 60, 120 and 180 min to record the amount of separated water.

2.3 Instrumentation

Identification of fractions in crude oils was evaluated using GC DHA (Agilent). The change of demulsifier's weight that takes place during heating at constant rate were performed by thermogravimetric analysis (TA Instrument).

3. Results and discussion

3.1. Identification of fractions in crude oils

Figure 1 shows several fractions in petroleum from different oil fields from East Kalimantan. Crude oil from field A exhibits higher content of aromatics yet lower content of parrafins and olefins compared to crude oil from field B. Chemical characterization of crude oil is a very challenging work due to its essentially complex nature [10]. Saturated and aromatic hydrocarbons are the most useful compounds discovered in petroleum to generate fuels and other profitable products. Some compounds or fractions have been reported to play a role in the formation of water in oil emulsions of crude oils, including paraffinic waxes, asphaltenes and resins and might be found in both the dissolved and particulate form

[4]. Paraffinic waxes might interact with asphaltenes and resins and form sub micrometer particles to stabilize the emulsions.

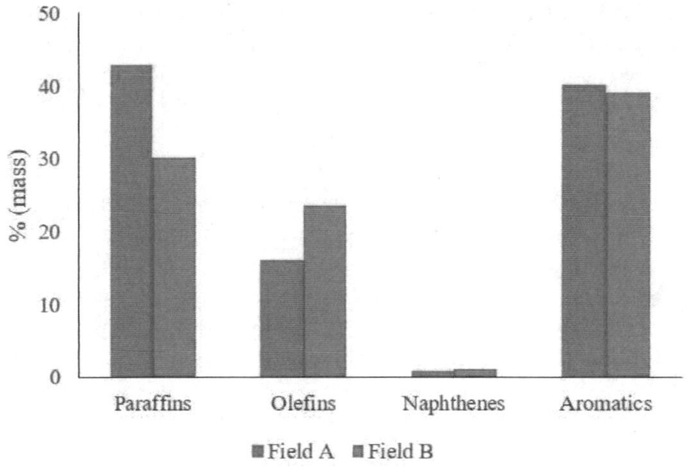

Figure 1. Chemical fractions in petroleum from different oil fields.

3.2. Identification of fractions in Pertadem series

Thermogravimetric analysis was employed in this work to evaluate components in demulsifier of Pertadem B-02 and Pertadem B-06 by monitoring the change of their weight that takes place during heating at constant rate of 20 °C/min (figure 2). The thermal decomposition of Pertadem B-02 started at 281.46 °C and finished at 582.25 °C, followed by major loss of its weight at temperature of 452.46 °C. Pertadem B-06 shows thermal decomposition that started at temperature of 182.00 °C and finished at 475.52 °C with major loss occured at temperature of 386.8 °C. By comparing peaks between Pertadem B-02 and Pertadem B-06, it can be noticed that they differ in chemical constituents which play a role in destabilizing the emulsions.

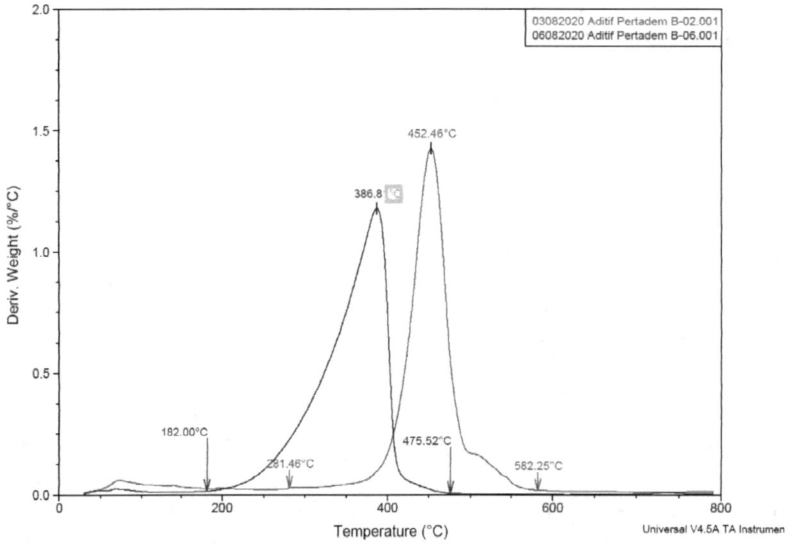

Figure 2. TGA profiles of Pertadem B-02 and Pertadem B-06 at heating rate of 20 °C/min.

3.3 Demulsification performance of tested demulsifiers

Demulsifying performances of Pertadem B-02 and Pertadem B-06 at 40 °C in 180 min were quite different for two crude oils as demonstrates in figure 3(a–b). Pertadem B-02 with final concentration of 18 ppm performed a remarkable demulsification efficiency of 100% for crude oil from field A in 10 min settling time and 97% for crude oil from field B in 15 min settling time. Maximum demulsification efficiency of Pertadem B-06 was 95% for crude oil from field A in 180 min settling time and 83% for crude oil from field B, both in 180 min settling time. It suggested that chemical constituents in Pertadem B-02 and Pertadem B-06 that play a role in breaking the emulsion stability until the two phases separate are differ (figure 4a–b). The presence of paraffinic waxes in those oils may generate the emulsion to be relatively stable and cannot be separated easily [11].

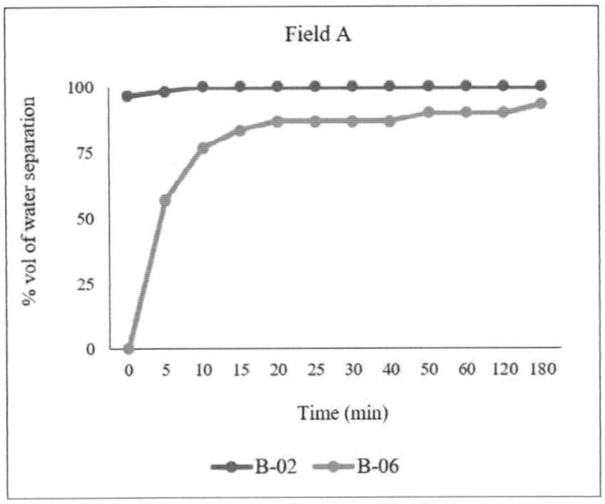

(a)

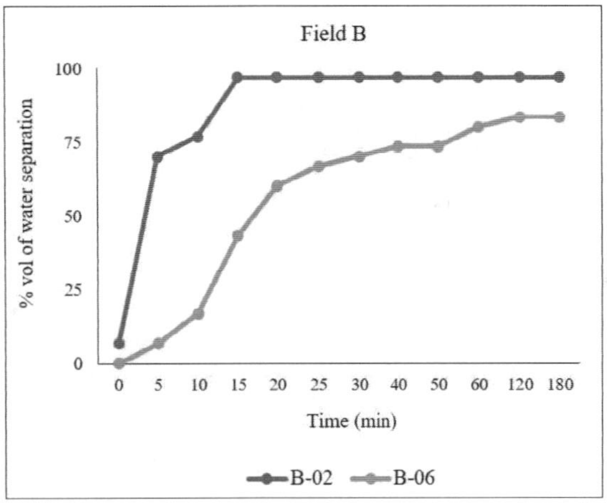

(b)

Figure 3. Emulsion formation by East Kalimantan's crude oils of: (a) field A and (b) field B with the addition of Pertadem B-02 and Pertadem B-06, respectively.

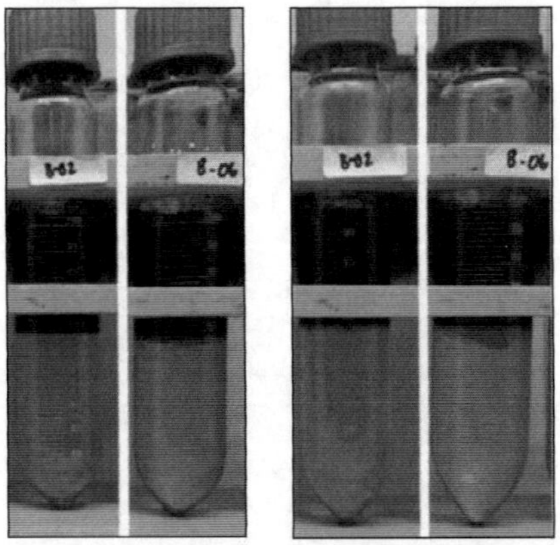

Figure 4. Appearance of the bottle test that used in in this work for crude oils from field A (a) and field B (b) at 40 °C in 180 min.

4. Conclusions

We have demonstrated the application of two PT Pertamina (Persero)'s commercial demulsifier series on destabilizing water in crude oil emulsions from different oil fields. Experimental data revealed that Pertadem B-02 had better dan faster demulsifying activity than Pertadem B-06. Fractions identifications showed emulsifying agents in crude oil of field A that promote very unstable emulsions compared to crude oil of field B from East Kalimantan. It should be noticed that specific demulsifier formulations can be modified to the type of crude oils and the nature and fractions of emulsifying agents in crude oils.

Acknowledgement

This study is supported and owned by PT Pertamina (Persero).

References

[1] Adewunmi A A and Kamal M S 2019 *J. Mol. Liq.* **279** 411
[2] Czarnecki J 2009 *Energy and Fuels* **23** 1253
[3] Peña A A, Hirasaki G J and Miller C A 2005 *Ind. Eng. Chem. Res.* **44** 1139
[4] Lee R F 1999 *Spill Sci. Technol. Bull.* **5** 117
[5] A Issaka S 2015 *J. Pet. Environ. Biotechnol.* **06**
[6] Raya S A, Mohd Saaid I, Abbas Ahmed A and Abubakar Umar A 2020 *J. Pet. Explor. Prod. Technol.* **10** 1711
[7] Zolfaghari R, Fakhru'l-Razi A, Abdullah L C, Elnashaie S S E H and Pendashteh A 2016 *Sep. Purif. Technol.* **170** 377
[8] Grenoble Z and Trabelsi S 2018 *Adv. Colloid Interface Sci.* **260** 32
[9] Hazrati N, Miran Beigi A A and Abdouss M 2018 *Fuel* **229** 126
[10] Niyonsaba E, Manheim J M, Yerabolu R and Kenttämaa H I 2019 *Anal. Chem.* **91** 156
[11] Zaki N, Schoriing P C and Rahimian I 2000 *Petrol. Sci. Technol.* **18** 945

IOP Conf. Series: Materials Science and Engineering **959** (2020) 012008 doi:10.1088/1757-899X/959/1/012008

The addition effect of *Pseudomonas aeruginosa* on biodegradation of methyl orange dye by brown-rot fungus *Fomitopsis pinicola*

A S Purnomo[1,*], D M Rahayu[1], R Nawfa[1] and S R Putra[1]
[1] Department of Chemistry, Faculty of Science and Data Analytics, Institut Teknologi Sepuluh Nopember, Kampus ITS Sukolilo, Surabaya, 60111, Indonesia.

[*] E-mail: adi_setyo@chem.its.ac.id; adi.spurnomo@yahoo.com

Abstract. The addition effect of *Pseudomonas aeruginosa* on methyl orange (MO) biodecolorization by *Fomitopsis pinicola* had been investigated. *P. aeruginosa* was added into *F. pinicola* culture at 2, 4, 6, 8, 10 mL (1 mL = 5.05×10^{12} CFU). The addition of 4 ml of *P. aeruginosa* showed the highest MO biodecolorization approximately 99.53%, while by *F. pinicola* only was 77.22% in potato dextrose broth (PDB) medium for 7 days incubation. Four metabolites were detected from MO degradation by mixed cultures such as 4-(4-(dimethylimino) hydroxy-cyclohexa-2,5-dien-1-ylidene) hydrazinyl) phenolate; 4-(hydroxy-4-iminio-cyclohexa-2,5-dien-1-ylidene) hydrazinyl) benzene sulfonate; 4-(4-(dimethylimino) methoxy-cyclohexa-2,5-dien-1-ylidene) hydrazinyl) hydroxy-benzene sulfonate; and 4-(hydroxy-4-iminio-cyclohexa-2,5-dien-1-ylidene) hydrazinil) hidroxy-benzene sulfonate. The mixed cultures transformed MO via three pathways: (1) desulfonylation, (2) demethylation, and (3) hydroxylation. These results indicated that *P. aeruginosa* can enhance MO biodecolorization by *F. pinicola*.

1. Introduction

The synthetic dye industry is currently showing fairly rapid development. One of the major uses of synthetic dyes is in the textile industry. In 2012, textile production increased rapidly as a result of the increasing demand for imported textile and textile products. Along with the development of the textile industry that continues to increase, textile industry waste production also increased. One type of waste generated from the textile industry is textile dye waste [1]. In textile dye waste, many synthetic dyes are dissolved in water, potentially causing pollution of the aquatic environment [2]. Azo dye is a dye that is often used in the textile industry which is 70% of the total dye [3]. Methyl orange (MO) is one of the most commonly used azo dyes. In the process of textile dyeing, 10-15% of MO dye is unused and thrown as waste [4]. Water-soluble MO dye as a textile industry waste disrupts the balance of the water ecosystem. In addition, the MO dye is quite stable because it is composed of aromatic compounds that cause MO difficult to be degraded in nature and harmful to the environment due to its carcinogenic.

Chemical-physical methods can be done to solve synthetic dye waste problems in industrial environments such as adsorption, precipitation, chemical degradation, and so on [5]. However, these methods generally relatively expensive and sometimes still produce toxic sludge. Therefore, another method is needed that a relatively cheaper and environmentally friendly method, that is through bioremediation [6–10]. In the bioremediation method, waste handling is done by utilizing the biological activity of microorganisms both fungi and bacteria in degrading textile dye waste. One of the fungi that can degrade dyes is the brown-rot fungi [1, 2, 11, 12].

Brown-rot fungi can degrade cellulose and hemicelluloses by using Fenton reactions by involving Fe(II) and H_2O_2 producing OH radicals [13]. *Fomitopsis pinicola* is one of brown-rot fungi that can degrade some organic pollutants such as DDT [14-21] and dyes [11]. *F. pinicola* degraded approximately 97% on liquid medium Potato Dextrose Broth (PDB) incubated for 14 days [11]. Although *F. pinicola* can degrade MO, but it is still required relatively long incubation times, thus, culture modification is needed to optimize *F. pinicola* ability to degrade MO.

In addition to using fungi, synthetic dye waste can also be degraded by bacteria [12, 22-24]. Besides, *P. aeruginosa* also degraded DDT in particular as well as in mixed cultures [19, 25, 26]. Therefore, the purpose of this study is to investigate the effect of addition of bacterium *P. aeruginosa* on the MO degradation process by brown-rot fungus *F. pinicola* including identification of metabolites and degradation pathway.

2. Materials and methods / another section of your paper (for review paper)

2.1. Chemicals

Methyl orange (MO, $C_{14}H_{14}N_3NaO_3S$), aqua DM, and ethanol 70% were purchased from PT. Sumber Ilmiah Persada Indonesia. Potato dextrose broth (PDB), potato dextrose agar (PDA), nutrient broth (NB), and nutrient agar (NA) were purchased from Merck, Germany.

2.2. Culture conditions

Culture bacterium *P. aeruginosa* NBRC 3080 (NITE Biological Resource Center, was incubated at 37 °C on NA plate. The colony was then inoculated into 60 mL NB, and then pre-incubated for 17 h at 37 °C with a shaker at a speed of 180 rpm [26, 27]. Culture fungus *F. pinicola* NBRC 8705 was maintained on PDA. This fungus was incubated at 30 °C. The mycelia were inoculated into 9 ml PDB then pre-incubated for 7 days at 30 °C [28].

2.3. Biodecolorization MO by F. pinicola

After pre-incubation, 10 mL of PDB medium was added into fungal cultures (final volume of 20 mL) and 1 mL MO 2000 ppm (final concentration of 100 ppm) was added to each *F. pinicola* inoculated flask. The cultures were incubated statically for 7 days at 30 °C. As a control, the cultures were killed by autoclave (121 °C, 20 min) after pre-incubation. The experiments were performed in duplicate [11].

2.4. Biodecolorization of MO by bacterium P. aeruginosa

After pre-incubation for 17 h, various volume of bacteria cultures at 2, 4, 6, 8, 10 mL (1 mL = 5.05×10^{12} CFU) were inoculated to PDB culture (final volume 20 mL). Each inoculated flask was added 1 mL MO 2000 ppm (final concentration of 100 ppm). The cultures were incubated for 7 days at 30 °C. As a control, the cultures were killed by autoclave (121 °C, 20 min) after pre-incubation. The experiments were performed in duplicate [29].

2.5. Biodecolorization of MO by mixed cultures of F. pinicola and P. aeruginosa

After pre-incubation (fungus for 7 days and bacteria 17 hours; section 2.2), various concentration of bacteria 2, 4, 6, 8, and 10 mL (1 mL = 5.05×10^{12} CFU) were separately added into *F. pinicola* cultures. PDB medium and MO (final concentration 100 ppm) were then added into the final volume of 20 mL. The headspace of each flask was sealed with a glass stopper. The cultures were incubated statically for 7 days at 30°C. As a control, the cultures were killed by autoclave (121°C, 20 min) after pre-incubation. The experiments were performed in duplicate [30, 31].

2.6. Analytical method and identification of metabolites
The cultures were centrifuged to separate the biomass. The supernatant absorbance was measured with a UV-Vis spectrophotometer at wavelength 350-600 nm, while the rest of supernatants were stored for metabolite product identification. The abiotic control was the cultures without *F. pinicola and P. aeruginosa.* The percentage of decolorization of MO was calculated at wavelength 465 nm by used the equation 1.

$$\% \text{ decolorization} = \frac{Ac - At}{Ac} \times 100\%, \quad (1)$$

where Ac = control absorbance and At = treatment absorbance.

Metabolites identification was carried out by analyzing the supernatants using LC-TOF MS, which ionization source with a mass range of 50-350. The sample was eluted by a gradient flow rate of 0.2 mL min^{-1} in 3 min, and then 0.4 mL min^{-1} in 7 min. The mobile phases were methanol and water at a ratio of 99:1 in 3 min, and then 61:39 in 7 min remaining. The column was TM RSLC Acclaim 120 C18 with 2.1x100 mm size which column temperature of 33 °C (11, 12, 32) .

3. Results and discussion

3.1. Biodecolorization of MO by F. pinicola
In this study, biodecolorization of MO was carried out in the PDB medium due to the most suitable medium for the growth of brown-rot fungi compared to the other liquid media [14]. The ability of *F. pinicola* to degrade MO was quantitatively determined by measuring the percentage of decolorization. figure 1 showed the profile absorbance of MO decolorization by *F. pinicola.* In the abiotic control (negative control), the maximum wavelength was detected at 465 nm. However, since *F. pinicola* produces some metabolites including organic acids, the pH medium became acid that caused the wavelength shifts to the right and detected at the 510 nm wavelength [12]. Profile absorbance showed that the absorbance of MO decreased when treated with *F. pinicola* (in this case was used as positive control) compared with abiotic control (negative control). Decreasing MO absorbance indicated that MO was degraded by *F. pinicola.* Percentages of decolorization of MO by *F. pinicola* were approximately 77.22%, for 7 days incubation periods. The ability of *F. pinicola* to degrade MO might be correlated with the ability of *F. pinicola* in producing hydroxyl radicals generated by Fenton reaction [14–17].

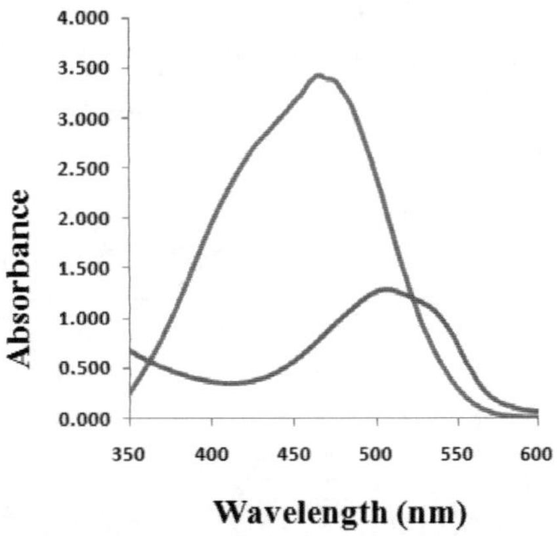

Figure 1. Absorbance profile of decolorization of MO by *F. pinicola* (red line) and abiotic control (blue line).

3.2. Biodecolorization of MO by P. aeruginosa

The absorbance profile of MO decolorization by *P. aeruginosa* was presented in figure 2. The highest peak was detected at 485 nm, due to the culture became acid. *P. aeruginosa* have been reported produces polyuronic acids such as mannuronis acid and guluronic acid [33]. The higher volume of bacteria resulted in decreasing of MO peak, indicating MO was further decolorized. The ability of the *P. aeruginosa* to decolorize MO is related to its ability to produce enzymes in its metabolic processes. Besides having the ability to produce degrading enzymes, *P. aeruginosa* is also capable of producing biosurfactant that also plays a role in biodecolorization process [34–36].

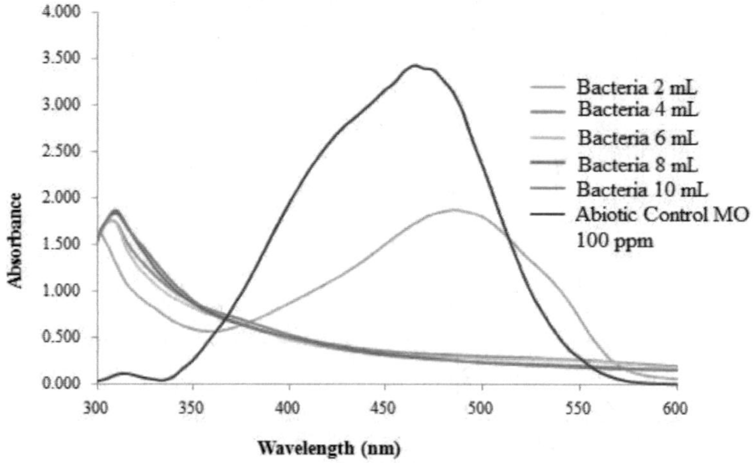

Figure 2. Absorbance profile of decolorization MO by *P. aeruginosa*.

The ability of *P. aeruginosa* to decolorize MO was also determined quantitatively based on the percentage of decolorization (table 1). Percentages of decolorization of MO by *P. aeruginosa* were approximately 49.59%, 90.27%, 91.06%, 91.36%, and 91.59% at 2, 4, 6, 8, and 10 mL of bacteria in PDB medium for 7 days incubation period, respectively. The results showed that the higher amount of *P. aeruginosa*, the higher percentage of MO biodecolorization was obtained, which the highest percentage of biodecolorization of MO was obtained approximately 91.59% at 10 mL of bacteria in PDB medium. However, in variation volume bacterium of 4-10 mL, MO decolorization was not significantly different approximately 91%. Since abundant population of bacteria might produce some toxic metabolites during the stationary phase, the competition of bacteria for surviving might occur rather than to decolorize MO [12].

Table 1. Percentage of biodecolorization MO by bacterium *P. aeruginosa*.

Amount of bacteria culture (mL)	Absorbance control	Absorbance sample	Decolorization (%)
2	3.424	1.726	49.59±0.09[a]
4	3.424	0.333	90.27±0.04[b]
6	3.424	0.306	91.06±0.03[b]
8	3.424	0.296	91.36±0.03[b]
10	3.424	0.288	91.59±0.02[b]

Analyses were conducted by spectrophotometer UV-VIS. Data are mean ± standard deviation (n=3). A 1 mL of bacteria ≈ 5.5×10^{12} CFU. Data followed by the different minor letter on each column indicates significantly different ($P < 0.05$).

3.3. Biodecolorization MO by mixed cultures

Figure 3 presented the absorbance profile of MO decolorization by mixed cultures of *F. pinicola* and *P. aeruginosa* in PDB medium for 7 days incubation. The addition of *P. aeruginosa* into *F. pinicola* culture can generally increase the MO decolorization compared to the abiotic control. In addition of 2 mL of *P. aeruginosa*, the absorbance was decreased, which indicated MO decolorization even in small amount due to the process of adaptation between fungus and bacterium. Among the mixed cultures, *F. pinicola* with 4 mL of *P. aeruginosa* showed the highest MO biodecolorization.

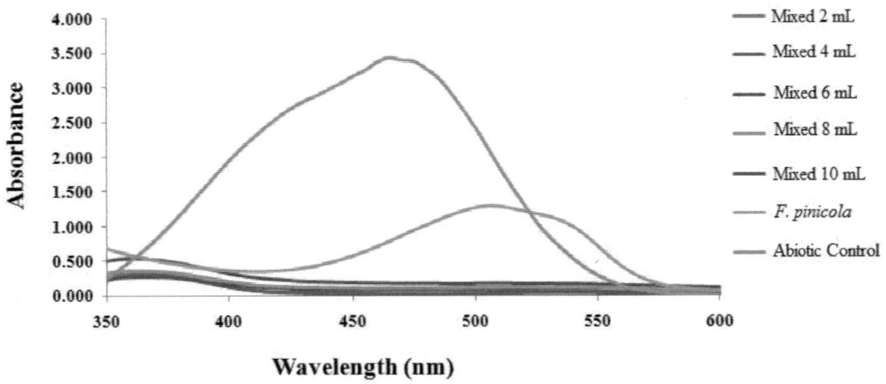

Figure 3. Absorbance profile of decolorization MO by mixed cultures of *F. pinicola* and *P. aeruginosa*.

Table 2. Percentage of biodecolorization MO by mixed culture of *G. trabeum* and *P. aeruginosa*.

Amount of bacteria culture (mL)	Absorbance control	Absorbance sample	Decolorization (%)
2	3.424	0.086	97.49±0.02[a]
4	3.424	0.016	99.53±0.02[b]
6	3.424	0.066	98.07±0.03[b]
8	3.424	0.108	96.85±0.05[b]
10	3.424	0.105	95.18±0.03[b]

Analyses were conducted by spectrophotometer UV-VIS. Data are mean ± standard deviation (n=3). A 1 mL of bacteria ≈ 5.5×10^{12} CFU. Data followed by the different minor letter on each column indicates significantly different ($P < 0.05$).

The ability of mixed cultures of *F. pinicola* and *P. aeruginosa* to decolorize MO was determined quantitatively based on percentage of decolorization (table 2). Percentages of decolorization of MO by mixed culture *F. pinicola* and *P. aeruginosa* were approximately 97.49%, 99.53%, 98.07%, 96.85%, and 95.18% at 2, 4, 6, 8, and 10 mL of bacteria, respectively, in PDB medium for 7 days incubation period. The addition of 2-4 mL of *P. aeruginosa* increased MO decolorization due to the enzyme dehydrogenase produced by *P. aeruginosa* might help to increase the decolorization of MO by *F. pinicola* [35]. However, in addition to 6, 8, and 10 mL, MO decolorization was decreased due to the competition between fungus and bacterium in getting the nutrients contained in the media [12]. The highest MO decolorization was obtained at mixed cultures of *F. pinicola* with the additional 4 mL of *P. aeruginosa*, which enhances the ability of decolorization by 22.31% (from 77.22% by *F. pinicola* culture only to 99,53% by mixed cultures). Mixed microbial cultures have more power to degrade pollutants because they produce more metabolites as well as enzymes [24].

3.4. The identification of metabolites
Based on characterization by LC-TOF/MS, four metabolites were detected at time retention 1.31; 2.36; 3.31; and 6.81mins (Figure 4). The identification of metabolites was determined based on the similarity between MS spectrum and time retention from the database.

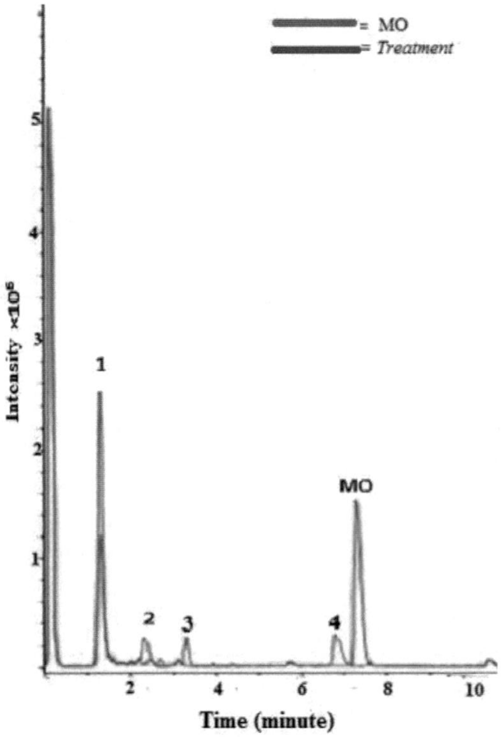

Figure 4. LC chromatogram profile of metabolites of MO degradation by mixed cultures (purple chromatogram is MO abiotic control; blue chromatogram is MO treatment.

Based on TOF/MS data (table 3), the peak at retention time 1.31 min has m/z 258 which is identified as 4-(4-(dimethylimino) hydroxy-cyclohexa-2,5-dien-1-ylidene) hydrazinyl) phenolate. The compound was reported previously which had fragments in m/z 118, which is in accordance with previous research on the degradation of azo dyes by laccase enzyme from WRF [37, 38]. At a peak of 2.36 mins, metabolite was identified as 4-(hydroxy-4-iminio-cyclohexa-2,5-dien-1-ylidene) hydrazinyl) benzene sulfonate with m/z 294. This compound was supported by the presence of fragments in MS data with m/z 214, fragment m/z 136 in accordance with research on metabolites produced by *F. pinicola* in MO (11). The peak of 3.31 mins with m/z 348 was identified as 4-(4-(dimethylimino) methoxy-cyclohexa-2,5-dien-1-ylidene) hydrazinyl) hydroxy-benzene sulfonate. This compound was supported by fragments m/z 213 [39]. The peak of 6.81 mins was identified as 4-(hydroxy-4-iminio-cyclohexa-2,5-dien-1-ylidene) hydrazinil) hydroxy-benzene sulfonate with m/z 311. This compound was supported by the presence of fragments in MS data with m/z 214 and m/z 108 in accordance with research on metabolites produced by *F. pinicola* in MO [11]. Based on the identification of metabolites, the MO decolorization pathway by mixed cultures of *F. pinicola* and *P. aeruginosa* cultures were proposed in figure 5.

TABLE 3. Metabolites of MO degradation by mixed cultures of *F. pinicola* and *P. aeruginosa*.

RT (min)	Mr	Name	Molecular formula	Structure
1.31	258	4-(4-(dimethylimino) 2-hydroxy cyclohexa-2,5-dien-1-ylidene) hydrazinyl) phenolate	$C_{14}H_{15}N_3O_2$	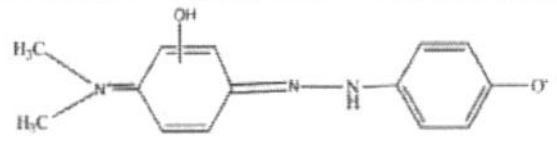
2.36	294	4-(2-hydroxy-4-iminio cyclohexa-2,5-dien-1-ylidene) hydrazinil) benzenesulfonate	$C_{12}H_{12}N_3O_4S$	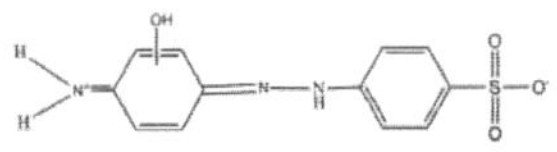
3.31	348	4-(4-(dimethylimino) methoxy cyclohexa-2,5-dien-1-ylidene) hydrazinyl) hydroxyl benzene sulfonate	$C_{14}H_{16}N_3NaO_4S$	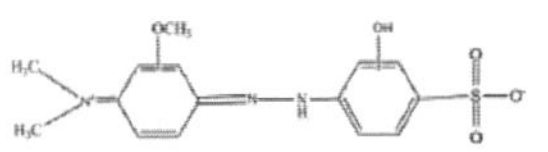
6.81	311	4-(2-hydrokxy-4-iminio cyclohexa-2,5-dien-1-ylidene) hydrazinil) 2-hidroxy- benzene sulfonate	$C_{12}H_{11}N_3O_5S$	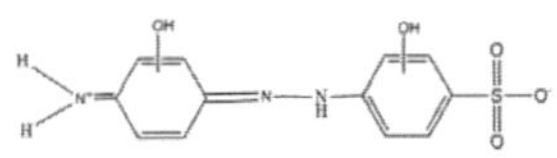

* The data were determined by LC-TOF/MS.

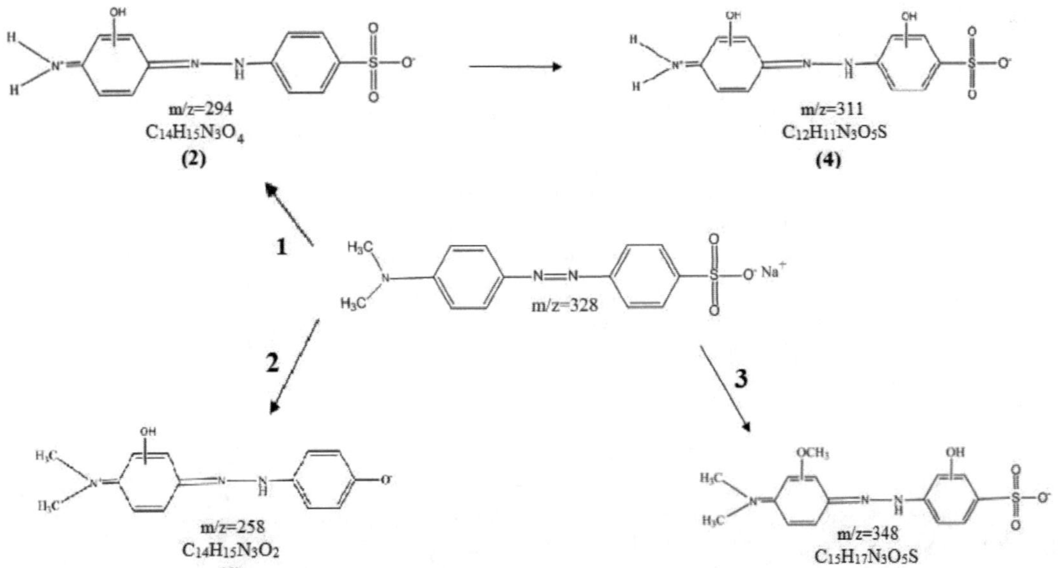

Figure 5. Proposed pathway degradation of MO by mixed cultures of *F. pinicola* and *P. aeruginosa*.

The MO degradation pathway by mixed cultures of *F. pinicola* and *P. aeruginosa* was proposed to divide into three ways. The first pathway: the MO was ionized by the media solution released Na^+, and then transformed into compound 2 by demethylation and hydroxylation, followed hydroxylation into compound 4. The second pathway: the ionized MO was transformed into compound 1 by desulfonylation, followed hydroxylation. The third pathway: ionized MO was dehydroxylated on both aromatic rings, followed methylated into compound 3. Hydroxylation occurs due to the presence of radical hydroxides that might be produced from Fenton reaction by *F. pinicola* that attack MO [19]. Previously, *F. pinicola* transformed MO via three pathways: (1) demethylation, followed by hydroxylation reactions; (2) hydroxylation, followed by demethylation; and (3) desulfonylation [11].

4. Conclusion

The addition of bacterium *P. aeruginosa* enhanced the ability of brown-rot fungus *F. pinicola* to decolorize and degrade MO. The highest MO decolorization was obtained in the addition of 4 mL of *P. aeruginosa* with decolorization percentages by 99.73%. The mixed cultures of *F. pinicola* and *P. aeruginosa* transformed MO into 4-(4-(dimethylimino) hydroxy-cyclohexa-2,5-dien-1-ylidene) hydrazinyl) phenolate; 4-(hydroxy-4-iminio-cyclohexa-2,5-dien-1-ylidene) hydrazinyl) benzene sulfonate; 4-(4-(dimethylimino) methoxy-cyclohexa-2,5-dien-1-ylidene) hydrazinyl) hydroxy-benzene sulfonate; and 4-(hydroxy-4-iminio-cyclohexa-2,5-dien-1-ylidene) hydrazinil) hidroxy-benzene sulfonate. Based on the identification of metabolic products, the mixed cultures transformed MO via three pathways: (1) demethylation, (2) hydroxylation, and (3) desulfonylation. This research indicated that the bacterium *P. aeruginosa* can increase the decolorization of the MO dye by the brown-rot fungus *F. pinicola*.

Acknowledgments

This study was funded under a fundamental research scheme by the Directorate of Research and Community Service, Directorate General of Strengthening Research and Development, Ministry of Research, Technology and Higher Education No. 5/E1/KP.PTNBH/2019.

References

[1] Rizqi H D and Purnomo A S 2017 *World J. Microbiol. Biotechnol.* **33** 92
[2] Purnomo A S, Andyani N E, Nawfa R and Putra S R 2020 *AIP Conf. Proc.* **2237** 020002
[3] Alamddine L and El Jamal M M 2009 *J. Chem. Technol. Metall.* **44** 127
[4] Selvam K, Swaminathan K and Chae K S 2003 *Bioresour. Technol.* **88** 115
[5] Moreira M T, Melgo I, Feijo G and Lema J M 2000 *Biotechnol. Lett.* **22** 1499
[6] Purnomo A S, Koyama F, Mori T and Kondo R 2010 *Chemosphere* **80** 619
[7] Purnomo A S, Mori T, Kamei I, Nishii T and Kondo R 2010 *Int. Biodeterior. Biodegrad.* **64** 397
[8] Purnomo A S, Mori T, Putra S R and Kondo R 2013 *Int. Biodeterior. Biodegrad.* **82** 40
[9] Purnomo A S, Putra S R, Shimizu K and Kondo R 2014 *Environ. Sci. Pollut. Res.* **21** 11305
[10] Purnomo A S, Nawfa R, Martak F, Shimizu K and Kamei I 2017 *Curr. Microbiol.* **74** 320
[11] Purnomo A S, Mauliddawati V T, Khoirudin M, Nafwa R and Putra S R 2019 *Int. J. Environ. Sci. Tech.* **16** 7555
[12] Purnomo A S and Mawaddah M O 2020 *Biodiversitas* **21** 2297
[13] Contreas D, Freer J and Rodriguez J 2005 *Int. Biodeterior. Biodegrad.* **57** 63
[14] Purnomo A S, Kamei I and Kondo R 2008 *J. Biosci. Bioeng.* **105** 614
[15] Purnomo A S, Mori T and Kondo R 2010 *Int. Biodeterior. Biodegrad.* **64** 560
[16] Purnomo A S, Mori T and Kondo R 2011 *Int. Biodeterior. Biodegrad.* **65** 691
[17] Purnomo A S, Mori T Kamei I, and Kondo R 2011 *Int. Biodeterior. Biodegrad.* **65** 921
[18] Sariwati A, Purnomo A S and Kamei I 2017 *Curr. Microbiol.* **74** 1068
[19] Sariwati A and Purnomo A S 2018 *Indon. J. Chem.* **18** 75
[20] Setyo P A, Dwi R H, Sri F, Sulistyo P H and Ichiro K 2018 *Res. J. Chem. Environ.* **22** 151
[21] Purnomo A S, Sariwati A and Kamei I 2020 *Heliyon* **6** e04027

[22] Asad S, Amoozegar M A, Poubaee A A, Sarbolouki M N and Dastgheib S M M 2007 *Bioresour. Technol.* **98** 2082
[23] Chen K C, Wu J Y, Liou D J and Hwang S C J 2003 *Biotechnology* **101** 57
[24] Moosvi S, Keharia H and Madamwar D 2005 *World J. Microbiol Biotechnol.* **21** 667
[25] Boelan E G and Purnomo AS 2019 *Hayati J. Biosci.* **26** 90
[26] Purnomo A S, Ashari K and Hermansyah F T 2017 *J. Microbiol. Biotechnol.* **27** 1306
[27] Wahyuni S, Suhartono M T, Khaeruni A, Purnomo A S, Asranudin, Holilah and Riupassa P A 2016 *Asian. J. Chem.* **28** 2731
[28] Purnomo A S 2017 Microbe-assisted degradation of aldrin and dieldrin *Microbe-Induced Degradation of Pesticides* ed S N Singh (Switzerland: Springer Nature) chapter 1 pp 1–22
[29] Purnomo A S, Rizqi H D and Rafi L H 2020 *J. Idn. Chem. Soc.* **3** 53
[30] Wahyuni S, Khaeruni A, Purnomo A S, Asranudin, Holilah and Fatahu 2017 *Asian J. Chem.* **29** 1119
[31] Grizca B E and Setyo P A 2018 *J. Physics Conf. Ser.* **1095** 102015
[32] Purnomo A S, Maulianawati D and Kamei I 2019 *J. Microbiol. Biotechnol.* **29** 1424
[33] Wang N, Tan F, Tsoi C and Fan X 2016 *Scientific Reports. Nature Publishing Group*, 1
[34] Li H, Zhang R, Tang L, Zhang J and Mao Z 2014 *Int. J. Environ. Sci.* **26** 1125
[35] Carlson D M and Matthews LW 1966 *Biochemistry* **5** 2817
[36] Nawfa R, Purnomo A S and Putro H S 2019 *Asian J. Chem.* **31** 2367
[37] Bidlan R 2003 *Studies on DDT degradatin by bacterial strains* Thesis (Central Food Technology Research Insitute)
[38] Baiocchi C, Marc G, Prevot A B, Palmisano L, Brussino M C and Pramauro E 2002 *Int. J. Mass. Spectrom.* **214** 247
[39] Prevot A B, Pramaro E, Pazzi M, Palmisano L, Marci G, Basso A, Baiocci C and Augugliaro V 2004 *Anal. Bioanal. Chem.* **378** 214

IOP Conf. Series: Materials Science and Engineering **959** (2020) 012009 doi:10.1088/1757-899X/959/1/012009

Polyelectrolyte Complex (PEC) film based on chitosan as potential edible films and their antibacterial activity test

N Ismillayli[1], I G A S Andayani[2], R Honiar[2], B Mariana[2], R K Sanjaya[3] and D Hermanto[1]

[1] Department of Chemistry, Faculty of Mathematics and Natural Science, University of Mataram, Mataram–West Nusa Tenggara 83125 Indonesia

[2] Laboratory of Analytical Chemistry, Faculty of Mathematics and Natural Science, University of Mataram, Mataram–West Nusa Tenggara 83125 Indonesia

[3] Department of Pharmacy, Faculty of Healthy Science, University of Kadiri, Kediri, East Java 64115 Indonesia

[*] E-mail: dhony.hermanto@unram.ac.id

Abstract. The synthesis of polyelectrolyte complexes (PEC) based on chitosan and determination of their antibacterial properties had been conducted. The microstructure of PECs obtained were described by SEM analysis, while inhibition activity of PECs against S. aureus and E. coli was determined by measuring inhibitory zone diameter. Characteristics of chitosan, alginate and κ-carrageenan as natural polymers which non-toxic, biodagradable and safe to eat meet the edible film criteria. Chitosan as a polycationic interacts with alginate and κ-carrageenan as polyanionic under the appropriate conditions to form PEC film. Based on FTIR spectra, it was found that interaction of chitosan and alginate as well as chitosan and κ-carrageenan was an electrostatic interaction. Microstructure study using SEM found that PECs have irregular and fibrous surface structure. Based on their inhibitory activity against S. aureus and E. coli, PECs have the strongest antibacterial activity compared to their original polymer. Therefore, PECs film could be excellent edible film for food coating that protect product from bacterial contamination.

1. Introduction

The use of packaging in food products is now an important requirement [1]. Packaging protects the product or foodstuff from physical impact and microbial contamination thereby extending its shelf life [2]. Petroleum-based synthetic packaging has been widely used today. Unfortunately, its non-degradability caused serious pollution problems [3]. Based on environmental safety considerations, edible film-based packaging materials have been developed [4,5].

Recently, there has been an increase in the development of bio-composite materials based on PEC that have the potential as edible films [6]. Besides its mechanical characteristics, edible films need to have resistance to food damage due to the presence of microorganisms. Chitosan has been known to be used as biodegradable film and food preservative that is resistant to microbes. The antibacterial properties of chitosan are obtained from a polymer structure that has a positively charged amine group [7–9]. However, chitosan has the disadvantage of leaching under acidic conditions and high water absorption. This weakness can be overcome through the formation of a PEC film [10,11].

In this work, a PEC as potential edible film based on chitosan (as polycationic) was made using carrageenan and alginate as polyanionic. Chitosan-carrageenan and chitosan-alginate polymers formed PEC through electrostatic interactions, without the use of crosslinking agents which are likely to be toxic [12]. The PEC film has the advantage of a dense structure that can inhibit the decay process by blocking oxygen inhalation [6]. The simple and easy manufacturing technique is suitable for large scale applications. The chitosan-carrageenan and chitosan-alginate films that were synthesized without any other toxic materials met the edible film criteria. The antibacterial properties of the formed films were also examined.

2. Materials and methods

2.1. Materials

Chitosan with 95 % degree of deacetylation (DDA) and low molecular mass (72 mPa.s, Mol. wt. 300–400 kDa) from crab shell, sodium alginate (45.3 mPa.s, Mol. wt. 10–20 kDa) from brown algae, κ-carrageenan (6.4 mPa.s, Mol. wt. 12-40 kDa) from red seaweed, S. aureus (ATCC 25923) and E. coli (ATCC 25922) were obtained from Sigma-Aldrich (UK), while glacial acetic acid (98%), sodium hydroxide, and hydrochloride acid (37%) were received from Merck (UK). Solutions were prepared with deionized water.

2.2. Chitosan-carrageenan film preparation

Chitosan-carrageenan film was made according to previous work by Carneiro *et al.* [13] by mixing at 1:1 molar ratio. Chitosan hydrosol was prepared by dissolving approximately 1.53 g of chitosan (0.38 % w/v) in 400 mL of 0.16 M HCl, while κ-carrageenan hydrosol was made by dissolution of 4.40 g of κ-carrageenan (0.63 % w/v) in 700 mL of deionized under stirring condition at 500 rpm (25 °C and 24 h). The addition of chitosan hydrosol to the κ-carrageenan hydrosol was done by using drop-wise fashion (approximately rate of 14 ml/min) in a 2 L Pyrex media bottle (Fisher Scientific, UK) under stirring at 500 rpm. Then, the mixture was cooled using a water bath at 25 °C for 15 min, followed by pH measurements using pH-meter (TOA *Electronic Ltd* model IM-20E). Polymer mixture was poured and molded into glass plate for drying at room temperature for 72 h.

2.3. Chitosan-alginate film preparation

The preparation of chitosan-alginate film was conducted by procedure described as previous work [10]. Briefly, hydrosol of chitosan at 1% (w/v) were prepared in 20 mL glacial acetic acid under stirring at 400 rpm using a magnetic stirrer (IKA Combimag REO Drehzahl Magnetic Stirrer Electronic, Germany), while alginate was dissolved in deionized water until hydrosol at concentration of 4% (w/v) was formed. Then, both alginate and chitosan hydrosols were left out overnight to remove air bubbles. The chitosan hydrosol (100 mL) was added to the alginate solution and added 8 mL of 32% HCl (100 mL) at 25°C, homogenized using a homogenizer (IKAT18Ultra Turrax, Germany) for 90 seconds. Following, NaOH 10% (w/v) was added till the pH of the suspension was elevated to 5.28 and homogenized again for 90 seconds. Mixture of polymer was poured and molded into glass plate for drying at room temperature for 72 h.

2.4. Antibacterial activity test

Antibacterial activity of PEC films against both E. coli and S. aureus was evaluated by using disc diffusion method as implemented by Kaya *et al.* [14]. As a reference standard 0.5 McFarland was used to adjust bacterial susceptibility. Suspension of bacterial culture was wiped onto a plate inside agar of Müeller-Hinton. On agar medium that has been inoculated with bacteria, paper discs were placed, and then loaded with 20 mL hydrosol film. It was incubated for 24 hours at 37°C. As positive control, it was used 10 mg gentamicin per disc and deionized water as a negative control. Inhibition zone diameter on a clear area of the paper disc was measured. All experiments were done in triplicate.

3. Results and discussions

3.1. Formation of PEC films

The formation of PEC films was confirmed by FTIR spectra. It was shown in Figure 1, along with their constituents polymers. Typical absorption peak of chitosan showed at 1643.0 and 1560.6 cm^{-1} while peak of κ-carrageenan was at 1231.1; 924.49; 846.48 cm^{-1}. Shifting of sulphate group absorption in κ-carrageenan from 1231.1 cm^{-1} to 1208.2 cm^{-1} was observed in chitosan- κ-carrageenan film. Shifts also occurred at 1647.6 cm^{-1} (κ-carrageenan) and 1643.0 cm^{-1} (chitosan) absorption which were then combined into absorption at 1652.2 cm^{-1} in PEC with a stronger intensity. The formation of the chitosan- κ-carrageenan PEC was indicated by absorption at 1556.1 and 1208.2 cm^{-1}. The loss of the absorption band at 1148.7 cm^{-1} as amine group characteristics indicated that the chitosan amine group has been protonated and interacted with the κ-carrageenan sulphate group. It was confirmed by the appearance of a peak in the area of 1556.1 cm^{-1} indicating the presence of the NH_3^+ ion. It was strengthened with absorption at 1208.2 cm^{-1} that indicated electrostatic interactions of chitosan- κ-carrageenan in molar ratio of 1:1. A pH 5 was chosen as the condition for chitosan- κ-carrageenan film synthesize. The amine groups of chitosan were protonated and interacted with the sulphate groups of κ-carrageenan under these condition. Chitosan is a weak base polymer with an intrinsic acid dissociation constant (pKa) of 6.5 [15], while κ-carrageenan consists of ionisable sulphate groups at pH above pKa of 2 [16]. It was found that optimal condition of chitosan-κ-carrageenan films formation at pH 5 with yield of 85% via ionic interaction.

The absorption at 3459 cm^{-1} indicated the presence of the -OH group of alginates and $-NH_2$ chitosan, 2924 and 2870 cm^{-1} are the absorption of CH (sp^3), 1578 cm^{-1} was the vibration of the $-COO^-$ group. The loss of amine group absorption at 1157 cm^{-1} indicated that the chitosan amine group has been protonated and interacted with the carboxylate group of alginate [17]. This was strengthened by the appearance of peaks in the area of 1550 cm^{-1} which showed the presence of protonated amine (–NH_3^+ ion). Peak intensity at wave number of 1398 cm^{-1} also showed electrostatic interactions on alginate-chitosan film, with alginate and chitosan mass ratio of 1:1. The presence of the absorption band mentioned above showed the existence of ionic bonds of protonated amine and the carboxylate groups. Chitosan consists of amine groups which have intrinsic pKa value of 6.5, while alginates consist of carboxyl groups that undergo ionization at pH of the solution above its pKa of 4.7 [18]. Therefore, at pH 5.28 chitosan presence at protonated form while alginate is in carboxylate form which is negatively charged. When both hydrosols were mixed in a 1:1 ratio, electrostatic interaction of protonated amine and carboxylate has occurred. Mechanism of PEC formation was described in Figure 2.

The surface morphology of chitosan, carrageenan, alginate, chitosan- κ-carrageenan and chitosan-alginate films was studied through Scanning Electron Microscopy (SEM) as shown in Figure 3. Here, the morphology of chitosan- κ-carrageenan films is less homogeneous than chitosan or κ-carrageenan alone. The chitosan film appeared microscopically smooth and uniform, while the κ-carrageenan and alginate films showed the clearest and smoothest surface. While, that surface structure of both PEC films was irregular and fibrous. The irregularity in the structure of the chitosan- κ-carrageenan and chitosan-alginate films indicated that complex aggregates were formed due to ionic interactions.

One of important parameters in edible film as a food coating is its physico-mechanical properties including tensile strength, elongation break, and modulus young. These characteristic affect oxygen diffusion elerates oxidative damage and food protection from physical impact during the transport and storage process. The physico-mechanical properties of PEC as edible film has been studied in previous work [6,11]. It has been found that PEC has better physico-mechanical properties than its constituent polymers due to its tighter microstructure.

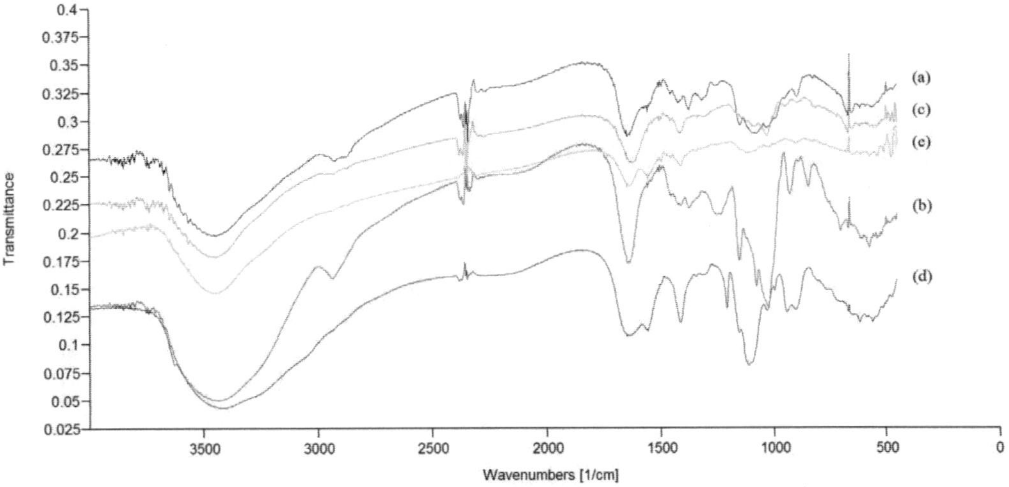

Figure 1. The FTIR spectra of (a) chitosan, (b) κ-carrageenan, (c) alginate, (d) chitosan-κ-carrageenan PEC and (e) chitosan-alginate PEC film

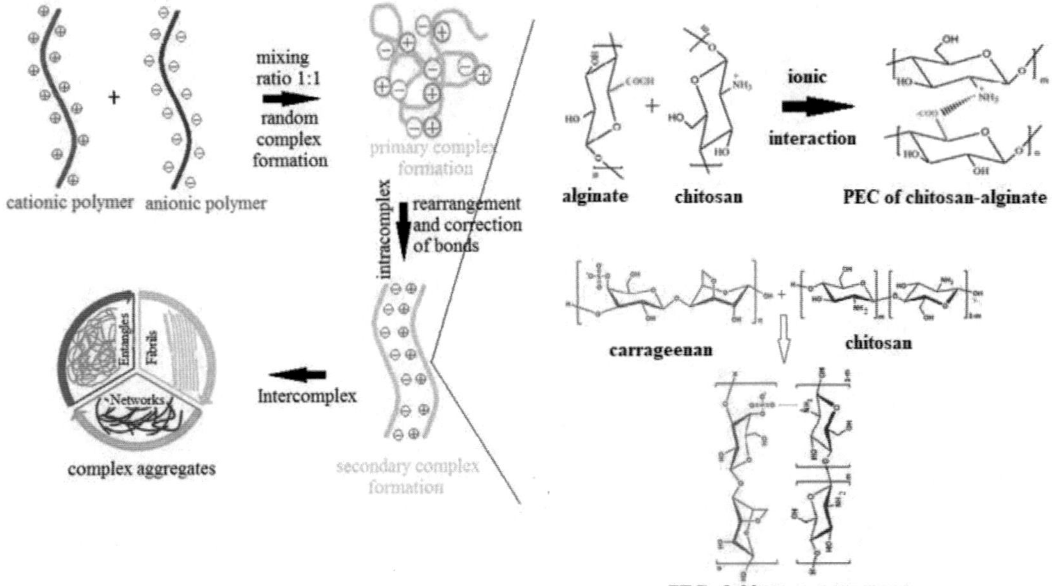

Figure 2. Mechanism of PEC formation [10,11].

(a) (b) (c) (d) (e)

Figure 3. SEM images of surface morphology of (a) chitosan, (b) κ-carrageenan, (c) alginate, (d) chitosan-carrageenan, (e) chitosan-alginate film with 3000x magnification.

3.2. Antibacterial activity test

Inhibitory activity of PECs and their constituent polymer against S. aureus and E. coli is shown in Figure 4. As shown in Figure 4, κ-carrageenan showed antibacterial activity against E. coli and S. aureus compared to the negative control of deionized water. The inhibitory effect on E. coli was higher than against S. aureus. The inhibitory effect by chitosan was found to be higher than that of κ-carrageenan in both types of bacteria. PECs have the strongest antibacterial activity compared to their original polymer. The antibacterial activity of the PECs can be due to the $-NH_3^+$ group which contributes a positive charge to PECs. Cationic charge can interact electrostatically with the major components of positive and negative bacteria, namely anionic phospholipid dipalmytol phosphatidylglyserol [19]. The interaction of protonated amine and sulfate in chitosan-carrageenan or carboxylate in chitosan-alginate is electrostatic that are not as strong as covalent bonds, allowed protonated amino groups to interact with anionic phospholipid of bacteria [6]. This interaction increases the permeability of the cell wall membranes of both bacteria, encourages release of nucleic acid, glucose, lactate dehydrogenase from cell as well as interfere transport of nutrient to cells causing bacterial death [20].

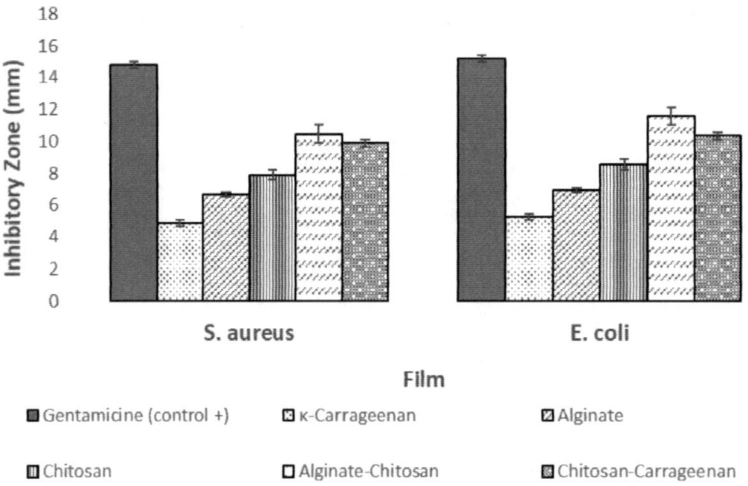

Figure 4. Antibacterial activity of films.

4. Conclusion

PECs based on chitosan, alginate and κ-carrageenan were prepared. PECs were formed by electrostatic interaction of chitosan-alginate as well as chitosan-κ-carrageenan. Surface morphology and structure of formed PECs was more irregular and fibrous structure than their native polymer. Antibacterial activity of PECs were the stronggest compare to their connstitients polymers. Hence, this biomaterial can be applied as active and safe coating for food product.

References

[1] Wyrwa J and Barska A 2017 *Procedia Eng.* **182** 770
[2] Han J, Garcia L R, Qian J P and Yang X T 2018 *Compr. Rev. Food Sci. Food Saf.* **17** 1
[3] Muller J, Martínez C G and Chiralt A 2017 *Materials* **10** 1
[4] Janjarasskul T and Krochta J M 2010 *Annu. Rev. Food Sci. Technol.* **1** 415
[5] Suput D Z, Lazić V L, Popović S Z and Hromiš N M 2015 *Food Feed Res.* 42 11
[6] Ismillayli N, Hadi S, Dharmayani N K T, Sanjaya R K and Hermanto D 2020 *IOP Conf. Ser. Mater. Sci. Eng.* **833** 012073
[7] Nouri A, Tavakkoli M, Ghorbanpour M, Agarwal S and Gupta V K 2018 nanocomposite *Int. J. Biol. Macromol.* **109** 1219

[8] Perinelli D R, Fagioli L, Lam J K W, Ba W, Filippo G, Casettari L and Bonacucina G 2018 *Eur. J. Pharm. Sci.* **117** 8
[9] Vo D and Lee C 2018 *Carbohydr. Polym.* **187** 1
[10] Hermanto D, Mudasir M, Siswanta D, Kuswandi B and Ismillayli N 2019 *J. Math. Fundam. Sci.* **51** 309
[11] Hermanto D, Mudasir M, Siswanta D, Kuswandi B and Ismillayli N 2020 *Molekul* **15** 40
[12] Hermanto D, Kuswandi B, Siswanta D and Mudasir M 2019 *Indones. J. Chem.* **19** 786
[13] Carneiro T N, Novaes D S, Rabelo R B, Celebi B, Chevallier P, Mantovani D, Beppu M M and Vieira R S 2013 *Macromol. Biosci.* **13** 1072
[14] Kaya M, Khadem S, Cakmak S, Mujtaba M, Ilk S, Akyuz L, Salaberria A M, Labidi J, Abdulqadir A H and Deligoz E 2018 *RSC Adv.* **8** 3941
[15] Das S N, Wagenknecht M, Nareddy P K, Bhuvanachandra B, Niddana R, Balamurugan R, Swamy M J, Moerschbacher B M and Podile A R 2016 *J. Biol. Chem.* **291** 18977
[16] Gu Y, Decker E A and Mcclements D J 2005 *Food Hydrocoll.* **19** 83
[17] Han J, Zhou Z, Yin R, Yang D and Nie J 2010 *Int. J. Biol. Macromol.* **46** 199
[18] Kulig D, Zimoch-Korzycka A, Jarmoluk A and Marycz K 2016 *Polym.* **8** 1
[19] Fernandes M M, Francesko A A, Torrent-Burgués J and Tzanov T 2013 *React. Funct. Polym.* **73** 1384
[20] Martins A F, Vlcek J, Wigmosta T, Reynolds M M, Popat K C and Kipper M J 2019 *Appl. Surf. Sci.* **136** 47877

IOP Conf. Series: Materials Science and Engineering **959** (2020) 012010 doi:10.1088/1757-899X/959/1/012010

Response surface methodology optimization of Congo Red dye adsorption onto MnFe-LDH adsorbent

L V Tan[1,2*], N T H Tham[3,4] and D T Sy[5]
[1] NTT Hi–Tech Institute, Nguyen Tat Thanh University, Ho Chi Minh City 700000, Vietnam.
[2] Ben Tre's Department of Science and Technology, Ben Tre Province 86000, Vietnam
[3] Center of Excellence for Green Energy and Environmental Nanomaterials, Nguyen Tat Thanh University, Ho Chi Minh City 700000, Vietnam
[4] Center of Excellence for Functional Polymers and NanoEngineering, Nguyen Tat Thanh University, Ho Chi Minh City 700000, Vietnam
[5] Institute of Chemistry, Vietnam Academy of Science and Technology, Ha Noi City, Vietnam

*E-mail: lvtan@ntt.edu.vn

Abstract. This study was carried out to optimize the adsorption process toward Congo Red (CR) dye over Mn-Fe LDH material in aqueous media. The effects of three variables including initial concentration, pH and reaction time were investigated with respect to maximal CR removal efficiency by utilizing response surface methodology (RSM). Through the variance of ANOVA, the second-order polynomial model was established in accordance with experimental data with high R^2 values (R^2=0.997), large F-value and small p-value, indicating that the proposed model is statistically significant. The estimated optimal conditions were validated by confirmation experiments. It was revealed that CR adsorption efficiency reached 53.21% at following optimal conditions: C_i = 48.09 mg/L; pH = 5.62; reaction time = 107 min. These results show that Mn-Fe LDH is used as an effective adsorbent for the removal of organic pigments in aqueous solution.

1. Introduction

Organic dyes, due to advantages such as resistant color, high water solubility and inexpensiveness, have been widely used in various industries including weaving, leather processing, paper, printing, plastics, pharmaceuticals, tanning, and others [1,2]. However, the discharge of wastewater containing organic dyes adversely affects human health and the environment [3,4]. Therefore, before being discharged into the environment, these dyes should be treated to minimize potential hazards [5–7]. In this study, Congo Red (CR) dyes have been evaluated such as a representative dye because its strong adsorption capacity on solids and its use in characterizing adsorptive materials [8,9].

In recent years, methods of treating environmental pollution are attracting much attention and adsorption has been shown to be a potential physicochemical method for removing dyes due to its advantages such as high efficiency, no toxic byproducts, applicability to regulatory systems, economic feasibility and ease of operation [10-14]. In particular, layered double hidroxides (LDH), due to advantages such as pore structure, large specific surface area and ion exchange capacity in the interlayer layer, has recently been proposed as a good candidate for manufacture of capable adsorbents [15–19].

In the previous studies, Mn-Fe LDH has been received much attention, which is a good candidate for the sensor, drug release [20]. The present study focuses on synthesizing Mn-Fe LDH materials and studying their adsorption capacity toward CR dye to expand the scope of applications. The conditions affecting adsorption performance including the initial concentration, pH value and reaction time were studied by surface response method (RSM).

2. Materials and method

2.1. Synthesis of Mn-FE LDH

Mn-Fe LDH was prepared by urea hydrolysis. The chloride containing Mn^{2+}, Fe^{3+} were mixed with urea solid (the molar ratios Mg^{2+} : Fe^{3+}: urea are 1:2:32), stirred for 15 min and then hydrothermally heated up to 100°C for 24h. Afterwards, the powder was filtered, washed with distilled water and ethanol for three times and then subjected to drying in at 90 °C for 24h [20].

2.2. Experimental

Mn-Fe LDH (0.04 g) was added to 100 mL CR solution with an appropriate concentration (40 mg/L to 60 mg/L) and pH (2 to 12). HCl 0.1M and NaOH 0.1M were used to adjust the pH. After the reaction had finished, 1 mL of the mixture was withdrawn and subjected to centrifugation at 3000 rpm for 10 min to remove the adsorbent. The remaining dye content was determined spectrophotometrically at λ = 497 nm. The adsorption efficiency was calculated based on the dye concentration before and after the process as follows

$$H\ (\%) = \left(1 - \frac{C_e}{C_o}\right) \times 100$$

where C_o and C_e are, respectively, initial and equilibrium dye concentrations (mg/L).

2.3. Experimental design with RSM

The CCD matrix was designed to evaluate the compatibility of experimental data and the response was selected as the CR elimination efficiency, as shown in table 1. A total of twenty experiments were set up by the central composite design with five levels including the low (–1), high (+1) and rotatable ($\pm\alpha$).

Table 1. Independent variables and their encoded levels.

				Levels				
No	Independent factors	Unit	Code	$-\alpha$	–1	0	+1	$+\alpha$
1	pH of solution (pH)	-	A	4.31	5	6	7	7.68
2	Concentration (C_o)	g/L	B	33.18	40	50	60	66.38
3	Time	min	C	86.36	100	120	140	153.36

3. Results and discussion

Through RSM method, 20 experiments were conducted to estimate the removal efficiency of CR and shown in table 2. The results of table 2 show that the effect of removing color is dominated by independent variables (initial concentration, pH and reaction time) and that adsorption performance is greatly determined by changes in surveyed parameters [21–25]. To select an appropriate model with statistically significant variables, the p-value threshold of the regressors is selected as 0.01, thus resulting in the following quadratic equation:

$$H\ (\%) = 50.34 - 3.87A - 6.24B - 2.20C + 0.7987AB - 2.31AC - 0.2227BC - 6.64A^2 - 3.4B^2 - 1.65C^2$$

Table 2. Matrix of observed and predicted values for CR adsorption capacity.

	Independent factors			CR adsorption			Independent factors			CR adsorption	
Run	pH (A)	C_o (B)	Time (C)	Actual (%)	Predicted (%)	Run	pH (A)	C_o (B)	Time (C)	Actual (%)	Predicted (%)
1	5	40	100	49.51	49.21	11	6	33.18	120	51.1	51.19
2	7	40	100	44.7	44.5	12	6	66.38	120	30.83	30.21
3	5	60	100	35.93	35.58	13	6	50	86.36	48.95	49.37
4	7	60	100	33.61	34.07	14	6	50	153.63	42.92	41.97
5	5	40	140	49.96	49.88	15	6	50	120	50.99	50.34
6	7	40	140	35.19	35.92	16	6	50	120	50.49	50.34
7	5	60	140	34.79	35.36	17	6	50	120	50.25	50.34
8	7	60	140	23.92	24.6	18	6	50	120	50.03	50.34
9	4.31	50	120	37.78	38.06	19	6	50	120	50.35	50.34
10	7.68	50	120	25.87	25.06	20	50	6	120	49.87	50.34

Table 3 displayed the ANOVA results that are used to evaluate the suitability of quadratic model and regression co-efficients. In general, the proposed model is statistically significant (95%) confirmed by large F-value and small p-values. Moreover, the coefficient of variance (CV), which represents the standard error of the estimate, achieved very low intensity (1.63%) and a high accuracy (AP = 354.42), which is indicative of high reproducibility of the model [26,27]. In particular, the correlation coefficient (R^2 = 0.997) shows that a significant proportion (99.70%) of the total variation can be explained by experimental models. Furthermore, an apparent random pattern of residuals could be observed in figure 1(a) and data points corresponding to the experimental and predicted values were distributed linearly on a straight line. Based on the result of lack of fit (p = 0.0561 > 0.05), the proposed model is consistent with experimental data.

Table 3. ANOVA for the CR removal models

Source	Sum of squares	Degree of freedom	Mean square	F-value	Prob. > F	Comment
Model	1604.20	9	178.24	373.11	$< 0.0001^s$	SD = 0.6912
A	204.26	1	204.26	427.57	$< 0.0001^s$	Mean = 42.35
B	531.62	1	531.62	1112.81	$< 0.0001^s$	CV(%) = 1.63
C	66.07	1	66.07	138.30	$< 0.0001^s$	Press = 31.92
AB	5.10	1	5.10	10.68	0.0085^n	R^2 = 0.997
AC	42.84	1	42.84	89.68	$< 0.0001^s$	$R^2_{(adj.)}$ = 0.994
BC	0.3969	1	0.3969	0.8308	0.3835^n	AP = 354.42
A^2	635.66	1	635.66	1330.59	$< 0.0001^s$	
B^2	167.54	1	167.54	350.69	$< 0.0001^s$	
C^2	39.41	1	39.41	82.50	$< 0.0001^s$	
Lack of fit	3.66	5	0.7317	4.75	0.0561	Not significant

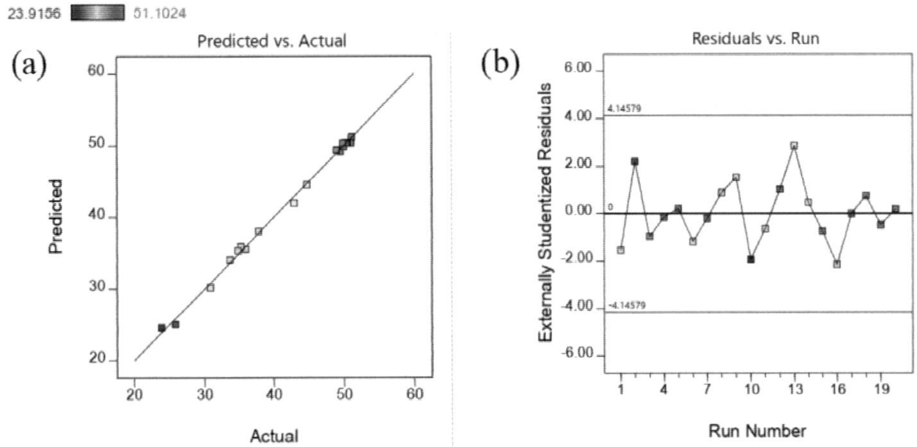

Figure 1. Actual versus predicted plot (a,b) for regression model of the percentage of CR removal.

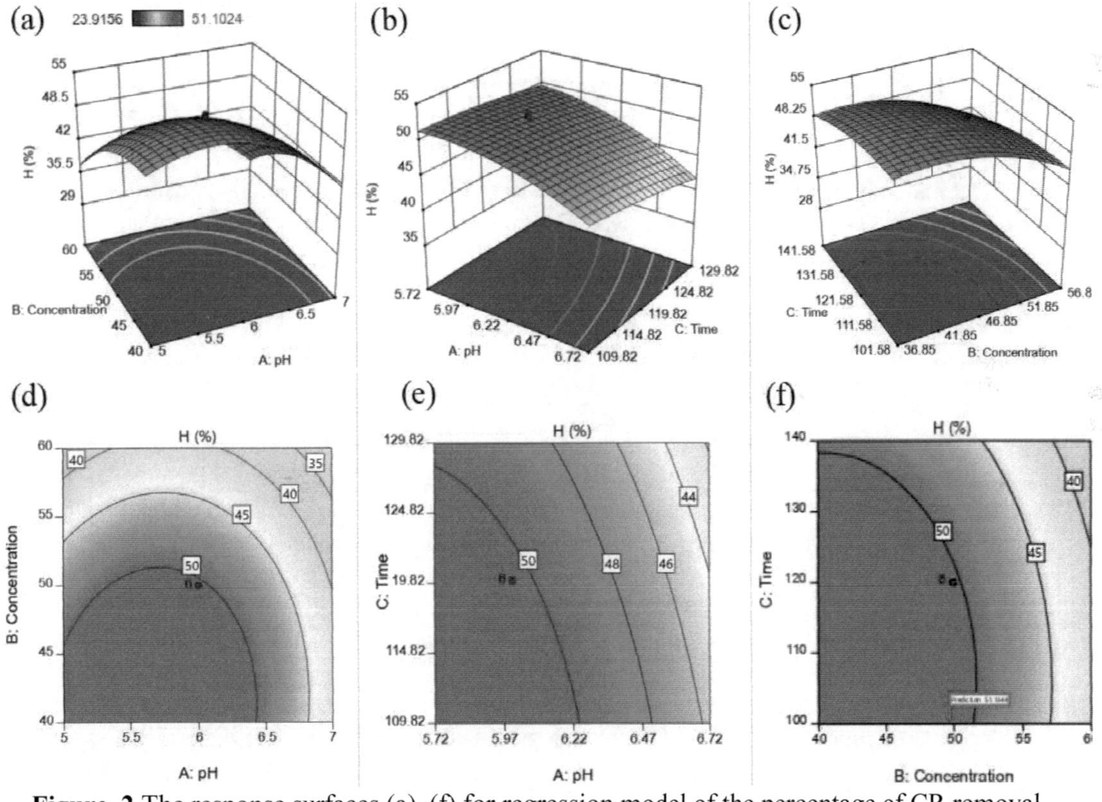

Figure. 2 The response surfaces (a)–(f) for regression model of the percentage of CR removal.

The removal efficiency of CR depends on the three variables, as shown in figures 2(a)–(f). The interaction between concentration and pH is illustrated in figures 2(a,d). It can be seen that reducing the concentration and pH value increases the adsorption capacity CR, as reflected by corresponding contour lines figures 2(a,d). Figures 2(b,e) suggest that the interaction between reaction time and pH for CR percent depends significantly on pH value. In a low pH aqueous solution, improved color removal efficiency could be observed. This is evidenced by figure 2(e) showing that convergent regions tended to be out of the investigation regions. Therefore, optimal conditions are likely to lie outside the survey area. Similarly, figures 2(c,f) show that the effect of concentration and reaction time on CR adsorption on Mn-Fe LDH. The optimal prediction results show that CR adsorption on Mn-Fe LDH depends significantly on the concentration and pH value. Attenuation of concentration and environmental pH leads to significant changes in adsorption performance [28–30]. In particular, the survey time reaches the optimal value in the survey conditions. The relationship between CR concentration and adsorption capacity could be explained by the lowered amount of adsorbed CR. In addition, increasing CR concentration also increased adsorption sites, leading to improved adsorption up to an equilibrium. On the other hand, changes in pH also are associated with the charge of adsorbent materials, leading to varying adsorption [25].

Solving the estimated equation with respect to maximal removal efficiency yielded following conditions: C_i = 48.09 mg/L; pH = 5.62; reaction time = 107 min. Actual experiment runs using those parameters resulted in the CR removal efficiency of 53.21%, approximately equal to the predicted value illustrated in the table 4. Those results demonstrate high compatibility of the proposed model with experimental data.

Table 4. Confirmation test.

Sample	pH (-)	Concentration (mg/L)	Time (min)	Adsorption capacity (%)			Desirability
				Predicted	Tested	Error	
Mn-Fe LDH	5.62	48.09	107	52.34	53.21	0.87	1.0000

4. Conclusion

Mn-Fe LDH material synthesized by hydrothermal method is used as an effective adsorbent to remove CR from aqueous solution. By adopting RSM optimization procedures, we found that optimal CR removal efficiency in this adsorption process was approximately 53.21%, corresponding to following optimal conditions: C_i = 48.09 mg/L; pH = 5.62; reaction time = 107 min. Further studies should contemplate the use of this promising adsorbent toward other dyes such as Methylene Blue or Rhodamine B to justify its applications in new wastewater treatment processes.

References

[1] Dai Y, Zhang N, Xing C, Cui Q and Sun Q 2019 *Chemosphere* **223** 12
[2] Shalla A H, Bhat M A and Yaseen Z 2018 *J. Environ. Chem. Eng.* 6 5938
[3] Alharbi O M L, Basheer A A, Khattab R A and Ali I 2018 *J. Mol. Liq.* **263** 442
[4] Ephraim E K, Anyama C A, Ayi A A and Onwuka J C 2018 *Adv. Mater. Sci. Eng.* **2018** 1
[5] Bilal M, Rasheed T, Iqbal H M N and Yan Y 2018 *Sci. Total Environ.* **644** 1
[6] Yang Y, Zhou T, Qiao Q and Chen S 2007 *J. China Univ. Min. Technol.* **17** 96
[7] Dastkhoon M, Ghaedi M, Asfaram A, Ahmadi Azqhandi M H and Purkait M K 2017 *Chem. Eng. Res. Des.* **124** 222
[8] Ma H T, Ho V T T, Pham N B, Bach L G and Phan T D 2018 *Appl. Mech. Mater.* **876** 91
[9] Karen A, Godínez L A, Espejel F, Ramírez-zamora R and Robles I 2019 *Waste Manag.* **85** 202
[10] Bach L G, Van Tran T, Nguyen T D, Van Pham T and Do S T 2018 *Res. Chem. Intermed.* **44** 1661
[11] Matouq M, Jildeh N, Qtaishat M, Hindiyeh M and Al Syouf M Q 2015 *J. Environ. Chem. Eng.* **3** 775
[12] Pi M, Lei C, Zhang F, Kuang P and Guo Y 2017 *J. Colloid Interface Sci.* **496** 158
[13] Veselská V, Hudcová B, Filip J, Komárek M and Číhalová S 2016 *Chemosphere* **168** 539
[14] Goh K H, Lim T T and Dong Z 2008 *Water Res.* **42** 1343
[15] Zubair M, Daud M, McKay G, Shehzad F and Al-Harthi M A 2017 *Appl. Clay Sci.* **143** 279
[16] Gupta V K and Suhas 2009 *J. Environ. Manage.* **90** 2313

[17] Mishra G, Dash B and Pandey S 2018 *Appl. Clay Sci.* **153** 172
[18] Sarfraz M and Shakir I 2017 *J. Energy Storage* **13** 103
[19] Mahjoubi F Z, Khalidi A, Abdennouri M and Barka N 2015 *J. Taibah Univ. Sci.* **11** 90
[20] Kang D, Yu X, Tong S, Ge M, Zuo J, Cao C and Song W 2013 *Chem. Eng. J.* **228** 731
[21] Wan S, Wang S, Li Y and Gao B 2017 *J. Ind. Eng. Chem.* **47** 246
[22] Wang W Z and Bian J M 2017 *Key Eng. Mater.* **730** 200
[23] Starukh G 2017 *Nanoscale Res. Lett.* **12**
[24] Huang G, Zhang K-L, Chen S, Li S-H, Wang L-L, Wang L-P, Liu R, Gao J and Yang H-H 2017 *J. Mater. Chem. B* **5** 3629
[25] Ahmed M A, brick A A and Mohamed A A A 2017 *Chemosphere* **174** 280
[26] Zhu C, Chen Z, Gao S, Goh B L, Samsudin L B, Lwe K W, Wu Y, Wu C and Su X 2019 *Prog. Nat. Sci. Mater. Int.* **2019** 628
[27] Chen K, Jiang S, Hong Y, Li Z, Wu Y L and Wu C 2019 *Prog. Nat. Sci. Mater. Int.* **6** 617
[28] Ding, Geng N, Zhou D S, Liu L, Yu Z L and Zong A S 2019 *Prog. Nat. Sci. Mater. Int.* **29** 672
[29] Dinh V P, Tran N Q, Tran Q H, Nguyen D T 2019 *Prog. Nat. Sci. Mater. Int.* **29** 648-654
[30] Lin M, Ye E and Zhang J 2019 3D *Prog. Nat. Sci. Mater. Int.* **29** 685-689

IOP Conf. Series: Materials Science and Engineering **959** (2020) 012011 doi:10.1088/1757-899X/959/1/012011

Identification and simulation model of mixed solutions of diethanolamine (DEA) and methyl diethanolamine (MDEA) for removal of acid gases

D Febriantini, S Pebriani, Usman*, B Purnomo, A K Rianto, A S Nugraha and N Khoirunnisa
Downstream Research & Technology Innovation, Innovation & New Ventures
PT Pertamina (Persero)
Jl. Raya Bekasi KM 20, Pulo Gadung, Jakarta 13920, Indonesia

* E-mail: usman01@pertamina.com

Abstract. Acid gases are often found in industrial and crude natural gas and represent technological problems to the environment and bad effect to the human health. The high-energy demand and processing restrictions make the amine absorption operation sensitive to any change in conditions. The mixture of diethanolamine (DEA) and methyl diethanolamine (MDEA) was developed as the removal of acid gases. The physical properties and the characteristics of the mixture were examined using FTIR, TGA and SEM-EDS and were described in this work.The HYSYS program were conducted in this work to reasonably predict removal amounts of acid gases from natural gas using the mixture diethanolamine (DEA) and methyl diethanolamine (MDEA) mixture as absorbents based on the plant specifications data from PT Pertamina (Persero)'s refinery unit. The correlations between the characteristics of the mixture and simulation model were revealed. The model predictions were in excellent agreement with the experimental data and the results were useful to represent the acid gas absorption process using alkanolamine solutions.

1. Introduction

Acid gases (e.g. H_2S and CO_2) are undesirable gases that are often found in industrial and crude natural gas and are severe challenges in energy, processing and chemical industries [1]. They may also carry other impurities of short-chain hydrocarbons (C1—C4), benzene, toluene and xylene (BTX), N_2, NH_3, CS_2 and COS that show negative impacts to the environment by greenhouse gas emissions and the human health due to their highly toxic combustion by products [2]. In the petroleum industry, sulfur compounds generate corrosion problems on the refinery units and catalyst poisoning employed to upgrade the fuel quality [3,4]. Therefore, a study on the practical and economical clean up of acid gas becomes essential.

There are many methodologies to decontaminate acid gases from natural gas. Alkanolamine or amine solutions such as monoethanolamine (MEA), diethanolamine (DEA), and methyldiethanolamine (MDEA) have been reported as valuable methods to absorb carbon dioxide and hydrogen sulfide from natural gas due to their reusability as absorbents and capability to hold acid gases in a large amount [5,6]. Two functional groups in these amines play significant roles i.e. the amino group supplies alkalinity required to increase the absorption rate of the acidic gasses; the hydroxyl group decreases the vapor pressure of the acid gases in the aqueous solution in order to improve their solubility [4,7,8].

IOP Conf. Series: Materials Science and Engineering **959** (2020) 012011 doi:10.1088/1757-899X/959/1/012011

Mixed amine systems commonly contain a tertiary amine that reacts spontaneously with H_2S by proton transfer yet relatively slowly with CO_2 compared to a primary or secondary amine [9–11]. Therefore, modifying CO_2 removal in a selective H_2S absorbtions system by varying the composition of amines is a beneficial feature. A blend of a secondary amine of diethanolamine (DEA) and tertiary amine of methyldiethanolamine (MDEA) is the most commonly employed constituents in the gas processing experiments [3].

Despite their advantages, it should be noticed that very few of the experimental data of the mixture of amines as a capture of acid gases were reported. The objective of the present work was to evaluate the properties of diethanolamine (DEA) and methyldiethanolamine (MDEA) mixture and to investigate their ability to capture acid gases in the simulation model by using the HYSYS program. The accurate information of diethanolamine (DEA) and methyldiethanolamine (MDEA) mixture's properties is essential before any modeling activity.

2. Materials & methods

2.1 Materials

Aqueous amines mixtures for density and kinematic viscosity analysis were prepared with double-distilled water. The solutes of DEA and DEA were Merck reagents with a purity of > 98 %. All solutions were formulated by mass employing a Matrix ESJ210-4B analytical balance with a precision of 0.0001 g.

2.2 Instrumentation

The density of amines solutions was evaluated using an Anton Paar DSA 4500 density meter at a temperature of 20 °C, while the kinematic viscosity of amines solutions was measured based on the transit time of the liquid meniscus through a glass capillary provided by Schott equipped by an electronic stopwatch for calculating efflux times. The viscometer was immersed in a thermostatic bath at 20 °C. Scanning electron micrograph (SEM) images were obtained using a Sigma Zeiss microscope with an attached camera. An elemental analysis of the corrosion effect of the amines solution was evaluated using energy-dispersive X-ray spectroscopy (EDX) spectra recorded on Amatek. Fourier transform infrared spectroscopy (FTIR) spectra were analyzed in the wavenumber of 4000-400 cm^{-1} region, on a Shimadzu infrared spectrophotometer. The thermal stability of the mixture observed during heating at a constant rate of 20 °C/min were performed by thermogravimetric analyzer (TA Instrument).

2.3 Simulated process as acid gas absorbents using HYSYS *program*

The purification process of acid gases by amines solutions was simulated by using the HYSYS program based on the plant specifications data from PT Pertamina (Persero)'s refinery unit. Various acid gas streams were mixed, compressed and cooled to the final condition of pressure of 5.917 barA and temperature of 40 °C then went into the LP Contactor. The LP contactor is a 16-trayed column where the mixed acid gas was contacted by an amine solvent which enters from the top of column. The treated gas would be sent to other unit for further treatment, while rich amine solutions were sent to amine solvent regeneration to remove dissolved H_2S and CO_2. Before going into the regenerator, rich amine solutions were flashed down to 2.386 barA in the flash drum. The regenerator is a trayed column with 32 trays. The heated rich amine solutions were fed to the regenerator on tray 28 (tray 1 is the bottom tray in the column). A horizontal thermosiphon regenerator reboiler provided the stripping vapor rising up the column which stripped the H_2S and CO_2 from the rich amine solutions. The hot lean amine solvent from bottom regenerator was pumped to the lean amine surge tank. Prior to reaching the surge tank, the lean amine solutions were cooled to 40 °C and filtered. The lean amine from surge tank then pumped to the LP Contactor to absorb H_2S and CO_2 from another mixed acid gas. Detailed simulation process is shown in table 1.

Table 1. Simulation data parameter.

Parameter	MDEA Solutions	DEA+MDEA Solutions
Lean amine concentration, % wt	30 (30% MDEA + 0% DEA)	30 (21% MDEA + 9% DEA)
Rich amine concentration, % wt	29.16 (29.16% MDEA + 0% DEA)	29.14 (20.4% MDEA + 8.74% DEA)
Circulation rate, kg/h	81,250	Same
Acid gas flow, kg/h	24,090	Same
H_2S inlet, % vol	10.61	Same
CO_2 inlet, % vol	0.34	Same

3. Result and discussion

3.1. Density and viscosity of amines solutions

The density and kinematic viscosity data for aqueous MDEA and mixture of DEA+MDEA solutions are presented in table 2. The data shows that density of mixture DEA+MDEA of 1.0287 g/mL was lower than MDEA solutions. The mixture of DEA+MDEA also showed a lower kinematic viscosity of 4.800 mPa·s than the solutions containing only MDEA of 3.677 mPa·s. Hence, the density and viscosity of aqueous amines measured in this work were affected by the addition of DEA and were in a good agreement with the study reported by Alavarez et al. (2006) [12]. It should be noticed that the transport properties of the fluids such as phase density, surface tension, viscosity and diffusion coefficients might affect the mass transfer coefficients and the interfacial contact area between fluids and gases [13].

Table 2. The experimental data of density and kinematic viscosity of amines solutions.

No.	Solutions	Density (g/mL)	Kinematic viscosity (mPa·s)
1	MDEA	1.0349	4.800
2	DEA+MDEA	1.0287	3.677

3.2. Characterization of the amines solutions

An experimental evidence of the functional group in a mixture of DEA+MDEA solutions was obtained through the FT-IR spectrum (figure 1). The spectrum represented the common characteristic stretching vibrations of O–H at approximately 3365 cm^{-1}. The associated with C–H stretching vibration bands were shown at 2941cm^{-1}. The band at 1447 cm^{-1} was caused by the C–O stretching vibration of the primary alcoholic groups in DEA and MDEA.

The thermal stability of amines solutions was investigated by recording the weight change during heating at a constant rate of 20 °C/min using a thermo-gravimetric analysis (figure 2a–b). Figure 2a shows the thermal decomposition of MDEA solutions that started at 30.41 °C and finished at 226.02 °C, while significant loss of its weight occurred at a temperature of 190.65 °C. However, mixture of DEA+MDEA solutions exhibited a more excellent stability compared to the solutions containing only MDEA. The thermal decomposition of the mixture started at temperature of 30.41 °C and finished at 229.25 °C with the major loss occurred at a temperature of 207.95 °C. By comparing peaks between MDEA solutions and mixture DEA+MDEA solutions, it can be noticed that they differ in chemical constituents which promote in stabilizing the matter. Addition of DEA into the solutions might improve the thermal stability of absorbents to capture acid gases.

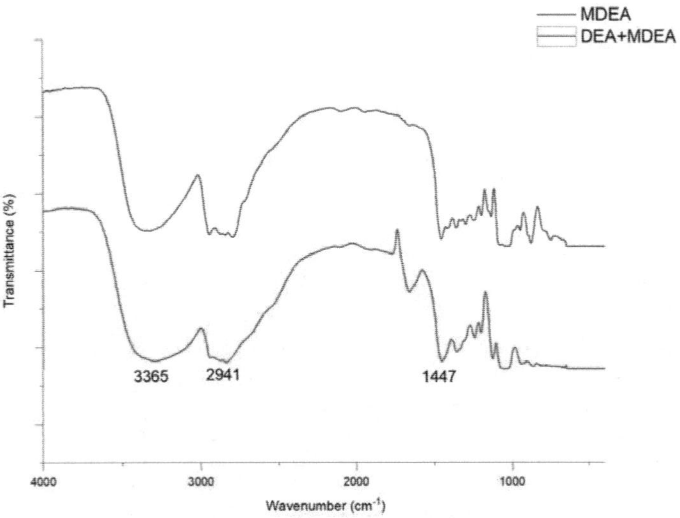

Figure 1. The FT-IR spectra of amines solutions.

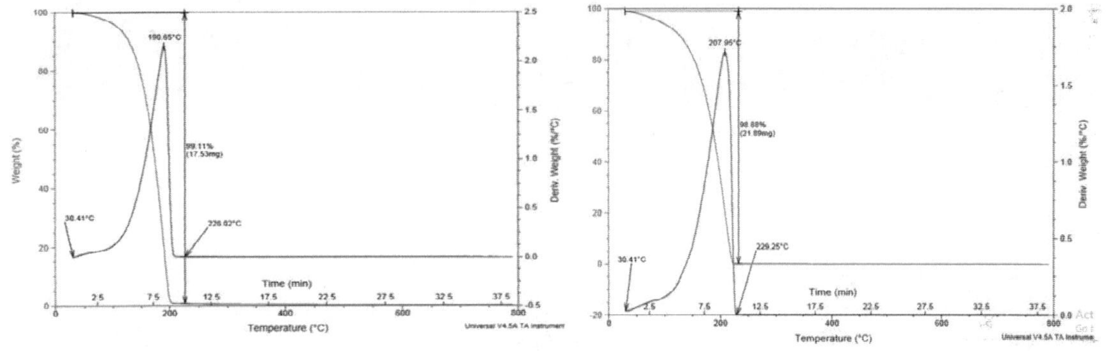
(a) (b)

Figure 2. Plots for thermal decomposition as weight percent loss (W %) vs temperature (T/°C) of: (a) MDEA and (b) mixture DEA+MDEA.

3.3 Corrosion effect of amines solutions

It has been reported that amine solutions might induce corrosion to the refinery equipment due to acidic and basic compounds that produce by thermal and chemical degradation [14], while some studies observed the activity of MDEA as corrosion inhibitors [8,9,15]. Table 3 describes the corrosion rate of carbon steel as materialsin a 4 h settling time in amines solutions at a maximum temperature of heating mantel (110–120 °C). The DEA+MDEA mixture showed a slower corrosion rate of carbon steel of 0.0269 mm/y compared to MDEA solutions (0.0737mm/y). These data were well agreement with the morphology and the surface structure of the carbon steel used as experimental materials and monitored by a scanning electron microscope (the images are shown in figure 4). The SEM images revealed and confirmed the difference of the morphology of the carbon steel after treatment with amines solutions in a 4 h settling time. The surface of the carbon steel altered to be rougher and inconsistent after treatment with the MDEA solutions compared to that with the DEA+MDEA mixture. The elemental analysis using energy-dispersive X-ray spectroscopy (EDX) revealed that the carbon steel after treatment with solutions containing only MDEA had more elements than the other one. These findings were also in an excellent agreement with our experimental data of thermal stability evaluated by a thermo-gravimetric analysis in which addition of DEA may improve the thermal stability of solutions to absorb acid gases.

Table 3. The experimental data of the corrosion effects of the amines solutions.

Solutions	Materials	Settling time (h)	Corrosion rate (mm/y)
MDEA	Carbon steel	4	0.0737
DEA+MDEA	Carbon steel	4	0.0269

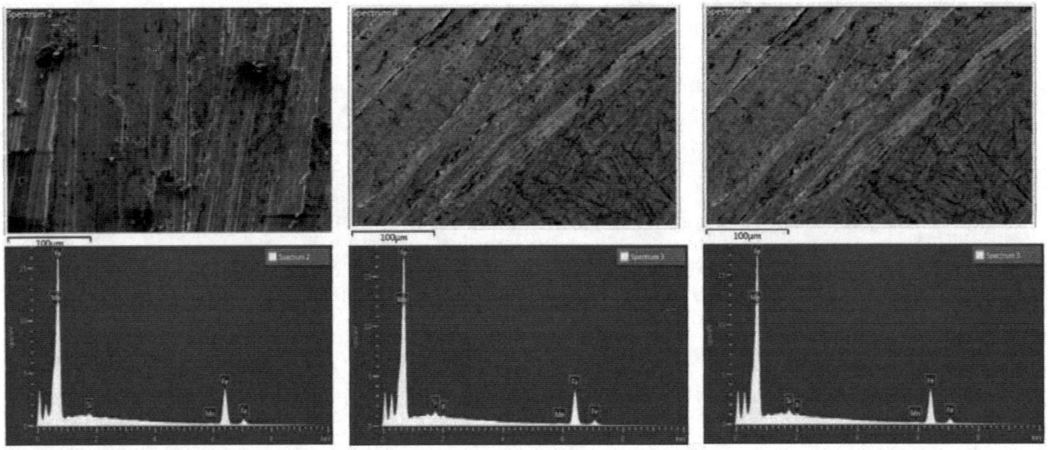

(a) (b) (c)

Figure 3. SEM-EDS images of carbon steel: (a) before treatment, (b) after treatment with MDEA solutions, and (c) after treatment with DEA+MDEA solutions.

3.4 Simulation as acid gases absorbents

The purification process of acid gases by amines solutions was simulated using the HYSYS program based on the plant specification data from PT Pertamina (Persero)'s refinery unit. The simulation results indicated that both solutions were powerful absorbents to reduce H_2S and CO_2 contents (table 4). The content of H_2S decreased extensively after treatment with DEA+MDEA solutions compared to MDEA solutions. By comparing the two amines solutions, the mixture of DEA+MDEA was not only sensitive to capture H_2S but also CO_2. The final content of CO_2 after treatment with DEA+MDEA solutions was 2.5 times lower than treatment with solutions containing MDEA only. These results confirmed the simulated process revealed by Naji and Abd (2019) [5]. They also reported that the plant using the DEA solutions required a lower energy than the plants using MDEA as solvents due to their lower vapor pressure. On our energy performance simulated data, the mixture of DEA+MDEA showed a lower reboiler energy of 9.95 Gcal/h compared to MDEA solutions (12.13 Gcal/h). The data also revealed that energy consumption using mixture of DEA+MDEA was more effective than that using MDEA at the regeneration system.

Table 4. The simulated data of amines solutions as an acid gases absorbent.

Acid gas	Initial content (% vol)	Final content of acid gases after treatment with different amines solutions (% vol)	
		MDEA	DEA+MDEA
H_2S	10.61	0.00179	0.00032
CO_2	0.3395	0.2175	0.0855
Reboiler energy (Gcal/h)		12.13	9.95

4. Conclusions

We have demonstrated the identification and lower corrosion rate of carbon steel after treatment with the mixture of diethanolamine (DEA) and methyl diethanolamine (MDEA) in removing acid gases such as H_2S and CO_2. The analysis of physical properties and characteristics of the mixture using FTIR, TGA and SEM-EDS confirmed the remarkable activity in absorbsing acid gases and excellent stability simulated by using HYSYS program based on the plant specification data from PT Pertamina (Persero)'s refinery unit. The model predictions were in excellent agreement with the experimental data and the results were useful to represent acid gas absorbtions process using alkanolamine solutions.

Acknowledgement

This study have received a support from PT Pertamina (Persero). The data and intellectual property related to this study is owned by PT Pertamina (Persero).

References

[1] Choi Y J, Sivanesan D, Lee J, Youn M H, Park K T, Kim H J, Grace A N, Kim I H and Jeong S K 2016 *J. Ind. Eng. Chem.* **34** 76
[2] Rufford T E, Smart S, Watson G C Y, Graham B F, Boxall J, Diniz da Costa J C and May E F 2012 *J. Pet. Sci. Eng.* **94–95** 123
[3] Al-Rashed O A and Ali S H 2012 *Sep. Purif. Technol.* **94** 71
[4] Gupta A K, Ibrahim S and Al Shoaibi A 2016 *Prog. Energy Combust. Sci.* **54** 65
[5] Naji S Z and Abd A A 2019 *Front. Energy* **13** 317
[6] Abd A A, Naji S Z and Barifcani A 2020 *Chinese J. Chem. Eng.*
[7] Akbar M M, Chemat F, Arunagiri A and Murugesan T 2016 *J. Therm. Anal. Calorim.* **123** 785
[8] Haron N, Sairi N A and Lee V S 2016 *Mol. Simul.* **42** 655
[9] Karunarathne S S, Eimer D A, Jens K J and øi L E 2020 *Fluids* **5**
[10] Akachuku A, Osei P A, Decardi-Nelson B, Srisang W, Pouryousefi F, Ibrahim H and Idem R 2019 *Energy* **179** 475
[11] Pal P, AbuKashabeh A, Al-Asheh S and Banat F 2015 *J. Nat. Gas Sci. Eng.* **24** 124
[12] Álvarez E, Gómez-Díaz D, La Rubia M D and Navaza J M 2006 *J. Chem. Eng. Data* **51** 955
[13] Weiland R H, Sivasubramanian M S and Dingman J C 2003 *53rd Annu. Laurance Reid Gas Cond. Conf. Norman, OK, 24 Febr. 2003* 79
[14] Bord N, Crétier G, Rocca J L, Bailly C and Souchez J P 2004 *Anal. Bioanal. Chem.* **380** 325
[15] Alef K and Barifcani A 2020 *J. Pet. Sci. Eng.* **184** 106522

Study of final temperature and heating rate variation to pyrolysis of Acacia (*Acacia mangium W.*) wood waste

Pranoto[1,*], K D Nugrahaningtyas[2] and R N O Putri[3]
Chemistry Department, Faculty of Mathematic and Science,
University of Sebelas Maret, Jl. Ir. Sutami 36A Kentingan, Surakarta, Indonesia 57126

[*] E-mail: pak_pran@yahoo.com

Abstract. The research pyrolysis of acacia wood waste has been carried out. This research was conducted to determine the optimal conditions that can be used for the pyrolysis process of acacia sawdust and the effect of the pyrolysis products. The pyrolysis process was carried out on the principle of slow pyrolysis using a batch type reactor. Pyrolysis was carried out by varying the final temperature of pyrolysis 375°C, 475°C, and 575°C and variations of heating rates 5°C/min, 10°C/min, and 15°C/min with a holding time of 20 minutes. The content of cellulose, hemicellulose and lignin at raw material was determined with the van soest method and the values are 32.75%, 35.16%, and 16.81%, respectively. The result of the pyrolysis process in the form of char at the final temperature of 575°C with a heating rate of 5°C/min produces a calorific value of 6.816 cal/gram. In the tar (oil) tested using GCMS, the acacia tar contains quite high phenol, acetic acid, 2 propanone. The results of pyrolysis can be seen that the highest activation energy value is found in the pyrolysis process with a variation of the heating rate of 15°C/min and the final temperature of the pyrolysis 575°C which is 49.56 kJ/mol.

1. Introduction

Population in the world from year to year was continues to increase. According to ESDM, consumption of energy in Indonesia reached 1.23 billion Barrels Oil Equivalent (BOE), which means an increase of 9% from the previous year. Energy consumption which continues to increase every year is not balanced with the existence of energy reserves in Indonesia, so it is necessary to renew energy. Renewing energy can be done by utilizing a biomass. Biomass can come from plants, animals, and industrial waste from agriculture, estate, forestry, farm and fisheries.

One of the biomass was potentially in Indonesia and has not been utilized properly is waste from the wood processing industry. The wood processing was produces 12–15% sawdust, 25–35% small ends, and 5–10% defective wood, with a total waste of 50.8% of the total raw material used [1]. Acacia (*Acacia mangium W.).* is wood that is widely used by wood industry in Indonesia. Acacia sawdust is a lignocellulotic waste containing about 32.8% cellulose, 34.32% hemicellulose and 14.65% lignin [2]. These compounds are compounds that have the potential to be converted into degraded biomass.

Energy conversion from biomass can be carried out by biochemical and thermochemical processes. Pyrolysis is one of the thermochemical biomass conversions. Pyrolysis is considered as the easiest technology to use for biomass conversion because pyrolysis has a high conversion ratio and easier control of the conversion process. Pyrolysis produces which have a high energy content [3].

Pyrolysis is a heating process that degrades biomass into char (charcoal), tar (oil) and gas [4] Factors that influence the pyrolysis process include the final temperature of the pyrolysis, heating rate, holding time, particle size, and particle weight [5]. Majedi *et al.* [6] have conducted research on mahogany sawdust pyrolysis by varying the final temperature of pyrolysis. The pyrolysis process is carried out with temperature variations of 350°C, 450°C, 500°C, and 600°C at a heating rate of 6°C/min and a heating rate of 1°C/min. The results showed that the percentage of mass reduction at a heating rate of 14°C/min with a final temperature of 600°C experienced a mass reduction of 75.24% with a little char product. In this study, it is known that variations in heating rate and final temperature affect pyrolysis products, but it is not explained what percentage of pyrolysis products and the character of pyrolysis products are formed so that it needs to be studied further about the character of pyrolysis products. Saddawi *et al.* [7] said that one of the things that affects the conversion of biomass using the pyrolysis method is the kinetics of the thermal decomposition reaction. In this study, it is said that the activation energy produced by a pyrolysis process can be seen from the mass degradation during the combustion process. Activation energy can be calculated using the Arrhenius equation, the Murray and White equation, Doyle, Senum and Yang, and Flynn Wall Ozawa and Criado using the reaction order of one. This research describes the calculation of the activation energy that occurs in the pyrolysis process but does not explain the effect of thermal degradation on the kinetics that occurs in the pyrolysis process.

Based on the problems regarding biomass waste, this research was carried out to determine the effect of the final temperature and heating rate of the pyrolysis process based on the yield of the resulting product, the decrease in mass during the pyrolysis process and the character of the pyrolysis products produced so that it can be used as energy conversion of acacia sawdust.

2. Research method

2.1 Equipment
The equipments used in the research are container raw material, *furnace* diameter 20 cm, electronic scale *A&D GF-300*, ADAM *thermoreader*, *thermocontroller*, *thermocouple tipe K*, condenser, container of the tar product, stopwatch, sets of *Fourier Transform-Infra Red (FTIR)* (Shimadzu type FTIR-8201 PC), set of *Gas Chromatography-Mass Spectrometer* (GC-MS).

2.2 Materials
The materials of the research is acacia sawdust obtained from Kedungombo, Sragen (7°25'32.8"S 110°48'48.9"E).

2.3 Research procedure

2.3.1 Preparation of the materials
Acacia sawdust is sieved to pass 20 mesh.

2.3.2 Pyrolysis of materials
The pyrolysis process of acacia sawdust is carried out by weighing 200 grams of material. The material laid on the furnace. Pyrolysis is carried out by varying the heating rate and final temperature. Pyrolysis of acacia sawdust was carried out by varying the final temperature were 375°C, 475°C, and 575°C. The heating rates used for acacia sawdust pyrolysis were 5°C/min, 10°C/min and 15°C/min. The holding time was carried out for 20 minutes. The addition of heating temperature is done every minute using a thermocontroller.

2.3.3 Testing the raw materials
The raw material to be used for the research is the van soest test (use to check the contens of cellulose, hemicellulose and lignin) and the proximate test to determine the chemical content of the raw material.

IOP Conf. Series: Materials Science and Engineering **959** (2020) 012012 doi:10.1088/1757-899X/959/1/012012

Raw materials are also tested using FTIR to determine the functional groups contained in the raw materials.

2.3.4 Testing the pyrolysis product
The pyrolysis process produces products in the form of char (charcoal) and tar (tar). In the two pyrolysis products, several tests were carried out. Char products are subjected to proximate test, calorific value test and characterization of functional group by FTIR. Meanwhile, the tar product was tested by FTIR and GC-MS characterization.

3. Result and discussion

3.1 Identification of acacia sawdust

3.1.1 Van Soest analysis
Based on the results of the tests that have been carried out, it is known that the acacia sawdust used has high levels of cellulose, hemicellulose, and lignin. According to Yang et al (2007), a biomass can be said to be good if it has a hemicellulose content of 20-30% of the total mass, so that acacia sawdust used in this study is biomass which is very good for energy conversion raw materials.

Table 1. Van Soest analysis result of sawdust Akasia.

Contents	Presentase (%)
Cellulose	32.75
Hemicelulose	35.16
Lignin	16.81

3.1.2 Proximate analysis
The proximate test results showed that the acacia sawdust before pyrolysis contained high levels of fixed carbon and volatile matter. High volatile matter will result in a significant reduction in mass during the pyrolysis process [4]. The water content of acacia sawdust indicates that acacia sawdust has a relatively low water content. High water content will reduce the heating value of a biomass. The water content of acacia wood is low so that acacia wood has a high calorific value. The test results of the calorific value of the acacia sawdust sample showed a low value. Based on SNI 01-6325-2000, biomass should have a heating value of at least 5000.00 cal/g. From the results of testing the calorific value of acacia sawdust waste before pyrolysis, it only has a calorific value of 4727.56 cal/g.

Table 2. The contens of the acacia sawdust.

Sample	Dust (%)	Water (%)	Volatile matter (%)	Fixed carbon (%)	Calor value (kal/g)
Acacia sawdust	1.68	12.81	60.36	25.52	4727.56

3.1.3 Analysis of functional group
The acacia sawdust that will be used as raw material for the pyrolysis process is carried out by FTIR test to determine the functional groups and chemical bonds contained in the sample. FTIR analysis can be seen in the FTIR spectra in figure 1. In the FTIR spectra, it can be seen that the raw material for acacia sawdust contains functional groups as in table 3.

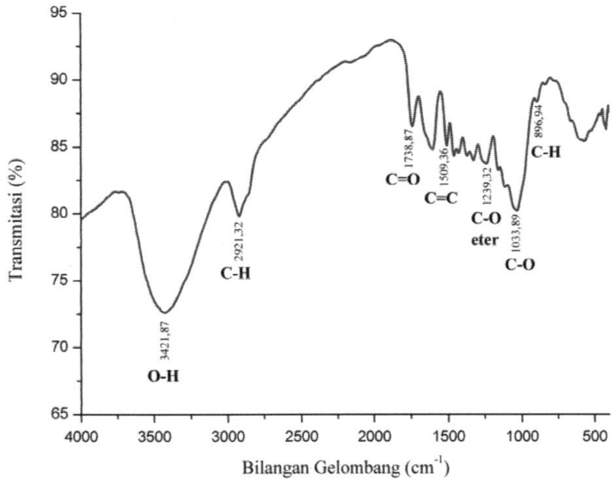

Figure 1. Spectra FTIR of Acacia sawdust.

Table 3. Analysis result of acacia sawdust by FTIR.

Functional group	Wave number reference(cm^{-1})	Wave number research (cm^{-1})
O-H	3550-3200	3421.87
C-H	2960-2850	2921.32
C=O	1750-1630	1738.87
C=C aromatic	1600-1500	1509.36
C-O eter	1275-1200	1239.32
C-O	1085-1030	1033.89
C-H aromatic	900-600	896.94

From table 3, it can be seen that acacia sawdust has O-H bonds which can be seen as the content of cellulose and hemicellulose. The known O-H bonds from the range of wave numbers are in the range 3550-3200 cm^{-1} [7]. The O-H bond can be said to be the water content in the raw material. In the FTIR spectra which can be seen in figure 1, it has a sufficiently wide O-H absorption so that it can be seen that the raw material has a high water content. In the range 2960-2850 cm^{-1}, it is known that there are C-H bonds which are cellulose and hemicellulose. In the 1750-1630 cm^{-1} bond there is a C=C bond that comes from hemicellulose. In the range of wave numbers 1600-1500 cm^{-1} there is a C=C bond which is known as the C=C bond of an aromatic compound which is known as a volatile substance in the raw material. At the wave number 1275-1200 cm^{-1}, it is known that there is a C-O bond which is an ether group which is included in the lignin content. In the range of wave numbers 1085-1030 cm^{-1}, it is known that there are C-O bonds derived from cellulose and hemicellulose, and at wave numbers 900-600 cm^{-1}, it is known that there are aromatic C-H bonds [7]. From the analysis carried out on acacia wood sawdust raw material, it can be seen that acacia wood dust originating from Kedungombo, Kalijambe, Sragen is a lignocellulotic biomass that can be used for energy conversion by the pyrolysis process.

3.2 Pyrolysis product

3.2.1 Yield product

The yield in pyrolysis products can be calculated using the mass balance principle. The yield calculation of the resulting pyrolysis products is by weighing the pyrolysis products in the form of solids (bio-char) and liquid (tar), while the products in the form of gases are calculated using the principle of mass balance. Based on the research that has been done, it can be concluded that the heating rate and final temperature affect the yield of the product from the pyrolysis process as depicted in figures 2, 3 and 4.

Based on the graphs in figures 2, 3 and 4, it can be seen that the heating rate and final temperature can affect the yield of the pyrolysis product. At the lower heating rate, the less mass decrease so that the mass of char formed is more. The mass decrease is also influenced by the final temperature, the final temperature was higher used in the pyrolysis process, was the mass decrease more. This results in fewer char products being formed. At high temperatures, a perfect heating process occurs so that many bonds in the cellulose and hemicellulose in the sample are damaged and make the sample more flammable, so that the product is less [8]. At the final high temperature, the decomposition reaction is very reactive, resulting in the conversion of biomass into liquid and gas products [9].

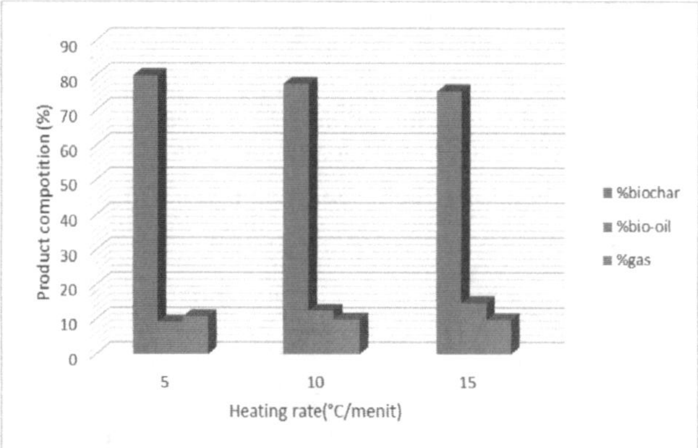

Figure 2. Yield graph of pyrolysis at final temperature 375°C.

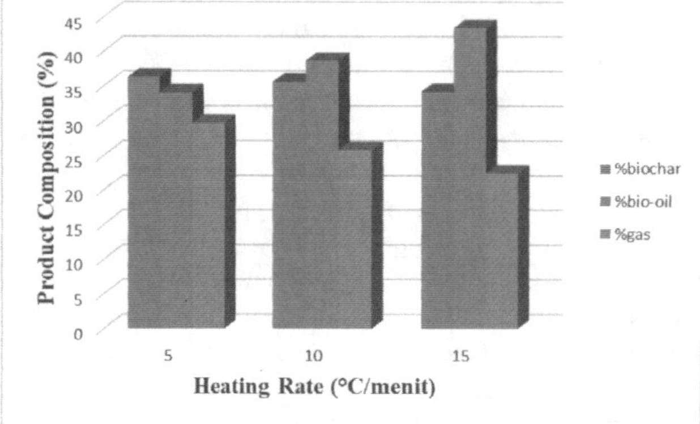

Figure 3. Yield graph of pyrolysis at final temperature 475°C.

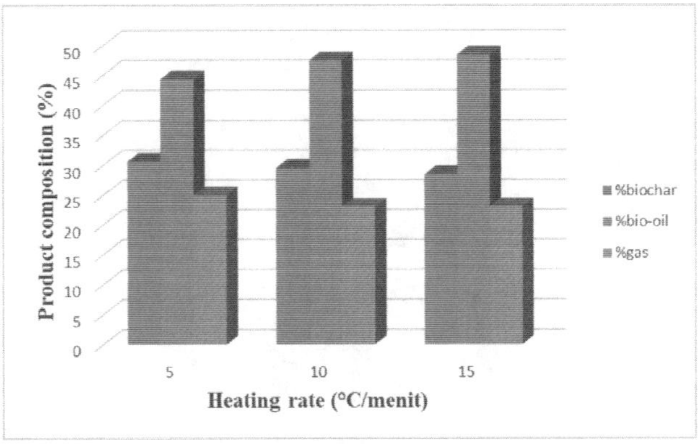

Figure 4. Yield graph of pyrolysis at final temperature 475°C.

3.2.2 Char (Charcoal)

Char (charcoal) is a product of pyrolysis in the form of solid. Char from the pyrolysis process has the main content in the form of carbon and contains some oxygen and hydrogen.

3.2.2.1 Calor value

The calorific value analysis is carried out on the solid product resulting from pyrolysis (char) which aims to obtain data on the heat energy released by a fuel due to the reaction or combustion process. The calorific value of raw acacia sawdust is 4727.56 cal/g. The calorific value of the pyrolysis results will be affected by the pyrolysis operating conditions such as the final temperature and heating rate. The results of the calorific value test on acacia sawdust pyrolysis products are shown in table 4.

Table 4. Calorific value result on char of acacia sawdust.

	Calorific value (cal/g)		
Heating rates (°C/min)	Final temperature 375 (°C)	Final temperature 475 (°C)	Final temperature 575 (°C)
5	5585.38	6467.72	6816.36
10	5558.02	6405.40	6742.76
15	5429.59	6281.83	6643.91

The heating rate affects the heating value of pyrolysis products. The higher the heating rate up to 15°C/min with a constant final temperature, the lower the calorific value. This happens because the calorific value is influenced by the moisture content and ash content in the pyrolysis products. At high heating rates, the chemical bonds are broken simultaneously while at low heating rates (<10°C/min) the weakest chemical bonds are broken. This can happen because there are still many chemical bonds that are still stable [10].

High heating rates can also predict high levels of volatile matter and a decrease in bound carbon content. The heating value will also be affected by the final temperature of the pyrolysis process. At the final temperature variation of 575°C the calorific value obtained will be higher [11].

3.2.2.2 Proximate analysis

Proximate analysis of acacia sawdust was carried out to determine the basic characteristics of the biomass after pyrolysis, such as the physical and chemical properties of the sawdust.

a. Proximate analysis on bio-char with variation of final temperature on same heating rate.

The proximate test was carried out on the bio-char product of the acacia wood sawdust pyrolysis process based on the variation of the final pyrolysis temperature (375°C, 475°C, 575°C) with a heating rate of 5°C/min can be seen in figure 5.

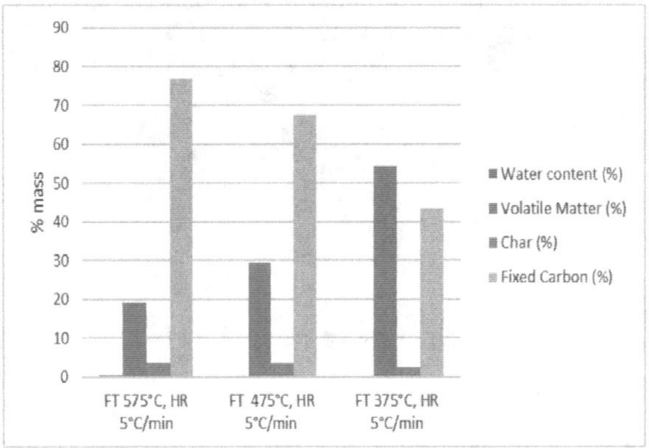

Figure 5. Graph of Proximate bio-char analysis with variation final temperature at the same heating rate.

b. Proximate analysis on bio-char with variation heating rate on same final temperature
The proximate test was carried out on the bio-char product of the acacia wood sawdust pyrolysis process based on variations in heating rate (5°C/min, 10°C/min, 15°C/min) with the final pyrolysis temperature of 575°C can be seen in figure 6.

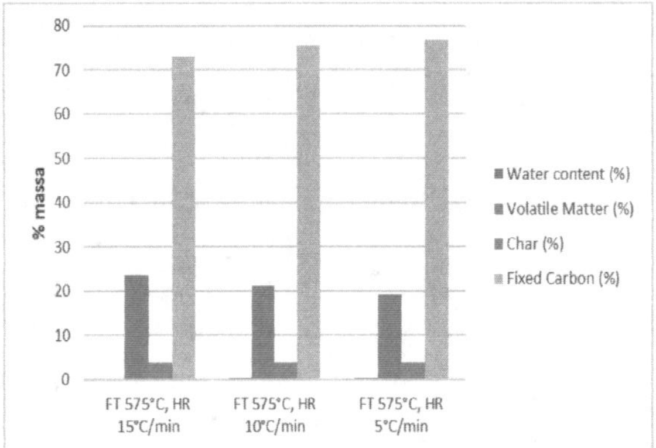

Figure 6. Graph of proximate bio-char analysis with variation heating rate at the same final temperature.

The proximate test results showed that the bio-char of acacia sawdust contained quite high fixed carbon and volatile matter even though with different heating rates. Fixed carbon increases with decreasing heating rate, this indicates that along with decreasing heating rate with high temperatures, the pyrolysis process will run perfectly. The ash content in bio-char corresponds to the content of fixed carbon. High levels of fixed carbon will increase the ash content because ash is a constituent of fixed carbon. Volatile matter levels decreased along with a decrease in the pyrolysis heating rate. This shows that at high heating rates the volatile matter is still high because the pyrolysis process occurs quickly so that the pyrolysis process has not occurred optimally. At low heating rates the pyrolysis process occurs slowly so as to maximize the evaporation process of volatile substances [6].

The water content of the proximate test results can also show how long the drying process takes on acacia sawdust pyrolysis. At a temperature of 575°C, it has a higher water content than the temperature below, so it shows that the pyrolysis process at this temperature takes a longer time. The water content of acacia sawdust indicates that the water content in this acacia sawdust is low. The water content of acacia wood is low, so this acacia wood has a high calorific value.

3.2.2.3 Functional group analysis

a. *Bio-char* functional group analysis with variation final temperature (375°C, 475°C, dan 575°C) at heating rate 5°C/min.
Analysis functional group with instrument FTIR on *bio-char* can be seen on spectra FTIR on figure 7. In the FTIR spectra, it can be seen that the bio-char from the pyrolysis of acacia sawdust contains functional groups as in table 5.

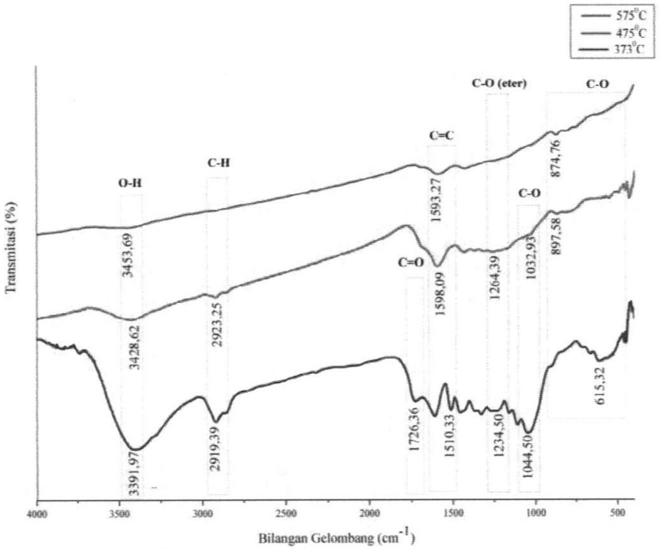

Figure 7. Spectra FTIR *bio-char* on same heating rate with variation final temperature pyrolysis.

Table 5. Result of functional group analysis of pyrolysis on same heating rate at variation final temperature.

Functional group	Wave number reference (cm^{-1})	Wave number research (cm^{-1}) 375°C	475°C	575°C
O-H	3550-3200	3391.97	3428.62	3453.69
C-H	2960-2850	2919.39	2923.25	-
C=O	1750-1630	1726.36	-	-
C=C aromatik	1600-1500	1510.33	1598.09	1593.27
C-O eter	1275-1200	1234.50	1264.39	-
C-O	1085-1030	1044.50	1032.93	-
C-H aromatik	900-600	615.32	897.58	874.76

From table 5, it can be seen that the bio-char from the pyrolysis of acacia sawdust has O-H bonds derived from cellulose and hemicellulose content. The O-H bond is known from the range of wave numbers, namely in the range 3550-3200 cm^{-1}[7]. The O-H functional group contained in this bio-char identifies that this bio-char still contains water. This O-H group will decrease in intensity with the increase in the final temperature of the pyrolysis process [7]. In figure 7, it can be seen that at low temperatures the absorption spectra of this O-H bond are wider than the spectra at high temperatures. In figure 7 it can be seen that the higher the final temperature of pyrolysis, the O-H absorption will be narrower. This is because the addition of the final pyrolysis temperature will reduce the water intensity in the bio-char. In the range 2960-2850 cm^{-1}, it is known that there are C-H bonds derived from cellulose and hemicellulose. In figure 7, it can be seen that the C-H group absorption appears to be narrower as the final temperature of pyrolysis increases. This is consistent with the theory that cellulose and hemicellulose will decompose in the temperature range 250-500°C [12]. In the 1750-1630 cm^{-1} bond there is a C=C bond that comes from hemicellulose.

In the range of wave numbers 1600-1500 cm^{-1} there is a C=C bond which is known as the C=C bond of an aromatic compound which is known as a volatile substance in the raw material. At the wave number 1275-1200 cm^{-1}, it is known that there is a C-O bond which is an ether group which is included in the lignin content. In figure 7, it can be seen that the C-O bond absorption which is the ether group of lignin looks narrower as the final temperature of pyrolysis increases. At 575°C, it can be seen that the absorption of this group is lost.

In the range of wave numbers 1085-1030 cm^{-1}, it is known that there are C-O bonds that come from cellulose and hemicellulose, and at wave numbers 900-890 cm^{-1}, it is known that there are aromatic C-H bonds. The aromatic C-H bond was only seen at 575°C and it is known that this compound appears at high temperatures.

Based on the spectral analysis mentioned above, it can be seen that cellulose and hemicellulose will be degraded along with the increase in the final temperature of pyrolysis and the water content in bio-char which is known from the absorption of O-H hydroxyl bonds decreases along with the increase in the final temperature of pyrolysis.

b. *Bio-char* functional group analysis with variation final temperature heating rate (5°C/menit , 10°C/menit, 15°C/menit) at the final temperature 575°C

FTIR analysis of the bio-char of pyrolysis products can be seen in the FTIR spectra in figure 8. In the FTIR spectra, it can be seen that the acacia sawdust sample contains functional groups as in table 6.

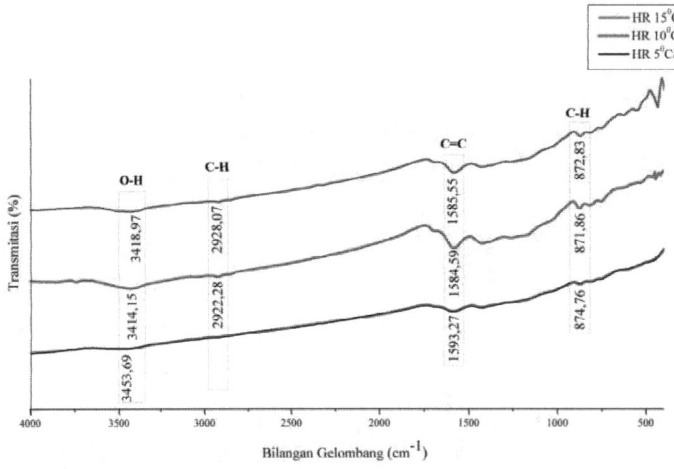

Figure 8. Spectra FTIR *bio-char* at final temperature 575°C with variation heating rate.

IOP Conf. Series: Materials Science and Engineering **959** (2020) 012012 doi:10.1088/1757-899X/959/1/012012

Table 6. FTIR bio-char analysis at the final pyrolysis temperature of 575oC with various heating rates.

Functional group	Wave number reference ($cm^{-1)}$	Wave number research (cm^{-1})		
		5°C/min	10°C/min	15°C/min
O-H	3550-3200	3453.69	3414.15	3418.97
C-H	2960-2850	-	2922.28	2928.07
C=C aromatik	1600-1500	1584.59	1584.59	1585.55
C-H aromatik	900-690	874.76	871.86	872.83

Based on the variation of heating rate with the same final pyrolysis temperature, it can be seen that this bio-char still contains water. As previously explained, at the high end temperature of pyrolysis, the O-H hydroxyl bond absorption at wave number 3550-3200 cm^{-1} is getting narrower, which indicates that there is still water content in the bio-char. At 1600-1500 cm^{-1} absorption, there is a C=C bond which indicates the presence of an aromatic group and at 900-690 cm^{-1} absorption there is an aromatic C-H bond. The presence of aromatic groups in the bio-char indicates that there are volatile substances from biochar which are formed at high temperatures in the pyrolysis heating process which is caused by damage to the volatiles of the bio-char material.

Based on the spectra analysis mentioned above, it can be seen that as the heating rate increases, the pyrolysis process occurs more rapidly, but the pyrolysis combustion process has not yet occurred completely. The faster heating rate also affects the hydroxyl group content, it can be seen that in figure 8, the absorption of hydroxyl groups at low heating rates looks narrow and will even disappear. This indicates that with a low heating rate a pyrolysis process will run perfectly because the water content will disappear over time.

3.2.3 *Tar*

Tar is a blackish liquid fuel produced from the conversion of biomass by pyrolysis. The largest organic components in tar are lignin derivatives, phenols, organic acids, alcohols, and carbonyl compounds such as aldehydes, ketones and esters [13]. Tar consists of molecules of different sizes derived from cellulose, hemicellulose and lignin. Tar is one of the alternative energy sources that is renewable and environmentally friendly. Tar is used for various industrial purposes, including combustion fuel and power generation to produce chemicals and can be mixed with diesel oil as fuel [14].

The tar composition varies depending on the nature of the biomass, pyrolysis temperature, heating rate, and environmental pyrolysis conditions. Generally tar consists of 51.2% water-soluble pyrolytic lignin, 25–30% water, 20–25% organic acids, 51% anhydrous sugar, 5–10% nonpolar hydrocarbons, and 10–25% other compounds. Tar has a pH between 2–4 which indicates that tar is acidic.

In this study, the tar obtained from the pyrolysis process was tested for its content using analysis by FTIR and GC-MS. This pyrolysis product tar was not subjected to heat testing and proximate test.

3.2.3.1 Functional group analysis

a. Funtional group analysis Bio-oil with variation final temperature (375°C, 475°C, dan 575°C) at heating rate 5°C/min
   Bio-oil, a product of acacia sawdust pyrolysis, is known to have brownish yellow and clear color. This bio-oil product is analyzed using FTIR to determine which chemical bonds and functional groups are present in this bio-oil. In the analysis using FTIR, it can be seen that the FTIR spectra in figure 9. In the FTIR spectra, it can be seen that the acacia sawdust sample contains functional groups as in table 7.

IOP Conf. Series: Materials Science and Engineering **959** (2020) 012012 doi:10.1088/1757-899X/959/1/012012

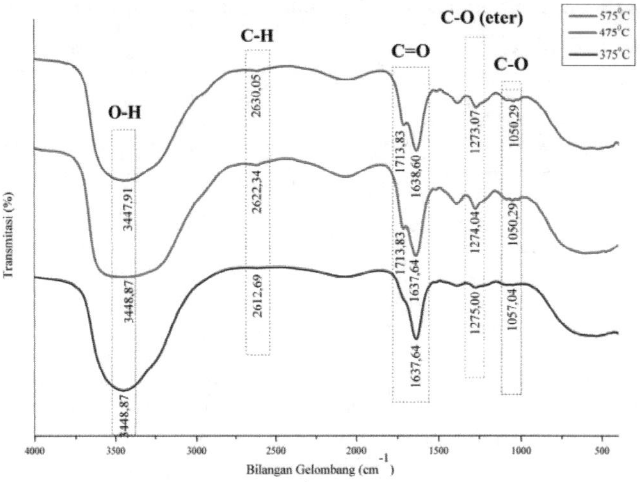

Figure 9. Spectra FTIR *tar* at heating rate 5°C/menit with variation final temperature.

Table 7. The results of the FTIR analysis of tar at the same heating rate as the final temperature variation.

Functional group	Wave number reference(cm^{-1})	Wave number research (cm^{-1})		
		375°C	475°C	575°C
O-H	3550-3200	3448.87	3448.87	3447.91
C-H	2900-2600	2612.69	2622.34	2630.05
C=O	1750-1630	1637.64	1713.83 & 1637.64	1713.83 & 1638.60
C-O eter	1275-1200	1275	1274.04	1273.07
C-O	1085-1010	1057.04	1050.29	1050.29

From table 7 it can be seen that the bio-oil resulting from pyrolysis of acacia sawdust with the same variations in the pyrolysis temperature and heating rate has O-H bonds derived from cellulose and hemicellulose content. This O-H bond is an O-H alcohol bond known from the wave number range, namely in the range 3550-3200 cm^{-1} [7]. The O-H functional group contained in this bio-oil is known to be a hydroxyl group which identifies that this bio-oil still contains water. This hydroxyl group will decrease in intensity as the final temperature of the pyrolysis process increases [7]. In figure 9, it can be seen that at low temperatures the spectra of the O-H bond are wider than the spectra at high temperatures. At the wave number 1750-1630 cm^{-1} there is a C=C bond that comes from hemicellulose. It is known that all the tars from the pyrolysis process of acacia sawdust contain cellulose. At temperatures between 300-500°C, it is known that cellulose and hemicellulose are differentiated.

In the range of wave numbers 1390-1340 cm-^{1}, it is known that there are C-H bonds which are known to be deformations of cellulose and hemicellulose. At low temperatures, it is known that the C-H bonds of hemicellulose are not too wide compared to high temperatures. This can occur because at high temperatures, namely above 400°C, cellulose and hemicellulose can be formed from the pyrolysis process.

At the wave number 1275-1200 cm^{-1}, it is known that there is a C-O bond which is an ether group which is included in the lignin content. In figure 14 it can be seen that the higher the final temperature of pyrolysis, the more visible the stretch in the bond spectra is. This shows that lignin is difficult to degrade at low temperatures. The higher the final temperature of the pyrolysis, the more visible the stretch in the wave number is. In the range of wave numbers 1085-1010 cm^{-1}, it is known that there are C-O bonds that come from cellulose and hemicellulose.

Based on the spectral analysis mentioned above, it can be seen that the higher the pyrolysis temperature is carried out, the more cellulose, hemicellulose, and lignin content is visible.

b. Funtional group analysis of Bio-oil with variation heating rate (5°C/min, 10°C/min, 15°C/min) at final temperature 575°C.
In the analysis using FTIR, it can be seen the FTIR spectra in figure 10. In the FTIR spectra, it can be seen that the acacia sawdust sample contains functional groups as in table 11.

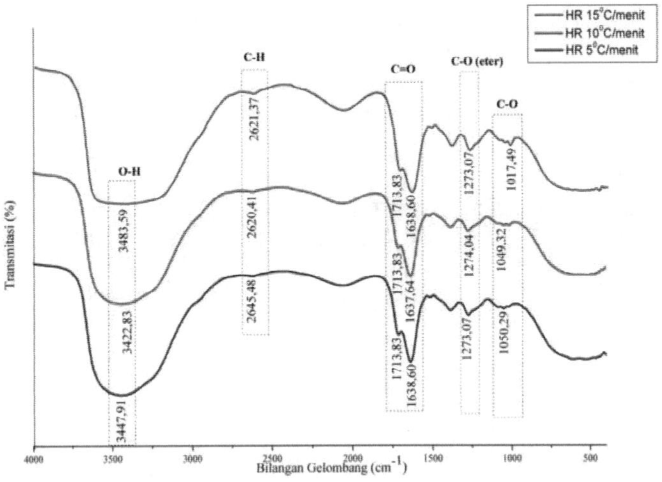

Figure 10. Spectra FTIR *tar* at final temperature 575°C with variation heating rate.

Table 8. Analysis FTIR *tar* at final temperature 575°C with variation heating rate.

Funtional group	Wave number reference(cm^{-1})	Wave number research (cm^{-1})		
		5°C/min	10°C/min	15°C/min
O-H	3550-3200	3347.91	3422.83	3483.59
C-H		2645.48	2620.41	2621.37
C=O	1750-1630	1713.83 & 1638.60	1713.83 & 1637.64	1713.83 & 1638.60
C-O eter	1275-1200	1273.07	1274.04	1273.07
C-O	1085-1010	1050.29	1049.32	1017.49

From table 8 it can be seen that the bio-oil resulting from pyrolysis of acacia sawdust with the same variation in heating rate and pyrolysis temperature has an O-H bond derived from cellulose and hemicellulose content. This O-H bond is an O-H alcohol bond known from the wave number range, namely in the range 3550-3200 cm^{-1} [7]. The O-H functional group contained in this bio-oil is known to be a hydroxyl group which identifies that this bio-oil still contains water. In figure 10, it can be seen that at a high heating rate of 15°C/min, it can be seen that the O-H absorption is wider than the O-H absorption at heating rates of 10°C/min and 5°C/min. It is known that at a fast heating rate the pyrolysis process has not yet occurred completely and has not been maximized. At a heating rate of 5°C/min, it can be seen that the absorption of the O-H group is narrower, which indicates that the hydrolysis group which is classified as water decreases more in the pyrolysis process and indicates that the pyrolysis process occurs optimally.

At the wave number 1750-1630 cm^{-1} there is a C = C bond that comes from hemicellulose. It is known that all bio-oils from the pyrolysis process of acacia sawdust contain cellulose. In the range of wave numbers 1390-1340 cm^{-1}, it is known that there are C-H bonds which are known to be deformations of cellulose and hemicellulose.

At the wave number 1275-1200 cm^{-1}, it is known that there is a C-O bond which is an ether group which is included in the lignin content. In the range of wave numbers 1085 -1010 cm^{-1}, it is known that there are C-O bonds derived from cellulose and hemicellulose.

Based on the spectra analysis mentioned above, it can be seen that the higher the heating rate, the pyrolysis process will occur faster and make the pyrolysis process not optimal. The pyrolysis process which is carried out at a slow heating rate will maximize the pyrolysis process and the absorption of functional groups of cellulose, hemicellulose and lignin can be seen.

3.2.3.2 Analysis GC-MS

For pyrolysis products in the form of liquid (bio-oil), chemical analysis in the form of gas chromatography - mass spectroscopy (GC-MS) is carried out to determine the content of any chemical compounds contained in the pyrolysis products. The products tested were heating rate of 5°C/min, where this variation had the most yield among the variations in the pyrolysis process carried out. GC-MS results can be seen in figure 11.

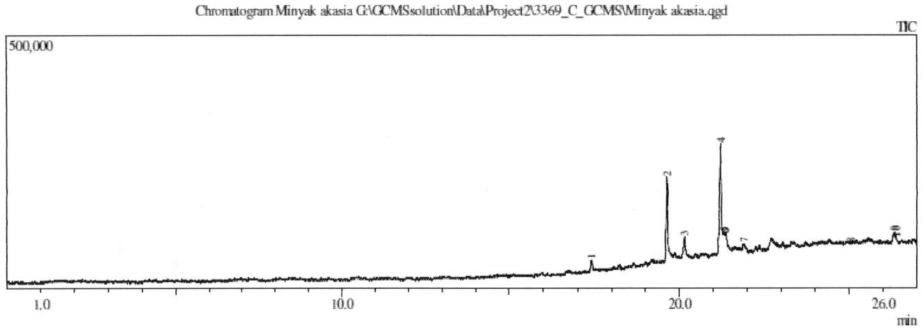

Figure 11. Spectra GC-MS on *tar* acacia sawdust wit heating rate 5°C/min at final temperature 575°C.

From the GC-MS results which can be seen in figure 11, there are 26 peaks of the number of compounds in tar. The seven most dominant chemical compounds include: acetic acid, phenol, CAS (hexane acid), 2-propanone, 2-amino, 1-propanol, propionic acid, 1-hydroxy-2-butanone, the seven dominant compounds are components in tar which is the decomposition of cellulose and hemicellulose. It can be seen that the pyrolysis product tar contains the components needed to become tar, as explained by Goyal et al (2006), where the organic components contained in tar, namely the contents of the acid, ester, alcohol, ketone groups. , aldehydes, alkenes, furans and other compounds which are characteristic of tar.

In this study it can be seen that the tar produced is acidic, this is because the pyrolysis temperature used is susceptible to 375°C–575°C. In this temperature range, cellulose, hemicellulose, and lignin compounds decompose. Hemicellulose decomposes at 200°C–275°C and cellulose decomposes at 240°C–350°C [15]. This tar has a phenol content of about 40–45%. This is because phenol from tar is obtained from lignin, where lignin can decompose at temperatures above 300°C explaining that the formation of phenol compounds in the pyrolysis process is the result of the decomposition of lignin in biomass at temperature range 400°C–500°C. The tar produced has a fairly large phenol content, so that the tar produced from the pyrolysis of this acacia sawdust can be used as a substitute fuel for petroleum..

3.3 Activation energy

Activation energy is defined as the energy that must be met to start a chemical reaction. The activation energy value of a sample indicates the ease with which the sample can react. The calculation of activation energy carried out in this study is based on a reaction kinetics formula using the first order which is often referred to as global kinetics (global kinetic). This study is adapted to the research of Himawanto [3] regarding the determination of the amount of activation energy in global kinetics using a graphical method where the formula used is based on the Arrhenius equation.

The Arhennius equation is a simple mathematical method and is widely used to obtain exponential factors and activation energies based on TGA experiments which depend on reaction rate constants [16]. Each variation is then plotted on a graph. The graph obtained from the Arrhenius equation is a graph of the relationship between ln(d/dt) and 1/T. The graph is then made linear regression to obtain the equation y = ax + c. The x coefficient or slope of the equation is multiplied by the gas constant (8.314 J/mol.K), so the activation energy value is obtained.

Table 9. The relationship of activation energy with variations in heating rate and final temperature

	Energi aktivasi (kJ/mol)		
Heating rate(°C/min)	Final Temperature 375°C (kJ/mol)	Final Temperature 475°C (kJ/mol)	Final Temperature 575°C (kJ/mol)
5	14.48	43.57	47.13
10	20.36	43.88	47.37
15	27.97	46.38	49.56

Table 9 shows the calculated results of the activation energy with variations in heating rate and final temperature. When viewed from the results of the calculation of the activation energy with variations in final temperature (375°C, 475°C and 575°C) and by varying the heating rate, the results obtained in table 9 show that the higher the temperature and heating rate, the resulting activation energy also increases. Increasing temperature causes the decomposition reaction of biomass to also increase. Increased decomposition reactions will require a large amount of energy to break the bonds contained in the biomass.

The lignocellulose content contained in biomass will be decomposed through several stages, where an increase in the rate of reaction that occurs in the pyrolysis process results in the biomass being decomposed slowly. The lignocellulose content contained in the sample of this study will differentiate the heating rate of a sample when it is treated thermally. If a material has a dominant lignin content, it requires a higher temperature to reach the devolatilization point. Due to the high temperature needed to break the bonds in lignin so that the energy required is also greater. In addition, the higher the heating rate, the acceleration of heat energy provided to break down the larger biomass compounds causes the decomposition rate of acacia sawdust to be faster and bigger.

3.4 Mass degradation

According to [6], the combustion process of solid materials is divided into three stages, namely the drying stage, the devolatilization stage, and the combustion stage. The drying stage is the stage where the sample will lose its moisture content, the devolatilization stage is the stage where the biomass conversion process chemically occurs. The biomass devolatilization stage generally occurs at temperatures around 200–500°C which is marked by a sharp decrease in the TG graph due to the decomposition of cellulose, hemicellulose and lignin [16]. Then the combustion stage on the TG graph shows a slight decrease, even closer to flat or straight because what happens is only the remaining lignin decomposition process which is getting slower. In this study, the thermal degradation analysis is divided into two variations, namely variations in heating rate and variations in final temperature respectively.

Pyrolysis in this study will be influenced by several factors. These factors include the final temperature and heating rate. The final temperatures used in this study were 375°C, 475°C and 575°C. The decomposition process of acacia sawdust with various heating rates at the final temperature of 375°C, 475°C and 575°C is depicted in figures 12, 13 and 14, respectively.

Figures 12, 13 and 14 show the influence caused by the heating rate in the biomass pyrolysis process. It can be seen from the acacia sawdust decomposition curve that the greater the heating rate, the more the graph will shift to the left, which means that the decomposition reaction occurs faster than the lower heating rate. The final temperature variations carried out in this study did not show any significant effect. This can happen because at the final temperature variation, pyrolysis takes place at the same rate as the example in figure 11 which is carried out at a rate of 5°C/min. So that the energy received by the biomass every one minute is the same.

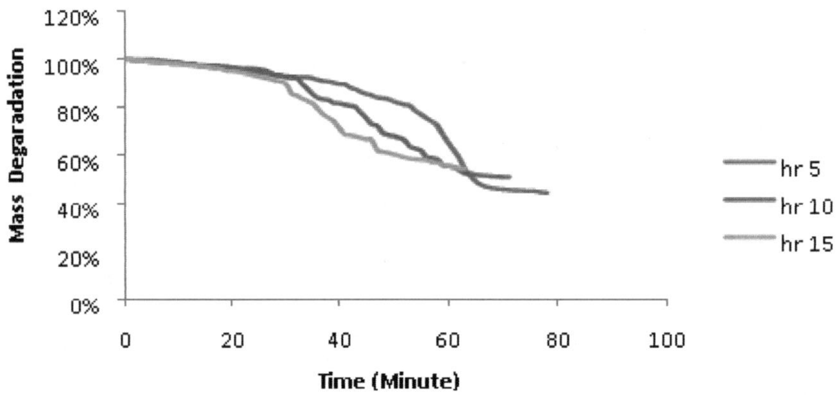

Figure 12. TG curve of acacia sawdust with variaton of heating rate at final temperature 375°C.

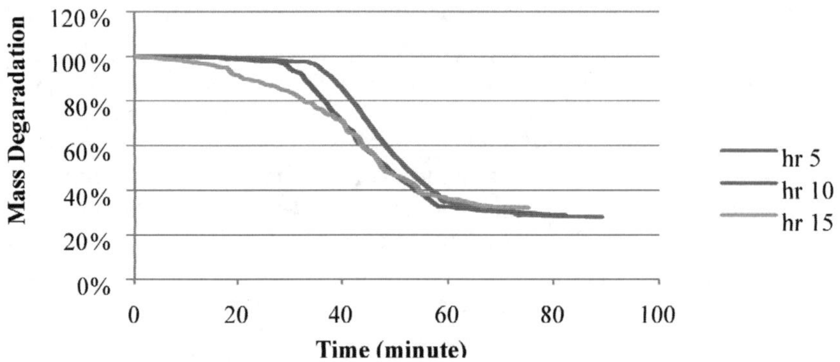

Figure 13. TG curve of acacia sawdust with variaton of heating rate at final temperature 475°C.

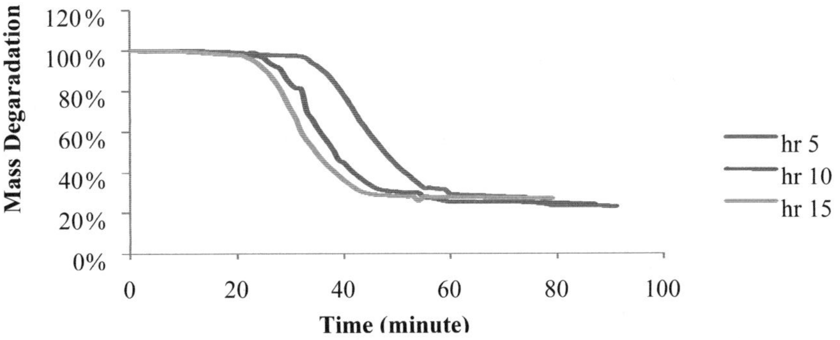

Figure 14. TG curve of acacia sawdust with variaton of heating rate at final temperature 575°C.

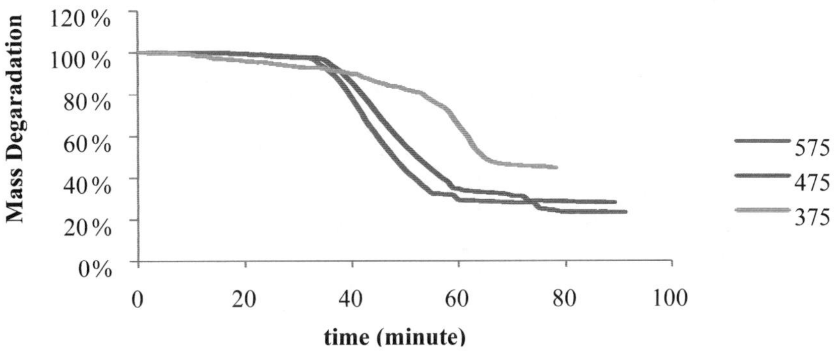

Figure 15. TG curve of acacia sawdust with variaton of final temperature at heating rate 5°C/min.

Figure 15 shows the decomposition curve of acacia sawdust with variations in the final pyrolysis temperature. In general, the variation in heating rate and final temperature is intended to determine which variation is best done for the pyrolysis process of acacia wood waste biomass. Based on the data table above, it can be seen that the best temperature for the pyrolysis process of acacia wood waste is at a temperature of 575°C and a heating rate of 5°C/min, this is indicated by a significant reduction in mass that occurs in the devolatilation process and the combustion process in the range temperature 500°C to 575.

4. Conclusions

The conclusions of this study are as follows. The final temperature of pyrolysis affects the amount of pyrolysis products produced. The mass reduction at a temperature of 575°C at a heating rate of 5°C/min reached an average of 76.34%. The calorific value produced by pyrolysis product with a final temperature of 575°C reaches 6816.36 cal/g. At the final temperature of 575°C, the highest activation value was 49.56 kJ / mol. The characters of char produced at high temperature are known to contain the functional groups O-H, C-H, C = C aromatic, and C-H aromatic. Pyrolysis heating rate affects the amount of pyrolysis products produced. The mass reduction at a low heating rate of 5°C/min at a temperature of 575°C reaches an average of 76.34%. The calorific value produced by pyrolysis products with a heating rate of 5°C/min reaches 6816.36 cal/g. At a heating rate of 15°C, the maximum activation value is 49.56 kJ/mol. The characters of char produced at high temperature are known to contain the functional groups O-H, C-H, C = C aromatic, and C-H aromatic.

References

[1] Rachman O and Malik J 2011 *Penggergajian dan Pemesinan Kayu Untuk Industri Perkayuan Indonesia* (Jakarta: Badan Penelitian dan Pengembangan Kehutanan)
[2] Sixta H 2006 *Handbook of Pulp* (Germany: Wiley-VCH Verlag GmbH & Co. KgaA)
[3] Himawanto D A, Dhewangga R D P, Indarto, Saptoadi H and Rohmat T A 2010 *J. Teknik Industri* **11** 127
[4] Demirbas A 2004 *J. Anal. Appl. Pyrolysis* **72** 97
[5] Naves D, Thunman H, Matos A, Tarelho L and Barea A G 2011 *Prog. Energy Combust. Sci.* **37** 611
[6] Majedi F, Wijayanti W and Hamidi N 2016 *Rotor* **9** 59
[7] Saddawi A, Jones J M, Williams A and Wojtowicz M A 2010 *Energy Fuels* **24** 1274
[8] Prvulovic S, Gluvakov Z, Tolmac J, Tolmac J, Matic M and Brkic M 2014 *Energy Fuels* **18** 2013.
[9] Azizi A, Erdawati A and Suhartono 2019 *J. Riset Sains Kimia Terapan* **8**
[10] Wardani S, Pranoto and Himawanto D A 2018 *AIP Conf. Proc.* **2049** 1
[11] Sridhar G 2007 Torrefaction of Bamboo *15th European Biomass Confference & Exhibition* (Berlin: German)
[12] Collard F and Blin J 2014 *Renew. Sustain. Energy Rev.* **38** 594
[13] Setiawan A, Andrio O and Coniwanti P 2012 *J. Teknik Kimia* **2** 9
[14] Raveendran K, Ganesh A and Khilar K C 1996 *J. Fuel* **75** 987
[15] Diebold J P and Bridgwater A V 1997 Overview of fast pyrolisis of biomass for the production of liquid fuels *Developments in thermochemical biomass conversion* eds A V Bridgwater and D G B Boocock (London: Springer)
[16] Yunanda I Y, Bahri S and Saputra E 2016 *Jom FTEKNIK* **3** 1

Carbon and zinc oxide synthesized by gelatin template as potential material for fight viruses covid-19: Future potential material

M Ulfa[1,*], D Prasetoko[2] and M Fajar[3]
[1] Chemistry Education Study Program, Faculty of Teacher Training and Education Sebelas Maret University, Jl. Ir. Sutami 36A, Surakarta 57126, Central Java Indonesia
[2] Department of Chemistry, Faculty of Science, Institut Teknologi Sepuluh Nopember, Keputih, Sukolilo, Surabaya, 60111, East Java, Indonesia
[3] Department of Chemistry, Faculty of Science, Gadjah Mada University55111, Centre of Java, Indonesia

[*] E-mail: ulfa.maria2015@gmail.com

Abstract. The COVID-19 pandemic that attacks the respiratory tract is spreading rapidly to all regions in Indonesia. In this pandemic situation, daily products such as sprays, tissue paper and masks need to be developed using nanomaterials. In this paper, graphite carbon and zinc oxide were synthesized using gelatin as a pore guide. The morphology and character of graphite carbon and zinc oxide were characterized by SEM, EDX and XRD to address appropriate active centers to counteract the protein spike interaction in covid-19 with ACE in the human body. Several illustrations are presented in this paper to understand the opportunity of carbon and zinc oxide materials as the basis for making antiviral devices to store viral molecules so that human-to-human transmission can be prevented. Keyword: graphitic carbon, zinc oxide, COVID-19, antiviruses.

1. Introduction

Corona 2 virus which causes severe respiratory infection or often called SARS-CoV-2 (The Severe Acute Respiratory Syndrome Corona Virus 2) causes the emergence of a new type of corona virus in 2019 called coronavirus disease or COVID-19 [1–2]. Transmission of COVID-19 took place very quickly with an indication in the 6 months since its appearance in the world has claimed 619,593 lives with a total number of cases reaching more than 15 million by mid-June 2020 [3]. When the transmembrane called Spike protein from SARS-CoV-2 interacts with a receptor called Angiotensin-Converting Enzyme 2 (ACE2) in the human body, ACE2 gives access to the virus entry to the cell surface of the organ marked by the virus that begins to multiply in the human body. [2]. That's why antiviral material is needed to prevent and combat the spread of COVID-19. Nanomaterials have received a lot of attention as antiviral materials for the prevention and control of COVID-19 because of several advantages of nanomaterial properties such as regular structure and antiviral properties in them [8–9]. One example is the application of immunomodulatory proteins encapsulated by nanoparticles which successfully counteract and eradicate SARS-CoV-2 [3–4].

The nanoscale material which is widely applied in the medical world is carbon and zinc oxide [5]. Graphitic carbon and zinc oxide material is hydrophobic with a two-dimensional structure. Graphitic carbon is in the form of a sp2-hybridized carbon layer [6]. However, naturally, graphitic carbon and zinc oxide has hydrophilic and electronic properties [7-8]. Even reduction of common carbon and graphitic carbon and zinc oxide by heating will add hydrophilic properties with low oxygen content [9]. Another uniqueness is that the layers on a graphitic carbon stack will interact through the π-π interaction with a special distance due to the presence of an oxygen layer in it [10] . The other material is zinc oxide which has been widely applied in the process of catalysis, drug control delivery, biosensors, biofuels, and uptake [1-3]. Zinc oxide is widely chosen as a catalytic material because it has many advantages, namely thermally stable, chemically inert, inexpensive and harmless [5–7]. The micropore category of zinc oxide is incapable of providing an entryway for applications involving large molecules [8–11].

The most interesting thing about not only carbon graphitic carbon and zinc oxide but also zinc oxide is that it has the ability to move electrons towards bacteria or viruses which is called migration [11–13]. The very negative charge on graphitic carbon and zinc oxide are advantageous because it causes the high affinity of the graphitic carbon and zinc oxide when it moves electrons towards viruses or bacteria which naturally have a very positive charge [14]. This movement or migration of electrons causes cytoplasmic efflux or cytoplasmic depletion in the cell. In addition, electron movement decreases cell metabolism, damages membrane lipids and induces oxidative stress which affects viral damage [4–5]. Besides, the movement of electrons to bacterial or viral cells produces reactive oxygen species that kill viruses and bacteria [6]. Research on the use of graphitic carbon and zinc oxide in feline viruses has shown that graphitic carbon and zinc oxide absorbs viral lipids well through hydrogen bonds and electrostatic interactions [7]]. Similar studies have noted that the interaction of graphitic carbon and zinc oxide and zinc oxide and virus loads causes the destruction of the viral membrane [8]. This is the basis of this review to predict graphitic carbon and zinc oxide that synthesized by gelatin template as a material to fight viruses, especially during this pandemic.

Gelatin is a potential organic molecule as a precursor to carbon. This is based on previous research that the carbon and nitrogen content in gelatin is 50% and 17% [15]. The carbon content in gelatin indicates the ability of gelatin to be converted into carbon. nitrogen in gelatin is a representation of the amine group as an active group that easily reacts with organic and inorganic substances through physical and chemical bonds [7]. This is evidenced by mixed gelatin with sodium silicate to synthesize graphitic carbon and zinc oxide followed by pyrolysis and silica release. The amine group (-NH2) in gelatin interacts strongly with silicate species which play a role in forming carbon pore walls or zinc oxide through hydrogen bonds. The interaction would stabilize the carbon skeleton even though the synthesis results showed many structural defects when observed by TEM (transmission electron microscopy). The previous research rationale in this study to synthesize graphitic carbon and zinc oxide from gelatin using different techniques in order to obtain carbon products with high regularity [15]

This study examines the synthesis and characterization of carbon graphite and zinc oxide from gelatin as a pore guide using self-assembly techniques. The characters of graphite carbon and zinc oxide were analyzed by SEM, EDX and XRD. Analysis of the potential use of carbon graphite and zinc oxide as a base material to fight Covid-19, achieved an illustrative approach adapted to the characterization results as suggestions for further research

2. Experimental

2.1 Material

The materials used in this study were gelatin, zinc sulfate, sucrose (Merck, 98%), sulfuric acid, hydrochloric acid, n-hexane, NaOH (Merck, 99%), hydrochloric acid, ethanol, methanol, acetone (Merck , 98%)

2.2 Synthesized graphitic carbon and zinc oxide

Graphite carbon and zinc oxide from gelatin were synthesized according to the procedure of Ulfa et al. (2016) using pore-directing gelatin. The ratio of carbon gelatin and Zn/gelatin on precursor: sulfuric acid: water was 1.00: 0.2: 100.00 (w/w). The successive synthesis stages are: gelatin infiltration to precursors, dehydration, pyrolysis and washing.

2.3 Characterization

Scanning electron microscopy, EDAX and X-Ray Diffraction were carried out on graphite carbon and zinc oxide. The study in this paper contains predictions that might occur if carbon and zinc oxide are used to restrain the interaction between the spike protein in Covid -19 and ACE-2 in the human body using the electromechanical properties of both materials. After characterization, an analysis will be carried out as outlined in the illustration of the use of carbon graphite and zinc oxide as materials for making equipment to prevent and fight COVID-19.

3. Result and discussion

Figure 1 is the EDAX result of a carbon graphite sample produced from a mixture of gelatin and sucrose which undergoes partial dehydration and carbonization at 100–160 ° C for 7 hours. The graphite carbon sample contains 70.5% carbon; 22.5% oxygen; 7.0% nitrogen. The nitrogen content indicates that there is an interaction that facilitates the bonding between the hydrogen in sucrose and the amine groups in the gelatin. According to Hsu et al. (2007) the involvement of hydrogen bonds in sucrose in interactions with gelatin will occur because gelatin has many amino groups that have a high affinity for strong interactions with sucrose. Only a small portion of the gelatin is converted to carbon because the energy used during dehydration and partial carbonization is not sufficient to carbonize all the gelatin molecules. The tentative conclusion, the character of the functional groups of the carbon source and the constituent elements of gelatin shows a strong interaction between gelatin and sucrose in producing graphite carbon.

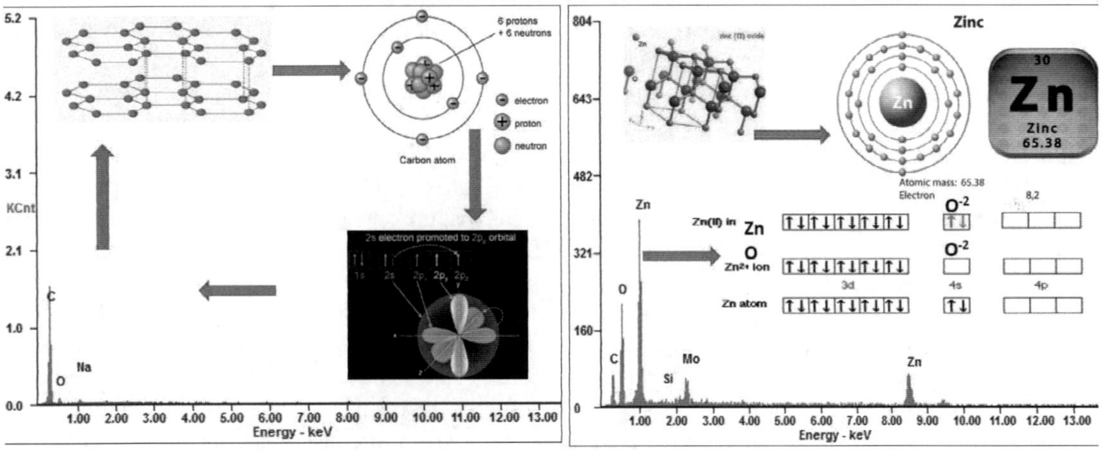

Figure 1. EDX of graphitic carbon and zinc oxide.

The zinc oxide in figure 1 (right) contains about 68% zinc and 22% oxygen. The carbon content of about 5% indicates that the gelatin is not completely decomposed at calcination temperature in a precurous mixture of zinc and gelatin as a pore guide. Zinc oxide samples also showed some impurities such as silica and molybdenum in small amounts of precursor impurities. The interaction between the carbon and amine groups in gelatin causes an electrostatic force so that the zinc element attaches to the gelatin micelles to form a morphology like thin glass shards (figure SEM). Electrons in graphite carbon and electrons in the outer orbitals of zinc can undergo migration and attack the virus so that the virus experiences deactivation. The EDX results show that both carbon graphite and zinc oxide are predicted to be capable of fighting viruses through the presence of electrons in the outer orbitals of each element.

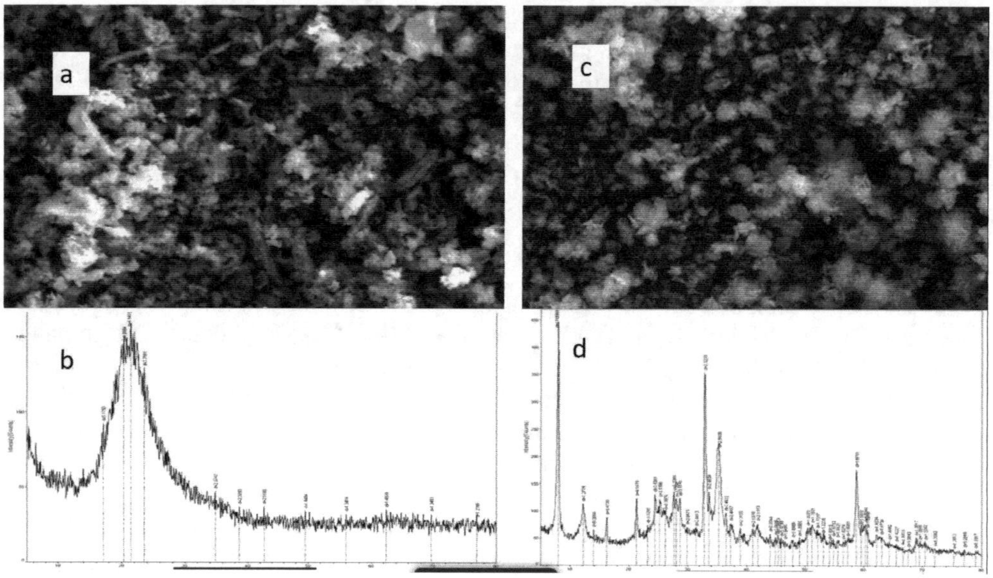

Figure 2. SEM and EDX of graphitic carbon and zinc oxide.

Figures 2a dan 2c shows the typical morphology of the synthesized graphitic carbon and zinc oxide from hydrothermal process. Pipe agglomerate shape and smooth surface can be observed in graphitic carbon. The morphology like thin glass broken-structure of zinc oxide shows also in figure 2c. This pipe agglomerate shape-like structure of carbon graphite and thin glass broken-structure-like of zinc oxide is generally a pattern in particle formation because gelatin acts as a molecular guide during infiltration by selecting the surface force with the smallest energy. The XRD patterns of graphite carbon and zinc oxide after high temperature heating are shown in figures 2b and 2d. There are two broad peaks at = 24.8 and 43.5 which are a reflection of plane (002) and plane (101). The vertices can be indexed to a hexagonal graphite grid. The widening of the peaks indicates the presence of an amorphous carbon phase in graphite carbon and zinc oxide [12]. This semi-hexagonal graphite structure takes place during the carbonization process at high temperatures. In the carbonization process, the carbon atoms in the graphite carbon are rearranged to form graphene sheets and partially collapse to produce a pore structure. This structure would be suitable for storage of substances having a molecular size smaller than micropores (<2 nm). Whereas in the XRD pattern of zinc oxide there are typical peaks of wurtzite and a little carbon which confirms the EDX data.

All the results of the characterization, especially the analysis of the elemental content, are a consideration to make an illustrative approach to the use of graphite carbon and zinc oxide as materials for making antiviral sprays, antiviral tissue paper and antiviral masks. The illustration in figures 3–5 is a prediction that can be used to determine whether carbon graphite and zinc oxide have virucidal potential so that in the future it is likely to prove to be an economical and efficient material to fight COVID-19 (figure 3). Figure 1 is an illustration for an antiviral spray. Considerations are made based on the nature of graphite carbon and zinc oxide having distinctive electrons in their molecular orbitals. This outer shell electron will be used as an antiviral center because of its ability to carry out movements that attack proteins in viruses. The affinity, porosity and stability of carbon graphite and zinc oxide have been processed using gelatin as the pore directing agent and functional groups. Gelatin functions to reduce the size of carbon and zinc oxide precursors to a nanometer level so that when used, it is easily dispersed on the surface of the virus. Automatically, the electrons in the outer shell will migrate so that the spike protein in Covid-19 will be damaged so that the virus will die.

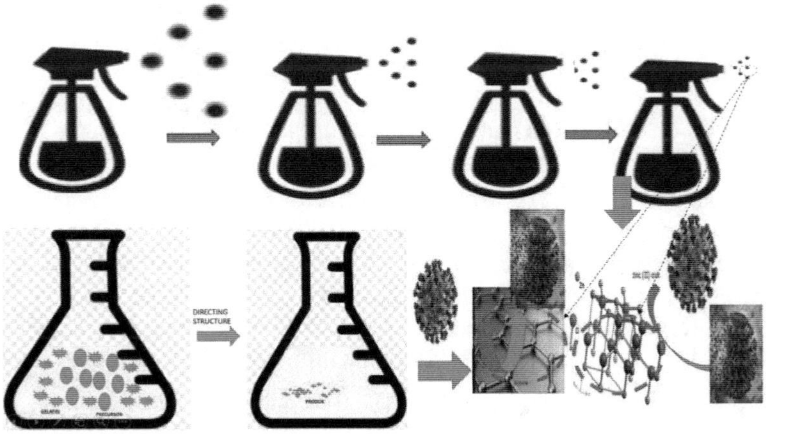

Figure 3. Illustration of graphitic carbon and zinc oxide using for spray antiviruses.

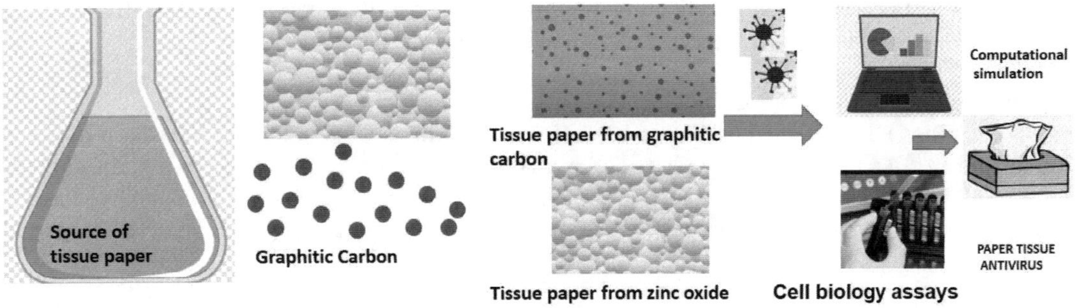

Figure 4. Illustration of graphitic carbon and zinc oxide using for paper tissue antiviruses

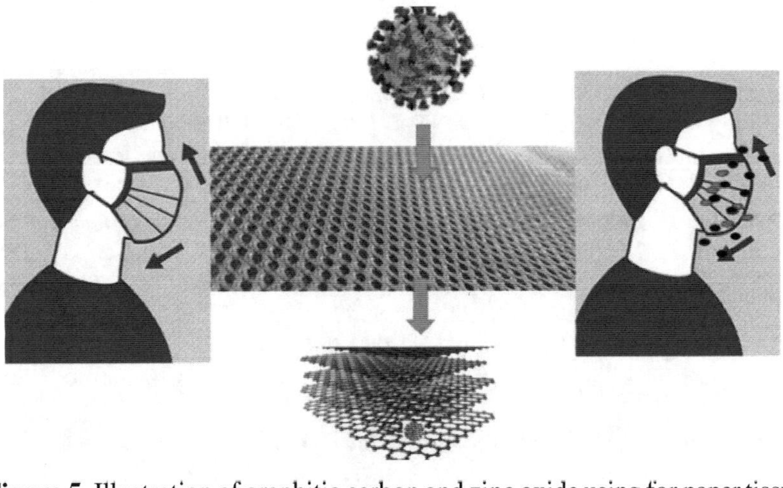

Figure 5. Illustration of graphitic carbon and zinc oxide using for paper tissue antiviruses.

Figure 4 is an illustration for creating antivirus tissue which is should be coated with increased hydrophobicity to keep tissue paper dry after washing that will prevent or repel the aerosol transmission of SARS-CoV-2. All these carbon and zinc-based composites would likely be facilitated to combat against the COVID-19 pandemic. The information comprised in this perspective provides a hypothetical approach that would likely help in preventing COVID-19 infection using carbon and zinc products with some development by considering the electro conductive and hydrophobic properties of graphitic carbon and zinc oxide nanomaterials, or their affinity with SARS-CoV-2. graphitic carbon and zinc oxide nanomaterials conjugated with antivirals can potentially trap and kill SARSCoV-2. Graphitic carbon and zinc oxide nanomaterials for SARS-CoV-2 can be computationally calculated using docking simulations and molecular dynamics. After that, molecular and cell biology tests can be carried out to determine the trend of interactions between SARS-CoV-2 and ACE2 molecules in living things. Antivirus tissue paper is expected to block the entry of coronavirus when consumers use it in their daily activities.

Figure 5 illustrates the inclusion of carbon and zinc oxide in an antiviral mask. Antiviral masks can be used to clean, protect the surface of the face as well as to maintain sanitation of human body. The presence of migrating electrons in the graphite carbon and zinc oxide in the mask is predicted to be able to block and mask the interaction of the SARS-CoV-2 Spike protein with ACE-2 in the human body. The antiviral mask is expected to be able to disinfect the infected area and prevent the virus from flying to other people. For healthy people, an antiviral mask will protect the face from viruses. Because it can be used repeatedly, this antivirus mask is more economical.

4. Conclusion

Graphite carbon and zinc oxide have a carbon atom and a zinc atom which have electron peculiarities. This second electron movement is predicted to be able to counter the interaction between the community of Covid-19 and ACE in the respiratory tract. These two materials are predicted to be the basic material as spray, paper tissue and masker antiviruses that have been blocked in the interaction between S-protein and human ACE2.

Acknowledgment

Authors acknowledges a research grant of Indonesian collaborative research program (PPKI 2020) funded by BOPTN of Sebelas Maret University, Gadjah Mada University and Institut Teknologi Sepuluh Nopember.

References

[1] Cui F and Zhou H S 2020 *Biosens. Bioelectron.* **165** 112349
[2] Aydemir D and Ulusu N N 2020 *Travel Med. Infect. Dis.* 101697
[3] Song Z, Wang X, Zhu G, Nian Q, Zhou H, Yang D, Qin C and Tang R 2015 *Small* **11** 1771
[4] Kong X *et al.* 2020 *Chem. Eng. J* **399** 125808
[5] Trikkaliotis D G, Mitropoulos A C and Kyzas G Z 2020 **600** *Chem. Eng. J* 124928
[6] Ward S, Lindsley A, Courter J and Assa'ad A 2020 *J. Allergy Clin. Immunol.* **146** 23
[7] Rahmani A R, Leili M, Azarian G and Poormohammadi A 2020 *Sci. Tot. Env.* **740** 140207
[8] Sun J, Bi H, Jia H, Su S, Dong H, Xie X and Sun L 2020 *Sci. Tot. Env.* **244** 118814
[9] Sharma N, Tripathi S K and Bhardwaj N K 2020 *Ind. Crops Prod.* **149** 112316
[10] Böger B, Fachi M M, Vilhena R O, Cobre A de F, Tonin F S and Pontarolo R 2020 *American. J. Infect. Control* **7** 11
[11] Sui L, Wang Y, Ji W, Kang H, Dong L and Yu L 2017 *Ind. Crops Prod.* **42** 29820
[12] Szczęśniak B, Phuriragpitikhon J, Choma J and Jaroniec M 2020 *J. Colloid Interface. Sci* **577** 163
[13] Raghav P K and Mohanty S 2020 *Med. Hypotheses* **144** 110031
[14] Mujawar M A, Gohel H, Bhardwaj S K, Srinivasan S and Hickman N 2020 *Mat. Today Chem.* **17** 100306

[15] Ravinayagam V and Rehman S 2020 *Saudi J. Bio. Sci.* **27** 1726
[16] Han H, Luo Q, Mo F, Long L and Zheng W 2020 *Lancet Infect. Dis.* **20** 655
[17] Ulfa M, Prasetyoko D, Mahadi A H and Bahruji H 2019 *Sci. Tot. Env* **711** 135066

IOP Conf. Series: Materials Science and Engineering **959** (2020) 012014 doi:10.1088/1757-899X/959/1/012014

Synthesis and characterizations of *C*-3-Nitrophenylcalix[4]resorcinarene as a potential chemosensor for La(III) ions

K T A Priyangga[1], Y S Kurniawan[1] and L Yuliati[1,2,*]

[1] Ma Chung Research Center for Photosynthetic Pigments, Universitas Ma Chung, Villa Puncak Tidar N-01, Malang 65151, East Java, Indonesia

[2] Department of Chemistry, Faculty of Science and Technology, Universitas Ma Chung, Villa Puncak Tidar N-01, Malang 65151, East Java, Indonesia

[*] E-mail: leny.yuliati@machung.ac.id

Abstract. *C*-3-nitrophenylcalix[4]resorcinarene (calix-3NO_2) is a macrocyclic compound with phenolic and nitrobenzene moieties and thus, promising to make a stable complex with a certain metal ion. Due to their high economic value for advanced materials, the detection of rare earth metals such as La(III) ions gains significant attention. In this work, we employed the calix-3NO_2 to establish a simple detection technique for La(III) ions using an ultraviolet (UV) spectrophotometer. The calix-3NO_2 was obtained in 40% yield through a condensation reaction between resorcinol and 3-nitrobenzaldehyde in acidic condition. It was confirmed that the calix-3NO_2 was able to detect La(III) ions in water:methanol media (4:1 v/v) by giving a significant additional peak at 268 nm due to charge transfer between ligand and metal ions. Furthermore, the calix-3NO_2 had selectively bound to La(III) ions compared with other metal ions such as Al(III), Ba(II), Ca(II), Cu(II), Mg(II), Mn(II), Ni(II), Zn(II), K(I), and Na(I). These findings are important in the design of an organic probe used for sensitive detection of La(III) ions using a simple UV spectroscopic technique.

1. Introduction

Due to its outstanding physicochemical properties, rare earth metals are considered as critical materials which are used in various applications, such as green and advanced technologies [1–3]. Lanthanum (La) is considered as one of the most important rare earth metals, which is widely used as a catalyst and energy-storage material [4]. In contrast to the important usage of La, the La ions are only found at the range of 10–300 mg kg^{-1} concentration in its mineral ore [1]. It was also reported that a trace amount of La ions (La(III)) is often present in the electronic wastes [5]. Considering those significant applications of La in our daily appliances, the recovery of La(III) ions from several sources has been evaluated over the past several years [4–6]. In the recovery purposes, it is necessary to ensure the presence of the La(III) ions in electronic waste. Therefore, the development of sensitive material to detect La(III) ions is critical prior to this recovery process.

Modified graphite materials were reported to be able to detect La(III) ions electrochemically [7]. Even though the successful detection of La(III) ions using the electrochemical method was achieved, it was not established using a UV-Vis spectrophotometer as it gave no significant signal response at 200–900 nm [7]. On the other hand, a spectroscopic method for the detection of metal ions has gained significant interests due to its simplicity and low-cost. It is known that an optical signal can be generated

IOP Conf. Series: Materials Science and Engineering **959** (2020) 012014 doi:10.1088/1757-899X/959/1/012014

after a suitable complexation phenomenon between organic compounds and certain metal ions [8]. Detection of La(III) ions using a fluorescent spectroscopic method has been successfully achieved using organic compounds, *i.e.* polydentate pyridyl-based quinoxaline derivative as the ligand [9]. In our previous works, we reported a sensitive detection of Fe(III) and Cu(II) ions using UV-Vis spectroscopic technique [10,11]. To the best of our knowledge, sensitive detection of La(III) ions using organic complexation agents by employing UV-Vis spectroscopic technique has not been developed yet.

As great organic complexation agents, calixarenes are supramolecular guest compounds that are easily synthesized through a one-step reaction from the phenolic compound and aldehydes derivatives [12]. The calixarenes-based compounds are well-known for complexation with certain metal ions [13-15]. As examples, calix[4]arene derivatives have been reported for their outstanding discrimination ability for metal ions complexation [12]. Similar to calix[4]arene framework, calix[4]resorcinarene framework was reported to be a suitable organic ligand for metal cations [16].

In this present work, we successfully synthesized one calix[4]resorcinarene derivative, which was *C*-3-nitrophenylcalix[4]resorcinarene (calix-3NO_2) from a condensation reaction between resorcinol and 3-nitrobenzaldehyde. The structure of the calix-3NO_2 is illustrated in figure 1. The sensing capability of the calix-3NO_2 was then evaluated against La(III), Al(III), Ba(II), Ca(II), Cu(II), Mg(II), Mn(II), Ni(II), Zn(II), K(I), and Na(I) metal ions using the UV spectroscopy.

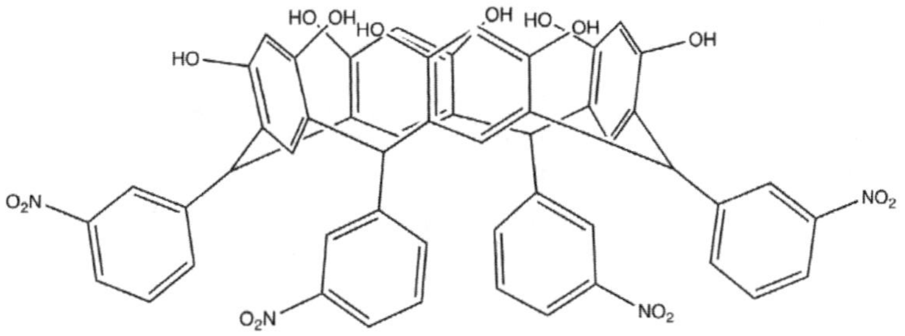

Figure 1. Chemical structure of *C*-3-nitrophenylcalix[4]resorcinarene.

2. Experimental section

2.1. Materials and apparatus

The material used in this work were 3-nitrobenzaldehyde ($C_7H_5NO_3$), resorcinol ($C_6H_6O_2$), hydrochloric acid (HCl) 37% v/v, ethanol (C_2H_5OH), methanol (CH_3OH), lanthanum chloride ($LaCl_3$), aluminum chloride hexahydrate ($AlCl_3 \cdot 6H_2O$), barium chloride ($BaCl_2$), calcium chloride ($CaCl_2$), copper(II) chloride dihydrate ($CuCl_2 \cdot 2H_2O$), magnesium chloride ($MgCl_2$), manganese(II) chloride tetrahydrate ($MnCl_2 \cdot 4H_2O$), nickel(II) chloride hexahydrate ($NiCl_2 \cdot 6H_2O$), zinc chloride ($ZnCl_2$), potassium chloride (KCl), and sodium chloride (NaCl), which were purchased from Merck in pro analytical grade. The instrumentations used in this work were Fourier transform infrared spectroscopy (FTIR, JASCO 6800), proton and carbon nuclear magnetic resonance spectrometer (^1H- and ^{13}C-NMR, JEOL JNM-ECZ500R/S1 500 MHz), and UV spectrophotometer (JASCO V-760).

2.2. Synthesis and characterizations of calix-3NO_2

The synthesis of calix-3NO_2 was conducted using a similar procedure as reported by Yamin *et al.* (2014) with several modifications [13]. In general, 3-nitrobenzaldehyde (4.53 g, 30 mmol) and resorcinol (3.30 g, 30 mmol) were dissolved and homogenized in absolute ethanol (75 mL) for 30 min. The HCl 37% v/v (3.0 mL) was then added dropwise into the mixture and the mixture was refluxed at 80 °C for 24 h. After the reaction was completed, the mixture was cooled at room temperature and then the mixture was added with cold distilled water (50 mL) to result in the formation of yellow precipitate. The solid product was washed with a mixture of cold ethanol and water until the filtrate was colorless. After drying at 80 °C for 15 h, the calix-3NO_2 product was obtained as a yellow powder.

IOP Conf. Series: Materials Science and Engineering **959** (2020) 012014 doi:10.1088/1757-899X/959/1/012014

The product was characterized using FTIR, ^{1}H-NMR, and ^{13}C-NMR spectrometries. The FTIR spectrum was obtained through an attenuated total reflectance (ATR) method, while the NMR spectrum was recorded in deuterated chloroform containing tetramethylsilane as the internal standard in the solvent.

2.3. Evaluation of calix-3NO$_2$ as a chemosensor agent for La(III) ions

At first, 20 mg L^{-1} of calix-3NO$_2$ was prepared as a stock solution by dissolving its solid product in a distilled water:methanol media (4:1 v/v). The chemosensing ability was evaluated by mixing calix-3NO$_2$ stock solution (1.5 mL) with a single of various metal ions (*i.e.* La(III), Al(III), Ba(II), Ca(II), Cu(II), Mg(II), Mn(II), Ni(II), Zn(II), K(I), and Na(I)) solution at a concentration of 20 mM in distilled water:methanol (4:1 v/v) media to reach a total volume of 3 mL. Afterward, the UV spectrum of each mixture was measured with a UV spectrophotometer at 200–400 nm. As a comparison purpose, the absorption spectrum of the calix-3NO$_2$ was recorded without the presence of metal ions.

The sensitivity test of La(III) ions detection was conducted similarly as mentioned above, but with various concentrations of La(III) ions. The stock solution of calix-3NO$_2$ (1.5 mL) was mixed with La(III) ions solution (1.5 mL) to give final La(III) ions concentration of 1.25, 2.5, 5, 10, and 20 mM) in distilled water:methanol (4:1 v/v) media. The UV spectrum of each mixture was recorded at 200–400 nm.

3. Results and discussion

3.1. Synthesis and characterizations of calix-3NO$_2$

The calix-3NO$_2$ was synthesized from resorcinol and 3-nitrobenzaldehyde compounds as the starting materials. Under the acidic condition, both starting materials were reacted to form a cyclic structure of calix [4]resorcinarenes as shown in figure 2. The calix-3NO$_2$ was successfully synthesized in 40% yield. The successful synthesis was indicated by the disappearance of the C–H aldehyde peaks of 3-nitrobenzaldehyde as the starting material at 2817 and 2713 cm^{-1} and the appearance of C–H methine peak of the *C*-arylcalix[4]resorcinarenes at 1428 cm^{-1} as shown in figure 3. The peaks at 1612 and 1519 cm^{-1} could be assigned to C=C aromatic stretching, while the N–O stretching and bending peaks were observed at 1351 and 830 cm^{-1}, respectively. The absorption peak at 1199 cm^{-1} corresponded to the C–O stretching, and the O–H stretching peak of calix-3NO$_2$ was detected as a broad peak at 3372 cm^{-1}.

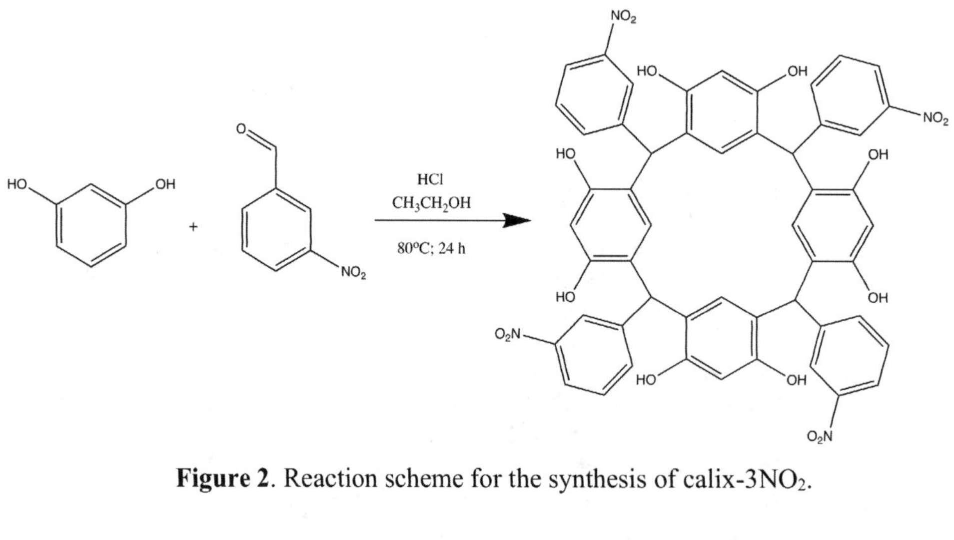

Figure 2. Reaction scheme for the synthesis of calix-3NO$_2$.

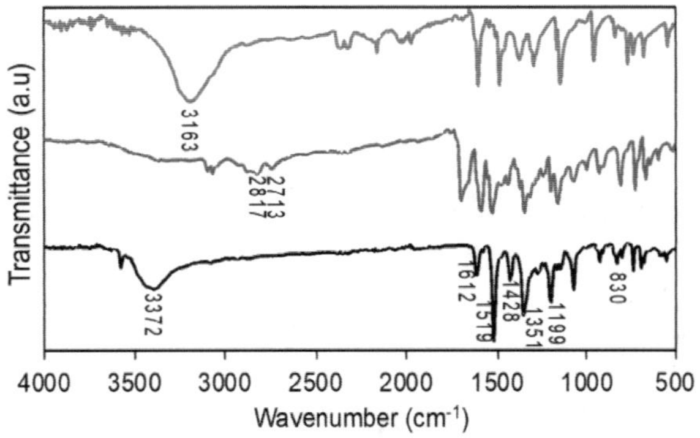

Figure 3. FTIR spectra of resorcinol (red line), 3-nitrobenzaldehyde (blue line), and calix-3NO$_2$ (black line).

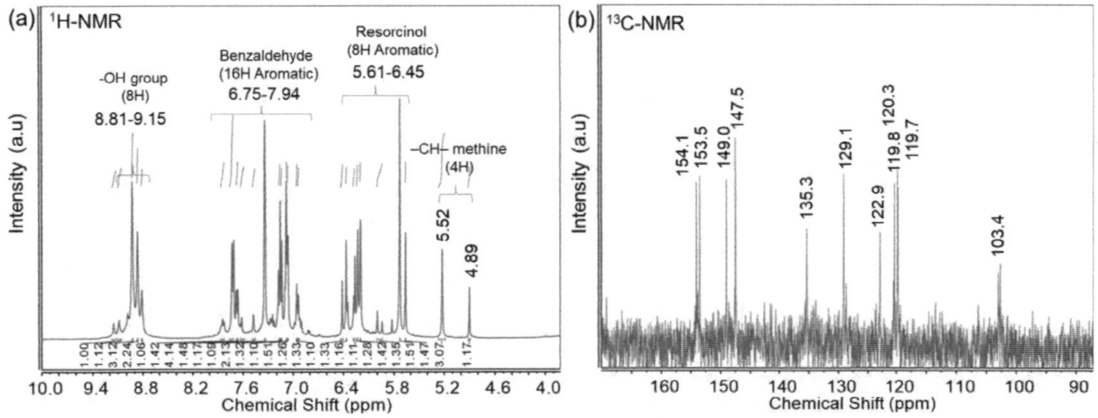

Figure 4. (a) ^{1}H-NMR and (b) ^{13}C-NMR spectra of calix-3NO$_2$.

The successful formation of calix-3NO$_2$ was also confirmed by NMR spectroscopy, in which spectra are shown in figure 4. The presence of C–H methine group in calix-3NO$_2$ was confirmed by a singlet peak at 5.22 ppm on the ^{1}H-NMR spectra and a peak at 103.4 ppm on the ^{13}C-NMR spectra. The detailed assignments of the NMR peaks are shown as followed. ^{1}H-NMR δ (ppm): 4.89+5.22 (s+s, 4H, –CH– methine), 5.61–6.45 (m, 8H, aromatic protons of resorcinol), 6.75–7.94 (m, 16H, aromatic protons of benzaldehyde), 8.81–9.15 (m, 8H, –OH). ^{13}C-NMR δ (ppm): 103.4 (–CH– methine), 119.7+119.8+120.3+122.9+129.1+135.3+ 147.5+149.0+153.5+154.1 (10 C aromatics). The FTIR and NMR spectra of the synthesized calix-3NO$_2$ were similar to the literature [13], indicating the successful synthesis of calix-3NO$_2$.

3.2. Evaluation of calix-3NO$_2$ as a chemosensor agent for La(III) ions

A good chemosensor agent should be able to give the distinguishable and significant response of the desired metal ion over the other metal ions. Therefore, the sensing ability of calix-3NO$_2$ was investigated for various metal ions, *i.e.* La(III), Al(III), Ba(II), Ca(II), Cu(II), Mg(II), Mn(II), Ni(II), Zn(II), K(I), and Na(I). These metal ions were selected to represent trivalent, divalent, and monovalent metal ions. The UV spectra of calix-3NO$_2$ without and with the addition of various single metal ions are shown in figure 5. While the addition of other metal ions only gave a slight spectrum difference as compared to the spectrum of the free ligand, it was confirmed that the addition of La(III) ions led to the increase in the absorbance intensity of the calix-3NO$_2$ at the UV region. This remarkable increase in the absorption intensity was also followed by a significant hypsochromic shift from 279 to 268 nm, which was only observed with the addition of La(III) ions. This shifted peak and increased absorption intensity could be generated from the charge-transfer between La(III) ions and the calix-3NO_2 ligand in the solution mixture through electrostatic and cation-π interactions [14,15]. In most cases, the absorbance of a calix compound is low since it can exist in various conformations such as cone, flattened partial cone, 1,2-alternate, 1,3-alternate, or a mixture of them. It was reported that the addition of certain metals could lead to a more stable and rigid conformation, which in turn gave a higher absorbance [17]. In this work, a similar phenomenon was observed where the addition of La(III) was shown to increase the absorbance of the calix-3NO_2. The formation of the La(III)-calix-3NO_2 complex resulted in a higher molar absorptivity due to the better stability and the more rigidity of the complex than the calix-3NO_2. Since such significant increased absorption intensity was not observed when another metal ion was added, this result clearly indicated that the calix-3NO_2 was selective for La(III) ions detection over the other metal ions, which would be a great benefit for La(III) ions chemosensing application.

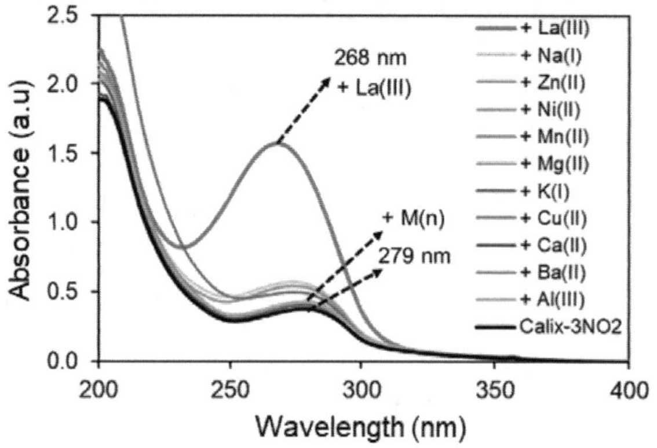

Figure 5. Optical spectra of calix-3NO_2 (10 mg L^{-1}) without and with the addition of each single metal ions solution at 20 mM concentration.

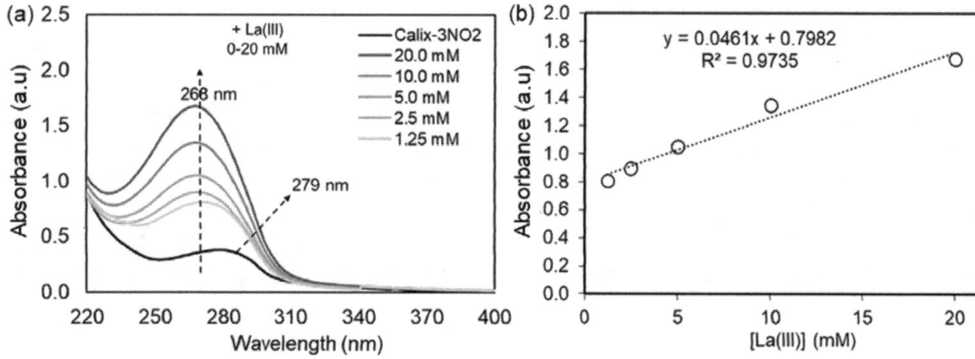

Figure 6. (a) Optical spectra of calix-3NO_2 (10 mg L^{-1}) without and with the addition of La(III) solution at 1.25–20 mM concentration, and (b) a linear plot of absorbance at 268 nm vs La(III) ion concentration.

Since the calix-3NO_2 was only responsive for La(III) ions, the sensitivity for La(III) ions was further evaluated. When the calix-3NO_2 solution was mixed with La(III) ions at a various concentration (1.25–20.0 mM), the optical spectrum of the calix-3NO_2 was changed as shown in figure 6(a). As expected, the higher concentration of La(III) ions led to the higher absorption peak intensity in the UV region of 230–320 nm. The absorption peak was blue-shifted from 279 nm to 268 nm with the increase of the

La(III) ion concentrations. When the absorbance value at 268 nm was plotted as a function against the La(III) ion concentration as shown in figure 6(b), a linear plot was obtained with the positive linear correlation (R^2) value of 0.9735. This result showed that the absorbance signal of calix-La(III) complexes at 268 nm was reliable enough to be used for the quantification of La(III) ion concentration in the range of 1.25–20 mM.

4. Conclusions

A simple and sensitive sensing technique for La(III) ion detection using the calix-3NO_2 in water:methanol media (4:1 v/v) was successfully carried out. The absorbance of calix-3NO_2 was increased and blue-shifted from 279 to 268 nm by the addition of La(III) ions. This could be observed as the indication of suitable complexation between the calix-3NO_2 and La(III) ions through electrostatic and cation-π interactions. In contrast, the addition of the other metal ions (*i.e.* Al(III), Ba(II), Ca(II), Cu(II), Mg(II), Mn(II), Ni(II), Zn(II), K(I), and Na(I)) did not give a significant difference in the absorbance spectrum of free calix-3NO_2, demonstrating that the calix-3NO_2 was a potential chemosensor for La(III) ion detection.

Acknowledgement

The authors thank for the financial support from Directorate General of Strengthening Research and Development, Ministry of Research, Technology and Higher Education of the Republic of Indonesia via the Fundamental Research (PD 2020, No. 041/SP2H/LT/MULTI/L7/2019 as revised in 041/SP2H/AMD/LT/MULTI/L7/2020 and No. 009/MACHUNG/LPPM/SP2H-LIT-MULTI/III/2020 as revised in 009/MACHUNG/LPPM/SP2H-LIT-MULTI/AMD/VI/2020).

References

[1] Jha M K, Kumari A, Panda R, Kumar J R, Yoo K and Lee J Y 2016 *Hydrometallurgy* **165** 2.
[2] Xie F, Zhang T A, Dreisinger D and Doyle F 2014 *Miner. Eng.* **56** 10.
[3] Bonificio W D and Clarke D R 2016 *Environ. Sci. Technol. Lett.* **3** 180.
[4] Cardoso C E D, Almeida J C, Lopes C B, Trindade T, Vale C and Pereira E 2019 *Nanomaterials* **9** 814.
[5] Zhao F, Repo E, Song Y, Yin D, Hammouda S B, Chen L, Kalliola S, Tang J, Tam K C and Sillanpaa M 2017 *Green Chem.* **19** 4816.
[6] Li C, Ma H, Venkateswaran S and Hsiao B S 2020 *Environ. Res.* **188** 109685.
[7] de Diego-Almeida R H, Monroy-Guzman F and Bustos E 2018 *J. Mater. Sci. Mater. Electron* **29** 15546.
[8] Hosseini H A, Nezhadali A and Darroudi M 2017 *Arab. J. Chem.* **10** S293.
[9] Zhao Q, Liu X M, Li H R, Zhang Y H and Bu X H 2016 *Dalton Trans.* **45** 10836.
[10] Fahmi M R G, Kurniawan Y S, Yuliati L and Lintang H O 2019 *AIP Conf. Proc.* **2175** 020062.
[11] Nugroho A A, Kurniawan Y S and Yuliati L 2020 *IOP Conference Series* **833** 012027.
[12] Kurniawan Y S, Sathuluri R R and Ohto K 2020 Droplet microfluidic device for rapid and efficient metals separation using host-guest chemistry *Advances in Microfluidic Technologies for Energy and Environmental Applications* ed Y Ren (IntechOpen) chapter 1 pp 1–19.
[13] Yamin B M, Abosadiya H M, Hasbullah S A and Jumina 2014 *Int. J. Adv. Sci. Eng. Inf. Technol.* **4** 125.
[14] Mahadevi A S and Sastry G N 2012 *Chem. Rev.* **113** 2100.
[15] Kumari H, Jin P, Teat S J, Barnes C L, Dalgarno S J and Atwood J L 2014 *J. Am. Chem. Soc.* **136** 17002.
[16] Budiana I G M, Jumina J and Anwar C 2016 *J. Appl. Chem. Sci.* **3** 289.
[17] Jumina J, Priastomo Y, Setiawan H R, Mutmainah M, Kurniawan Y S and Ohto K 2020 *J. Environ. Chem. Eng.* **8** 103971.

IOP Conf. Series: Materials Science and Engineering **959** (2020) 012015 doi:10.1088/1757-899X/959/1/012015

Synthesis and characterization of silver nanoparticles supported on highly ordered Biphenylene-Bridged Periodic Mesoporous Organosilica

A M Khikmah[1], Y K Krisnandi[1] and I Abdullah[1,*]
[1] Department of Chemistry, Faculty of Mathematics and Natural Sciences, Universitas Indonesia

[*] E-mail: iman.abdullah@sci.ui.ac.id

Abstract. Biphenylene-Bridged Periodic Mesoporous Organosilica (Bph-PMO) has been successfully synthesized using sol-gel method in the presence of Octadecyltrimethylammonium bromide (ODTMABr) surfactant as a template. Furthermore, Bph-PMO has been functionalized using amine (NH_2-Bph-PMO) in a two-step reaction, comprising nitration using HNO_3 65% / H_2SO_4 96% and reduction with $SnCl_2$ / HCl 37%. The white powder of Bph-PMO and the pale yellow powder of NH_2-BPh-PMO were characterized using FTIR, XRD, and TEM. Bph-PMO and NH_2-Bph-PMO showed long order periodicity between 2θ from 7.5-38.0 (7.5°, 14.9°, 22.5°, 30.1°, and 38.0°). The SEM images confirmed that all materials have uniform spherical morphology. Impregnation and reduction of silver nanoparticles in NH_2-Bph-PMO have been carried out using $AgNO_3$ as a precursor and $NaBH_4$ as a reducing agent. The TEM characterization showed that the structure of NH_2-Bph-PMO material is confirmed to have a hexagonal mesoporous structure with molecular periodicity. While in Ag/NH_2-Bph-PMO has homogeneous particle size distribution of 240-280 nm. The crystalline size of silver nanoparticles on the surface of NH_2-Bph-PMO calculated using the Debye-Scherrer equation is 8.05 nm. Nitrogen adsorption-desorption showed NH_2-Bph-PMO and Ag/NH_2-Bph-PMO were classified into mesoporous materials with pore diameters of 3.3 nm and 3.2 nm, respectively.

1. Introduction

Periodic Mesoporous Organosilica (PMO) as advanced porous hybrid material have been calling much attention due to the wide application in catalysis, adsorption, molecular recognition, sensor, applications in the electronic and optical fields [1]. The organic bridge of the PMO can be changed to obtain various properties and different interfacial interactions. Therefore, it can produce material for a variety of desired applications. In previous reports, the various organic bridges were incorporated into the framework of the PMO such as: ethane [2], benzene (-C_6H_4-) [3], and biphenylene (-C_6H_4-C_6H_4-) [4]. PMO is easily prepared and exhibits crystal-like walls, high specific surface areas, high pore volume, and high thermal stability. The presence of mesopores with high surface area and uniform diameter at PMO allows the effectiveness and ease to access to the active sites of catalyst [5].

In addition, the insertion of functional groups after synthesis has been widely reported. Ohashi et al. [3] reported 28% amination of the phenyl group PMO in a two-step chemical transformation procedure, the first functionalization process was carried out with the addition of a strong acid to incorporate a nitro group into the phenyl ring and followed by a reduction to the amine. The insertion of the amine functional groups into the organic bridge framework of PMO is expected to produce useful materials.

This is because the amine groups can be converted into various functional groups via amides (-NCO-R), imides (-NHC-R), and the formation of diazo urea (-NN-R).

The functionalized mesoporous silica surface can be grafted with various metal nanoparticles such as Pd [6], Pt [7], Ag [8], Ir [9], Ru [10]. Mesopore silica not only helps uniform dispersion of metal nanoparticles but also offers a high surface area for efficient catalysts [11]. Silver is a relatively inexpensive metal that is increasingly being used in various fields of the chemical industry because it has unique electrical, thermal, optical, and electrochemical properties. Silver nanoparticle grafting in catalytic support is a promising approach in designing heterogeneous catalysts for various variations of organic transformation [12]. Karimin et al. [13] added gold nanoparticles to the surface of phenylene PMO with the bottom impregnation and reduction methods for the Ulman aryl iodide coupling reaction. Zhang et al. [6] reported Pd impregnation at PMO for the same application.

In this report, we synthesized a highly ordered biphenylene-bridged PMO (Bph-PMO) and further functionalized the biphenylene bridge with the addition of amine functional group. Furthermore, we describe for the first time, modification of amine-functionalized Bph-PMO with silver nanoparticles through chemical impregnation followed by reduction. The material formed was evaluated for several characteristics such as surface morphology, crystallinity, surface area, pore size, and particle size. The results of these characterizations can then be used to see the potential utilization of Ag/NH_2-Bph-PMO in its application as a heterogeneous catalyst for organic reactions.

2. Materials and method

2.1. Materials

4,4'-Bis(triethoxysilyl)-1,1'-biphenyl, octadecyltrimethylammonium bromide (Aldrich), $SnCl_2$, and $NaBH_4$ were obtained by Sigma Aldrich. NaOH pellet, HNO_3 65%, isopropylamine, $AgNO_3$, and acetonitrile were provided by Merck. Ethanol and HCl 37% were purchased from Smart Lab, H_2SO_4 96% was obtained from Mallinckrodt.

2.2. Synthesis of Biphenylene Periodic Mesoporous Organosilica (Bph-PMO)

The Bph-PMO preparation was carried out by following method in the previous report[14] with modification. First, 4,4'-bis(triethoxysilyl)-1,1'-biphenyl (2.5 mmol) was added to a mixture of octadecyltrimethylammonium bromide (3.2 mmol), 6 M NaOH (30.4 mmol) and deionized water (3.3 mmol) under stirring at room temperature for 24 hours. The suspensions were moved to an autoclave and heated (100°C) for 24 hours. The materials were then filtered and washed with deionized water. The surfactant removal was done by extracting 1.5 g of as-prepared material in 150 mL ethanol with 6.2 g of 2 M HCl at ambient temperature for 8 hours.

2.3. Functionalization of Bph-PMO with amino functional group

Functionalization of Bph-PMO with amine group was conducted using a previously reported method for mesoporous phenylene-silica [3] through a two-step treatment. Nitration of Bph-PMO (1.05 g) was conducted in a mixed solution of 96% v/v H_2SO_4 (34.4 ml) and 65% v/v HNO_3 (13.3 ml). The suspension was stirred for 48 hours at ambient temperature. After the reaction, the suspension was poured into cold water (600 ml). The solid product was filtered and washed with deionized water. The reduction of nitro group was done by suspending the NO_2-Bph-PMO in a mixture of 37% v/v HCl (15 ml) and $SnCl_2$ (1.59 g). The mixture was stirred continuously at ambient temperature for 48 hours, then was poured into deionized water (600 ml), filtered and washed with deionized water, followed by isopropylamine (20 ml). The obtained NH_2-Bph-PMO is a pale yellow powder.

2.4. Impregnation and reduction of silver nanoparticles on NH_2-Bph-PMO

Impregnation and reduction of silver nanoparticles were performed by following the method in the previous report for impregnated silver nanoparticles on NH_2-SBA-15 [8] with solvent modification. 500 mg NH_2-Bph-PMO was dispersed in 11.6 ml acetonitrile and sonicated for 16 minutes. The suspension was then moved to another beaker containing silver nitrate solution (50 mg of $AgNO_3$ in 0.3 ml acetonitrile was added in dark condition, under magnetic stirring). The mixture was stirred for 24 hours at room temperature. The mixture had a pale brown color.

In a separate beaker, $NaBH_4$ solution was prepared by dissolving 23.75 mg $NaBH_4$ in 2.6 ml acetonitrile. To this solution, the suspension (silver ion adsorbed on NH_2-Bph-PMO in acetonitrile) was added, and stirred for 1 hour. Finally, the suspension was filtered and washed with an amount of acetonitrile. This product was named Ag/NH_2-Bph-PMO.

3. Results and discussion

Powder X-ray Diffraction (XRD) analysis of the parent Bph-PMO, NO_2-Bph-PMO, NH_2-Bph-PMO, and Ag/NH_2-Bph-PMO can be observed in figure 1a. The molecular-scale periodicity for Bph-PMO was observed through five sharp diffractions apparent at intermediate diffraction angle (7.5°, 14.9°, 22.5°, 30.1° and 38.0°). Similar diffraction pattern was observed for NO_2-Bph-PMO and NH_2-Bph-PMO. Both meso periodicity and molecular scale are maintained after nitration and nitro reduction into amine groups. The first strong low angle peak ($2\theta<5°$) assigned to the (100) reflection attributable to an ordered arrangement of meso channel [4] could not be observed due to the limitations of the instrument used.

After the impregnation with silver nanoparticles, the structure of NH_2-Bph-PMO was preserved. The sharps diffraction peaks at 38.1°, 44.2°, 64.5°, and 77.4° could be assigned to the Ag(111), (200), (220) and (311), which indicated the presence of silver nanoparticles with face center cubic nanostructure (JCPDS card no. 04-0783) at NH_2-Bph-PMO. Based on calculation using the Debye Scherer equation, the average crystal size of silver nanoparticles is 8.05 nm, respectively.

The nitration and amination reaction on Bph-PMO was characterized using FTIR (fgure 1b). The nitro groups in the biphenylene ring was confirmed by the presence of a C-N stretching band at 1262 cm^{-1} and by two strong bands at 1353 and 1534 cm^{-1} which corresponded to the N-O stretching band of the nitro group. The band at 1301 cm^{-1} originating from the C-N stretching vibration, indicates the covalent linkage of nitro groups (NO_2) to biphenylene moieties in Bph-PMO. The transformation of nitro to amine group was observed by the disappearance of the N-O stretching band and the presence of a N-H bending band at 1619 cm^{-1}. The stretching vibration of N-H from primary amine normally found at 3300-3400 cm^{-1}, could not be observed, since the vibration peak in this wave number area overlap with the strong and broad peak of the OH stretching of physisorbed water molecules [15]. The FTIR spectrum of Ag/NH_2-Bph-PMO is similar to that of NH_2-Bph-PMO, but appeared at a lower intensity due to the presence of silver nanoparticles in NH_2-BPh-PMO.

The nitrogen adsorption-desorption isotherm (figure 1c) confirm the type IV isotherms for NH_2-Bph-PMO and Ag/NH_2-Bph-PMO with small loop hysteresis each, indicating the occurrence of capillary condensation, which is typical of mesoporous material. The BET surface area and total pore volume of NH_2-Bph-PMO are 601 m^2/g and 0.36 cm^3/g, respectively. After the introduction of silver nanoparticles on NH_2-Bph-PMO, the surface area and pore volume decreased to 267 m^2/g and 0.17 cm^3/g, respectively. These results confirmed that the presence of silver nanoparticles on the surface fills the pore of NH_2-Bph-PMO. The Barrett, Joyner and Halenda (BJH) pore size distribution of NH_2-Bph-PMO and Ag/NH_2-Bph-PMO (figure 1d) are 3.3 nm and 3.2 nm, respectively. Therefore, both of them could be classified into mesoporous materials.

The Transmission Electrons Microscopy images (figure 2a and 2b) show the hexagonal mesoporous order of NH_2-Bph-PMO with pore diameter of 3.15 nm. Dark light lines show hexagonal pore channels with the periodicity of molecular scales. The morphology of NH_2-Bph-PMO and Ag/NH_2-Bph-PMO were observed by SEM-EDS analysis as shown in Figure 2c and 2d. Both materials have a uniform morphology with spherical in shape. Ag/NH_2-Bph-PMO shows a smoother surface compared to NH_2-Bph-PMO. Table 1 shows the presence of silver metal with loadings of 2.1%. It shows that NH_2-Bph-PMO have been successfully impregnated with silver nanoparticles. Compared with other supports of amine functionalized mesoporous silica materials such as NH_2-SBA-15, the impregnated silver nanoparticles had a larger crystallite size of 9-20 nm [16]. The presence of organic moieties with molecular periodicity of Bph-PMO results in uniform functionalization of the amines. The presence of amine group on the organic biphenylene bridge results in the formation of smaller and monodispersed silver nanoparticles [17].

4. Conclusion

Amine functionalized Bph-PMO was successfully synthesized by a two-step reaction. The presence of the amine functional groups in biphenylene bridge of the Bph-PMO was confirmed by FTIR analysis. Silver nanoparticles were also successfully embedded over the surface and pore of NH_2-Bph-PMO with loadings of 2.1%. Bph-PMO, NH_2-Bph-PMO, and Ag/NH_2-Bph-PMO exhibited a highly ordered mesostructured hexagonal. The TEM characterization confirms that amine-functionalized Bph-PMO had hexagonal mesostructures. All of the materials have spherical morphology from SEM images. In addition, calculation of the crystallite size using the Debye Scherrer equation provides the size of silver nanoparticles in the NH_2-Bph-PMO of 8.05 nm.

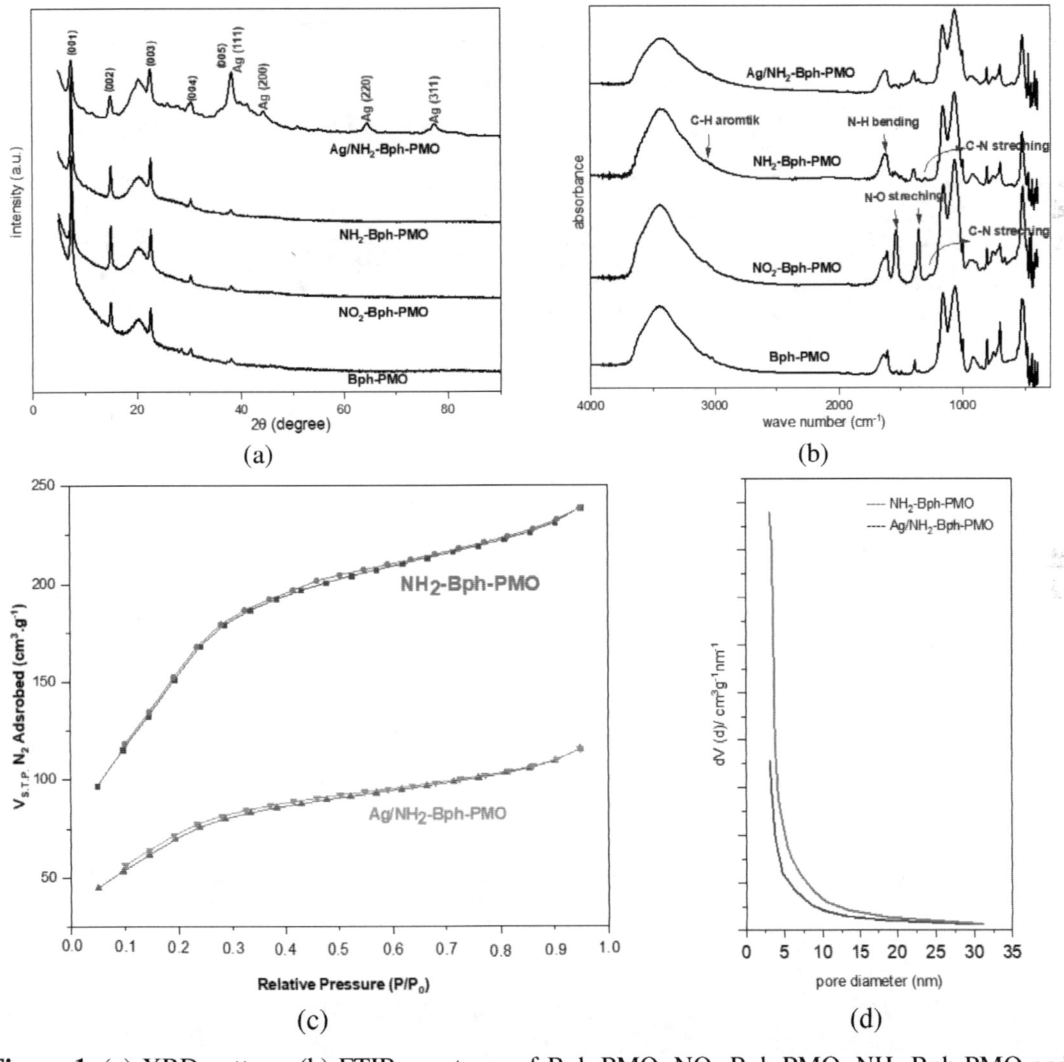

Figure 1. (a) XRD pattern, (b) FTIR spectrum of Bph-PMO, NO_2-Bph-PMO, NH_2-Bph-PMO and Ag/NH_2-Bph-PMO, (c) nitrogen adsopsion-desorption ishotherm and (d) pore size distribution of NH_2-Bph-PMO and Ag/NH_2-Bph-PMO.

IOP Conf. Series: Materials Science and Engineering **959** (2020) 012015 doi:10.1088/1757-899X/959/1/012015

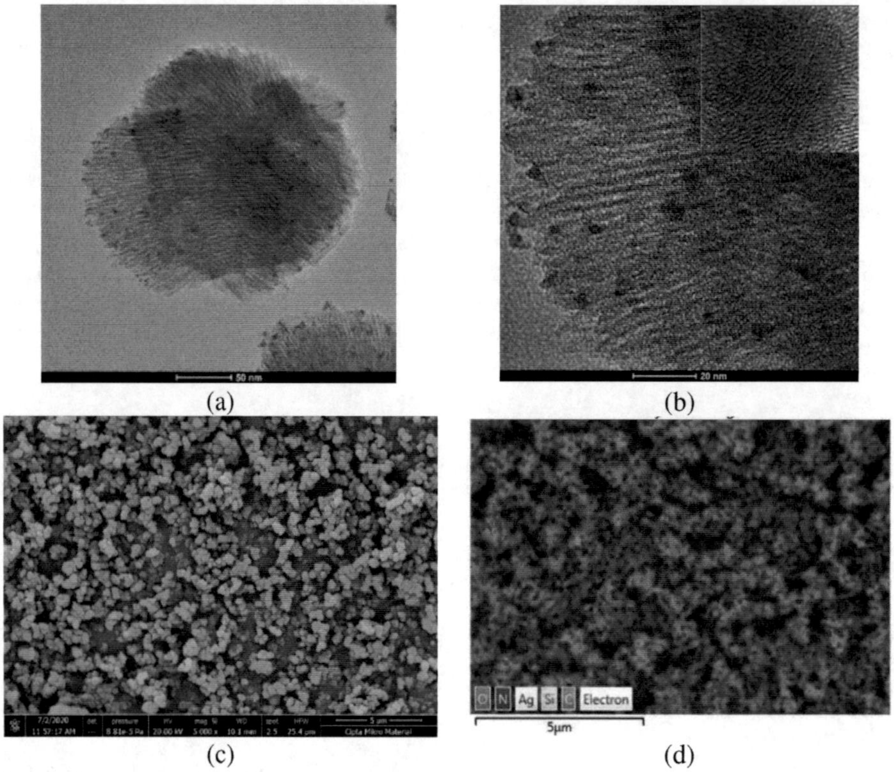

Figure 2. TEM images of NH_2-Bph-PMO (a) scale bar 50 nm (b) scale bar 20 nm, (c) SEM images of Ag/NH_2-Bph-PMO (5000x) and (d) mapping analysis of atoms in Ag/NH_2-Bph-PMO.

Table 1. Elemental analysis of Ag/NH_2-BPh-PMO.

Element	Wt. %
C	54.7
O	19.5
N	18.9
Si	4.8
Ag	2.1

Acknowledgements
The authors acknowledge Universitas Indonesia for funding this research through PUTI Prosiding research grant with contract No. NKB-1021/UN2.RST/HKP.05.00/2020.

References
[1] Inagaki S, Guan S, Fukushima Y, Tetsu O and Terasaki O 1999 *J. Am. Chem. Soc.* **121** 9611
[2] Asefa T, Ishii C Y, MacLachlan M J and Ozin G A 2000 *J. Mater. Chem.* **10** 1751
[3] Ohashi M, Kapoor M P and Inagaki S 2008 *Chem. Commun.* **7** 841
[4] Lourenço M A O, Mayoral A, Diaz I, Silva R A and Ferreira P 2015 *J. Micropor. Mesopor. Mat.* **217** 167
[5] Van Der Voort P, Esquivel D, De Canck E, Goethals F, Van Driessche I and Francisco J R S 2013*Chem. Soc. Rev.* **42** 3913

[6] Zhang F, Yin J, Chai W and Li H 2010 *Chem. Sus. Chem.* **3** 724
[7] Zhang X, Wang L J and Jiang W 2013 *Mater. Sci. Forum.* **745-756** 539
[8] Khan A Y and Bandyopadhyaya R 2014 *J. Elect. Chem.* **727** 184
[9] Gruning W R, Siddiqi G, Safonova O V and Coperet C 2014 *Adv. Synth. Catal.* **356** 673
[10] Haghshenas Kashani S, Moghadam M, Tangestaninejad S, Mirkhani V and Moh. Baltork I 2018 *Catal. Lett.* **148** 1110
[11] Wooh S, Huesmann H, Tahir M N, paven M, Wichmann K, Vollmer D, Tremel W, papadopoulos P and Butt H J 2015 *Adv. Matter.* **27** 7338
[12] Yan Z, Xu Z, Yu J and Jaroniec M 2017 *J. Colloid. Interf. Sci.* **501** 164
[13] Karimi B and Esfahani F K *Chem. Commun.*, **47** 10452
[14] K M P, Yang Q and Inagaki S 1999 *J. Am. Chem. Soc.* **124** 15176
[15] Suriyanon N, Punyapalakul P and Ngamcharussrivichai C 2015 *J. Matter. Chem. Phys.* **149** 701
[16] Kayal U, Mohanty B, Bhanja P, Chatterjee S, Chandra D, Hara M and Bhaumik A 2019 *Dalton Trans* **48** 2220
[17] Zhang W, Mei Y, Huang X, Wu P, Wu H and He M 2019 *J. Matter. Interfaces* **11** 44241

IOP Conf. Series: Materials Science and Engineering **959** (2020) 012016 doi:10.1088/1757-899X/959/1/012016

Synthesis of nanocomposites cellulose-Fe_3O_4/ZnO as novel green catalyst for biodiesel production from coconut oil

S N A R Azahra[1,*] and Helmiyati[1]
[1] Department of Chemistry, Faculty of Mathematics and Natural Sciences, University of Indonesia, Depok, 16424, Indonesia.

[*] E-mail: helmiyati@sci.ui.ac.id

Abstract. Cellulose-Fe_3O_4/ZnO nanocomposites have been successfully synthesized. This process was supported by characterization using Fourier Transform InfraRed (FTIR), X-ray Diffraction (XRD), Scanning Electron Microscopy (SEM), and Scanning Electron Microscopy-Mapping (SEM-Mapping) instruments. The successfully synthesized- nanocomposites then were tested for a catalytic activity in the transesterification reaction to form biodiesel. The results of the biodiesel production were characterized using Gas Chromatography - Mass Spectroscopy (GC-MS). The highest conversion of coconut oil into biodiesel was 90.6%, with a catalyst amount of 0.6% wt and a 120 minutes reaction time. The biodiesel production using GC-MS shows that the most formed-methyl esters are dodecanoic acid methyl esters (lauric acid methyl esters). The use of cellulose as a catalyst support by combining Fe_3O_4/ZnO composites is very promising to be a green catalyst in the reaction of biodiesel production.

1. Introduction

Recently, environmental pollution and global warming have been increasing along with the uplift in the use of fossil resources, so it is crucial to develop alternative energy sources that are clean and renewable [1, 2]. The total energy consumption worldwide has increased significantly due to overloaded-population, which cause huge impacts to the environment throughout the world. According to the projection of international energy use in 2016, the total of world energy consumption increased by 48% from 2012 to 2040. This results in the demand for fossil fuels facing two main challenges namely the scarcity of resources and their negative impacts on the environment [3]. Biodiesel is known as one of the most promising alternative fuels that is renewable, non-toxic and biodegradable, and a suitable fuel to replace diesel derived from petroleum. Biodiesel can be obtained from several renewable sources, such as vegetable oils and animal fats [4]. Biodiesel is also known as fatty acid methyl ester (FAME) which can be obtained by a transesterification reaction. It is a chemical reaction between triglycerides and alcohol in the presence of a catalyst to produce methyl esters (biodiesel) and glycerol [5, 6].

Vegetable oil is a potential candidate as a substitute source for fossil fuels. It is a natural oil containing a mixture of triacylglycerol, with several advantages, renewable, non-toxic, and a biodegradable source [7]. Coconut oil is one of the vegetable oils which is a raw material of compounds. It consists of various types of triglyceride fatty acids and around 90% of it is a saturated fatty acid compound, especially lauric acid and myristic acid [8].

In order to accelerate the process of biodiesel production, it is necessary to add both homogeneous and heterogeneous catalysts [9]. Heterogeneous catalysts are mostly used as an alternative of homogeneous catalysts in the chemical reactions. This is because the use of heterogeneous catalysts can minimize the impact of environmental damages, easy to separate from the reaction mixture, regenerable, and reusable [10, 11]. A lot of studies have been conducted using inorganic substances as catalysts such as Fe_3O_4, CaO, ZnO, Al_2O_3 and MgO. However, the use of inorganic catalysts is not modified and it may cause low thermal stability and release in water [12].

ZnO (metal oxide) is an inorganic catalyst that is commonly used and widely applied as a heterogeneous catalyst, because of its low cost, good stability, non-toxic and environmentally friendly properties [13]. By combining ZnO nanoparticles with other metal oxides, it can produce nanoparticles with a high surface area, a good dispersion, and smaller crystal sizes [14]. Besides that, the disadvantage of ZnO is it tends to form an agglomeration that has reduced its surface catalytic activity. In order to overcome this problem, a modification was carried out by magnetizing the ZnO nanoparticle catalyst with magnetite nanoparticles [15]. Combining ZnO with Fe_3O_4 can increase the overall catalytic activities compared to using only pure ZnO, because the addition of Fe_3O_4 can improve the stability, the catalyst surface area and is also easily separated using external magnets [16]. Iron (II) nanocatalysts which are doped with ZnO nano particles have been successfully synthesized as a promising catalyst for biodiesel production with 91% conversion [17]. CaO-Fe_3O_4 nanocatalysts have also been successfully synthesized for the transesterification reaction of palm kernel oil into methyl ester with a 69.7% conversion value [18].

It is interesting to develop metal oxides nano composites supported by biopolymers to be heterogenous catalysts. This is because biopolymer-based nanocomposites have a synergistic effect with inorganic materials to produce a good catalyst activity. The advantage of biopolymers is that they are abundant in nature, inexpensive, environmentally friendly, and biocompatible [19]. One of biopolymers like cellulose is an attractive compound for a catalyst support. Cellulose-Fe_3O_4 nanocomposite has been successfully synthesized which was applied as a heterogeneous catalyst for the synthesis of biodiesel using oleic acid with a conversion of 89.21% and a 5 hours reaction time [12] ZnO/SiO_2 nanocomposites with cellulose support derived from rice husk have been successfully synthesized and used as a biodiesel catalyst from coconut oil with 95% conversion during 270 minutes [20].

In this research, a new group of catalysts from a combination of ZnO, Fe_3O_4, and cellulose as the support is used as a catalyst for biodiesel synthesis through a transesterification process of coconut oil. The effect of the catalyst amount and the reaction time during the transesterification process on coconut oil are investigated.

2. Experimental

2.1. Materials
Pure cellulose powder (Merck) is used as a catalyst support, $FeCl_3$-$6H_2O$ (Merck), $FeCl_2$-$4H_2O$ (Merck) for the synthesis of nano Fe_3O_4, $Zn(CH_3COO)_2$ (Merck) as a precursor of ZnO nano, NaOH (Merck) as a pH regulator, also methanol (Merck) and coconut oil (local brand) for biodiesel production.

2.2. Synthesis of Fe_3O_4 nanoparticles
Nanomagnetite synthesis (Fe_3O_4) refers to a research conducted [15] through modifications. Mixed the $FeCl_2$-$4H_2O$ solution with $FeCl_3$-$6H_2O$ in a stoichiometric ratio of Fe^{2+}: Fe^{3+} (1:2), stired for 45 minutes at 80°C. Then, added it to 1M NaOH solution around pH $\leq$ 11, maintained the solution at 80°C for 1.5 hours until a black precipitate was obtained. The precipitate obtained then was rinsed with distilled water and ethanol, dried it in an oven at 103°C. The Fe_3O_4 nanoparticles obtained were characterized using FTIR, and XRD.

2.3. Synthesis of cellulose - Fe_3O_4/ZnO nanocomposites
Synthesis of cellulose-Fe_3O_4/ZnO nanocomposites was divided into two steps. The first step was the synthesis of Fe_3O_4/ZnO composites which refers to [21] through a modification by firstly combining ZnO and Fe_3O_4 nanoparticles. Fe_3O_4 nanoparticles were dispersed in 150 mL of distilled water and sonicated for 30 minutes, then added zinc acetate and stirred for 30 minutes, pH of the solution was

adjusted to 9 using 5M NaOH addition, and stired it for 30 minutes, then refluxed at 100°C for 1 hour. The obtained Fe_3O_4/ZnO composites were washed with distilled water and ethanol, dried it in an oven at 60°C for 24 hours, and characterized using FTIR, and XRD.

The second stage was that synthesis of cellulose-Fe_3O_4/ZnO nanocomposites, refers to [22]. The Fe_3O_4/ZnO composite was dispersed in 10 ml of 0.25 M NaOH and stirred for 1 hour at a room temperature. Then, the cellulose dispersion in distilled water was added slowly into Fe_3O_4/ZnO dispersion and stirred for 6 hours at a room temperature. Then the precipitate was filtered, rinsed, and dried in an oven at 60°C for one night. Cellulose-Fe_3O_4/ZnO nanocomposites were characterized using FTIR, XRD, and SEM.

2.4. Application of cellulose-Fe_3O_4/ZnO nanocomposites as a catalyst in biodiesel production

The catalytic activity of cellulose-Fe_3O_4/ZnO nanocomposites was evaluated for the transesterification reaction of coconut oil with methanol based on the research of [8, 12]. The reaction was conducted with various amount of catalyst (0.45%; 0.60%; 0.75% wt) and reaction times (60 minutes, 120 minutes, 180 minutes). The reaction experiment was conducted in a 500ml round neck flask containing coconut oil, methanol and cellulose-Fe_3O_4/ZnO nanocomposites as a catalyst. After that, the calculation of conversion percentage was done using equation (1) [23].

$$\text{Conversion (\%)} = \frac{\text{weight of biodiesel that obtained}}{\text{weight of oil that used in reaction}} \text{ x } 100\% \quad (1)$$

3. Result and discussion

3.1. Cellulose-Fe_3O_4/ZnO nanocomposites

Synthesis of cellulose-Fe_3O_4/ZnO in which cellulose acts as a catalyst support of Fe_3O_4/ZnO composites by combining Fe_3O_4/ZnO with cellulose biopolymer, it is hoped that Fe_3O_4/ZnO composites will be dispersed evenly on the surface of cellulose biopolymer. The illustration of cellulose-Fe_3O_4 / ZnO structure is shown in figure 1.

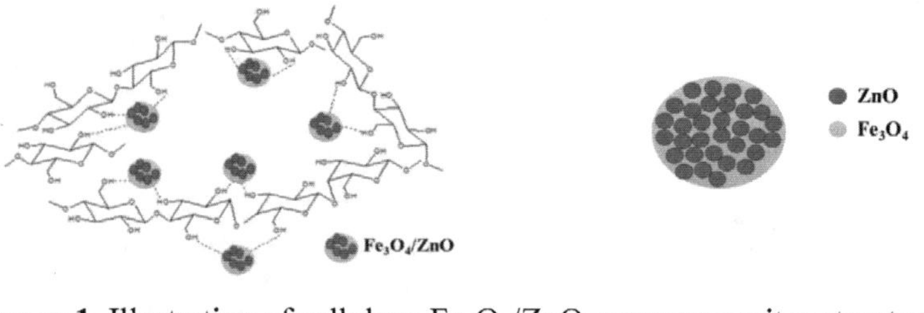

Figure 1. Illustration of cellulose-Fe_3O_4/ZnO nanocomposites structure.

3.2. Characterization of cellulose-Fe_3O_4/ZnO nanocomposites

The characterization using FTIR was carried out in an absorption area of 4000-400 cm^{-1} to determine the functional group of synthesized-cellulose-Fe_3O_4/ZnO. FTIR spectra from the result of cellulose-Fe_3O_4/ZnO synthesis can be seen in figure 2. Figure 2a shows Fe_3O_4 with a typical peak absorption at 577.83 cm^{-1} which indicates the presence of Fe-O functional group. Figure 2b shows that ZnO with a typical peak absorption at 432.24 cm^{-1} indicating the presence of Zn-O functional group. Figure 2c shows the FT-IR spectra of cellulose. A typical absorption at 800 cm^{-1} can be seen which is a typical absorption of the 1.4 β glycosidic bond. Figure 2d shows the cellulose-Fe_3O_4/ZnO nanocomposite spectra. It can be seen that the absorption peaks appear from the constituent compounds and there is a shift from each of the absorption peaks. It can be seen also, a typical peak arises from cellulose which is indicated by a 1.4 β glycosidic bond, a typical peak of Fe_3O_4 that indicates the presence of a Fe-O functional group, and a typical peak of ZnO that indicates the presence of a Zn-O functional group.

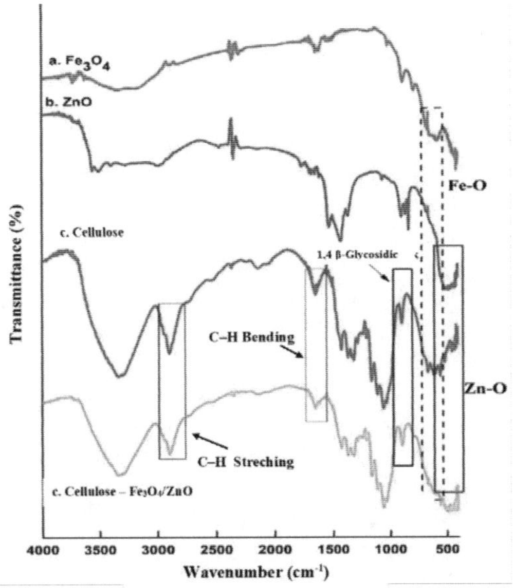

Figure 2. FTIR spectra synthesis result of: (a) Fe_3O_4, (b) ZnO, (c) cellulose, and (d) cellulose-Fe_3O_4/ZnO.

The characterization of Fe_3O_4/ZnO using XRD can be seen on the XRD diffractogram shown in figure 3. Figure 3a is a diffraction pattern of Fe_3O_4. The peaks can be observed at 2θ: 21.29°; 35.65°; 43.33°; 53.53°; and 63.04 which shows the typical peak of Fe_3O_4, this peak is in accordance with JCPDS No. 75-0033 [24]. Figure 3b is a diffraction pattern of ZnO nanoparticles, where peaks can be seen at 2θ: 31.92°; 34.56°; 36.39°; 47.68°; 56.75°; 62.97°; 66.53°; 68.09°; 69.24°, this peak is in accordance with JCPDS No. 36-1451 [25] which shows the typical peak of ZnO. Figure 3c is a diffraction pattern of composite Fe_3O_4/ZnO. The peaks can be seen at 2θ: 31.72°; 34.46°; 36.21°; 47.65°; 55.83°; 62.91°; 66.49°; 67.85°; and 69.19° which show the presence of ZnO compared to Figure 3b where there is a shift at peak 2θ. The presence of Fe_3O_4 peaks in Fe_3O_4/ZnO composites can be seen from 2θ: 36.21; 55.83°; and 62.91°, compared to Figure 3a where there is a shift at peak 2θ. This showed the successful synthesis of Fe_3O_4/ZnO.

The characterization of Cellulose-Fe_3O_4/ZnO using XRD can be seen on the XRD diffractogram shown in figure 4. Figure 4a, is a Fe_3O_4/ZnO composite diffractogram. Figure 4b shows the diffraction pattern of peak 2θ pure cellullose: 15.09°; 22.64°; and 34.71°, which is a typical characteristic of cellulose. Figure 4c shows the diffraction pattern of cellulose-Fe_3O_4/ZnO nanocomposites, from Figure 4c if it is compared with Figure 4a and Figure 4b it can be seen that the peaks of each cellulose and Fe_3O_4/ZnO remain exist and a shift in diffraction patterns occurs in the cellulose-Fe_3O_4/ZnO nanocomposite. This supports that cellulose-Fe_3O_4/ZnO nanocomposite had been successfully synthesized.

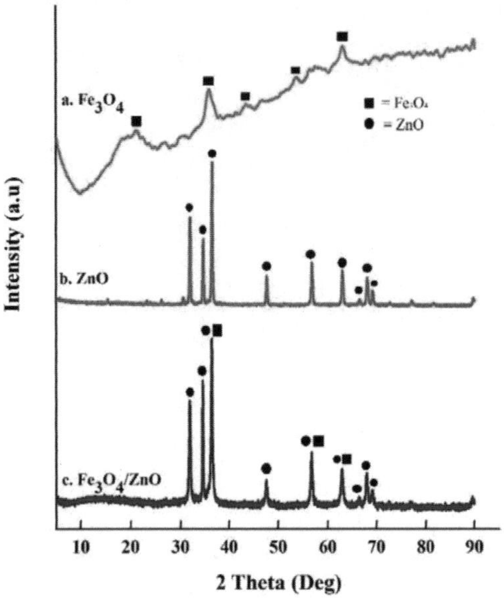

Figure 3. Diffractogram of: (a) Fe_3O_4, (b) ZnO, and (c) Fe_3O_4/ZnO.

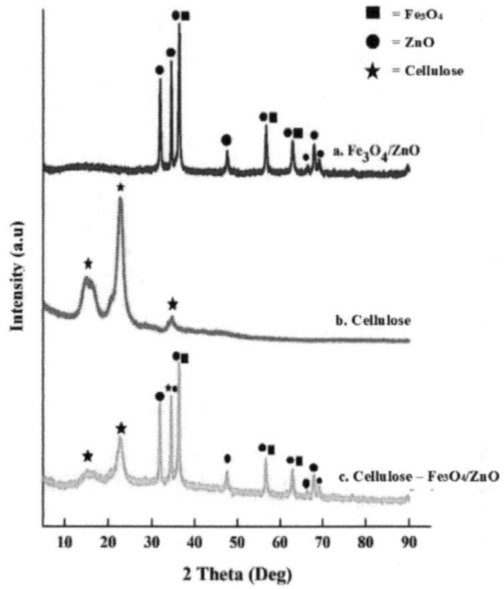

Figure 4. Diffractogram of: (a) Fe_3O_4/ZnO, (b) Cellulose, and (c) Cellulose-Fe_3O_4/ZnO.

The characterization using SEM was used to analyze the morphology surface of synthesized-cellulose-Fe_3O_4/ZnO nanocomposites. The results of cellulose-Fe_3O_4/ZnO nanocomposite characterization using SEM can be seen in figure 5. Figure 5a shows the SEM result of cellulose-Fe_3O_4/ZnO nanocomposite with 3000 times magnification. There were white fine and even granules which were Fe_3O_4/ZnO composites on the surface of cellulose in the form of fiber. Figure 5b is the SEM result of cellulose-Fe_3O_4/ZnO nanocomposite with 6000 times magnification. It was clearly visible that there was an even distribution of Fe_3O_4/ZnO composites on the surface of cellulose which was a catalyst support. The morphology of cellulose- Fe_3O_4/ZnO nanocomposites supported by SEM-Mapping can be seen in figure 6. SEM-Mapping was carried out aiming to determine the distribution of the elements in cellulose-Fe_3O_4/ZnO nanocomposites.

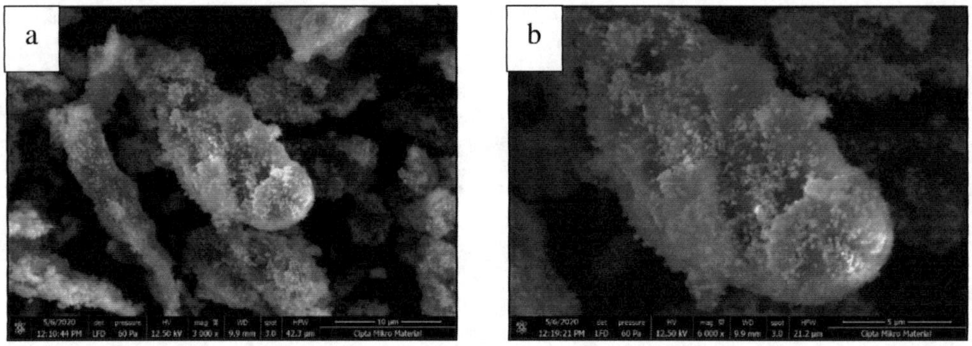

Figure 5. SEM results of cellulose- Fe_3O_4/ZnO with magnification: (a) 3000x and (b) 6000x.

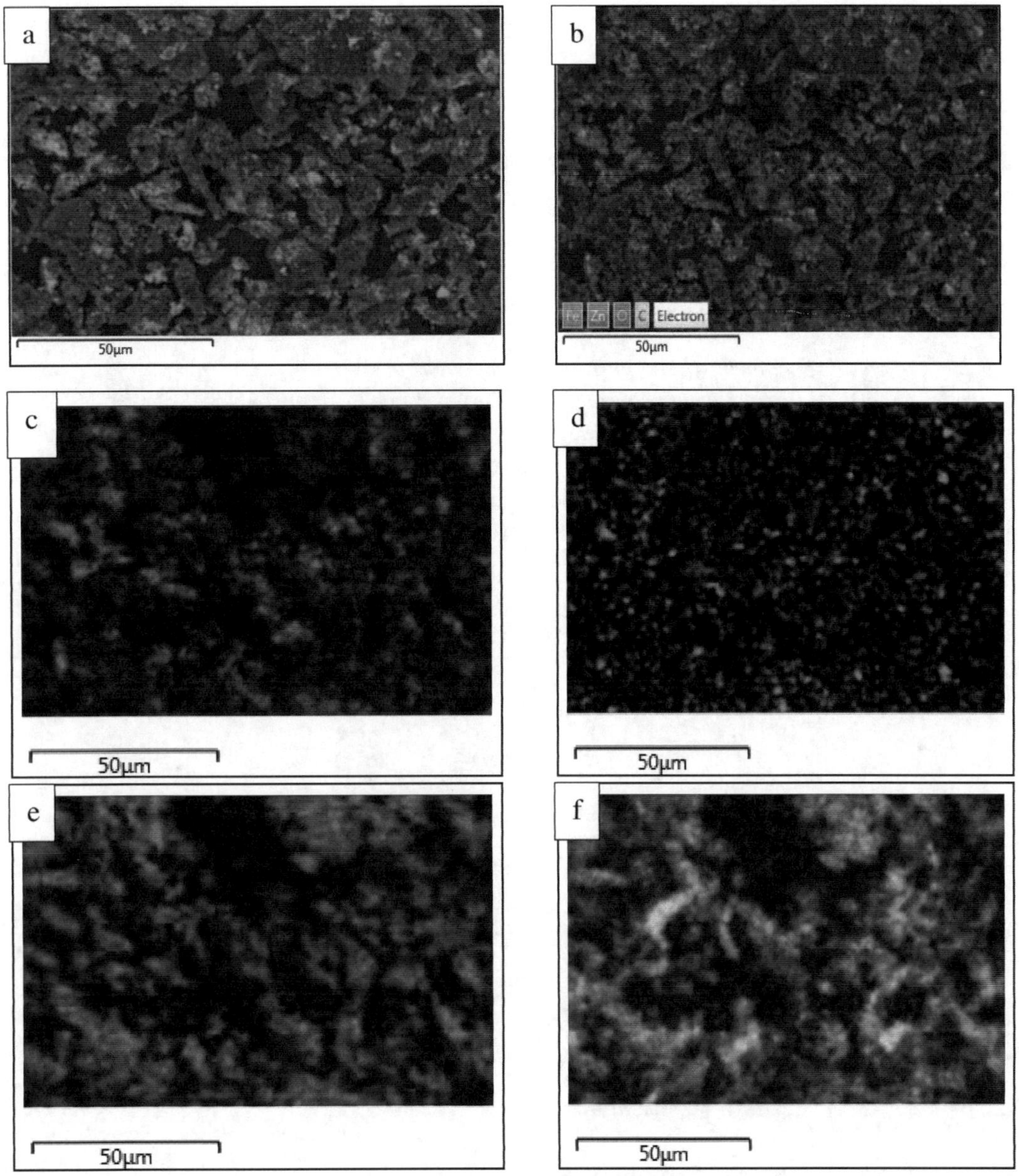

Figure 6. Morphology of cellulose-Fe_3O_4/ZnO nanocomposites with: (a) 1000x magnification, (b) distribution of whole elements, (c) element Zn, (d) element Fe, (e) element O, and (f) element C on nanocomposites.

In figure 6a can be seen that on the surface of the cellulose, in the form of smooth fibers and on its surface, there are small white granules that were evenly distributed as Fe_3O_4/ZnO composites. Figure 6b shows the overall distribution of nanocomposite elements. It was clearly seen the distribution of each element and on the surface there exists elements of Fe and Zn. Figure 6c shows the color distribution of Zn elements in blue and it can be explained that the presence of Zn elements comes from dispersed ZnO on the surface of cellulose evenly. In addition, the distribution of Fe elements derived from Fe_3O_4 is also seen scattered on the surface of cellulose which can be seen in figure 6d in a color of purple. Figure 6e shows the distribution of O elements in red which is distributed quite a lot and evenly, because the amount of O elements comes from cellulose and Fe_3O_4/ZnO composites. Figure 6f is the distribution of

IOP Conf. Series: Materials Science and Engineering **959** (2020) 012016 doi:10.1088/1757-899X/959/1/012016

element C in a uniformly distributed green color derived from cellulose. The results of SEM-Mapping analysis supported the success of cellulose-Fe_3O_4/ZnO nanocomposite synthesis and it can be observed that cellulose had successfully become a support for Fe_3O_4/ZnO, this result was strengthened by the results of FTIR and XRD characterizations.

3.3. Cellulose-Fe_3O_4/ZnO nanocomposite activity as a catalyst in biodiesel production

Cellulose-Fe_3O_4/ZnO nanocomposites that have been successfully synthesized are used as heterogeneous catalysts for the conversion of coconut oil into biodiesel through transesterification reaction using methanol. The use of catalysts in the transesterification reaction aiming to accelerate the conversion of triglycerides into biodiesel with a by-product in the form of glycerol.

3.3.1. Effects of catalysts concentration on biodiesel production

In this study, the effects of catalyst concentration in biodiesel production (0.45%; 0.6%; and 0.75% wt) can be seen in figure 7. The increased amount of catalyst also result in the escalation of biodiesel conversion results due to an increase in the active side derived from Fe_3O_4/ZnO composites. Futhermore, cellulose as a support can increase the adsorption of reactants on the catalyst so the reaction becomes more effective and the transesterification reaction for biodiesel production can be accelerated. The optimum amount of catalyst was obtained at 0.60% wt with 90.6% conversion and the addition of the catalyst amount do not show any significant differences in conversion. This result was better than our previous study where the obtained conversion was 89.21% with a weight of 1.5% wt catalyst using a cellulose-magnetite nanocomposite catalyst [12].

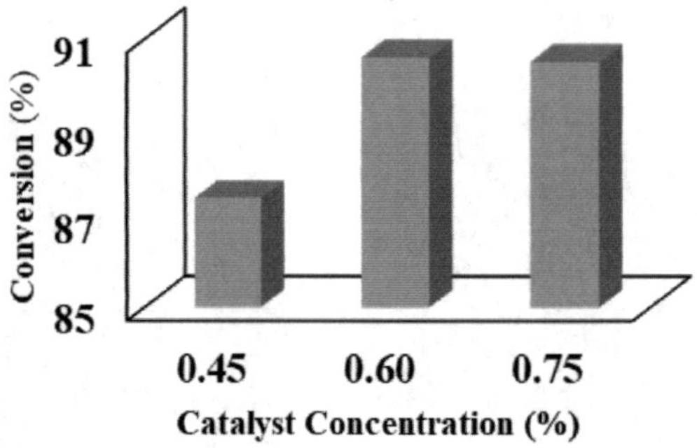

Figure 7. Biodiesel conversion with effect of catalysts concentration.

3.3.2. Effect of reaction time on biodiesel production

In this study the effect of reaction time on biodiesel production (60 minutes, 120 minutes, and 180 minutes) can be seen in figure 8. Based on figure 8, it can be seen that with an increase in the reaction time from 60 minutes to 120 minutes, a rise in conversion percentage from 83.55 % to 90.60% was obtained. This is because the reaction time increases the collisions between particles, the interaction of reactants, and catalysts. Thus the biodiesel production also increases. When, the reaction time is increased to 180 minutes, there is a decrease in biodiesel conversion results, this is because the conditions have reached equilibrium, it is known that the reaction occurs was a reversible reaction that was likely to react back towards the reactants [26].

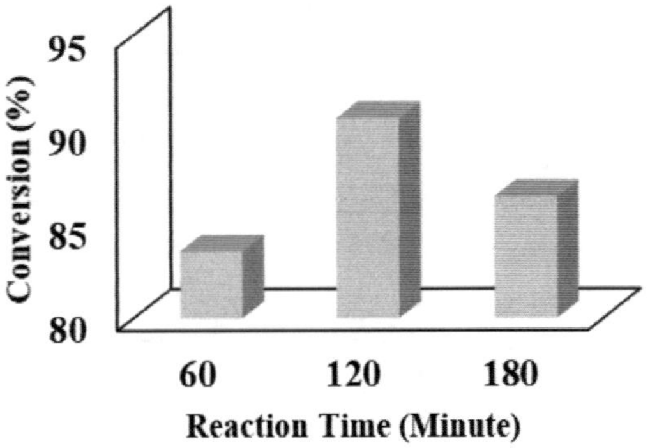

Figure 8. Biodiesel conversion with effect of reaction time.

3.4. Characterization of biodiesel products with GC-MS

The biodiesel products obtained then were characterized using GC-MS instrument which can be seen in figure 9. The GC-MS characterization is carried out to determine biodiesel produced from the conversion of coconut oil with methanol using a cellulose-Fe_3O_4/ZnO nanocomposite based-catalyst. Figure 9 shows the chromatogram of the product resulting from the transesterification reaction of coconut oil using a cellulose-Fe_3O_4/ZnO nanocomposite catalyst. The result was that the highest area percentage of biodiesel appeared at the retention time of 12,653 minutes and dodecanoic acid methyl esters biodiesel was obtained with the highest area percentage of 23.89%. In addition, there were also other types of biodiesel that have been successfully produced from coconut oil using a cellulose-Fe_3O_4/ZnO nanocomposite catalyst, which is shown in table 1. Based on table 1, an abundant biodiesel type gained is myristic acid methyl ester with an area percentage of 17.12%. This is because coconut oil is a triglyceride that contains mainly lauric acid and myristic acid [8].

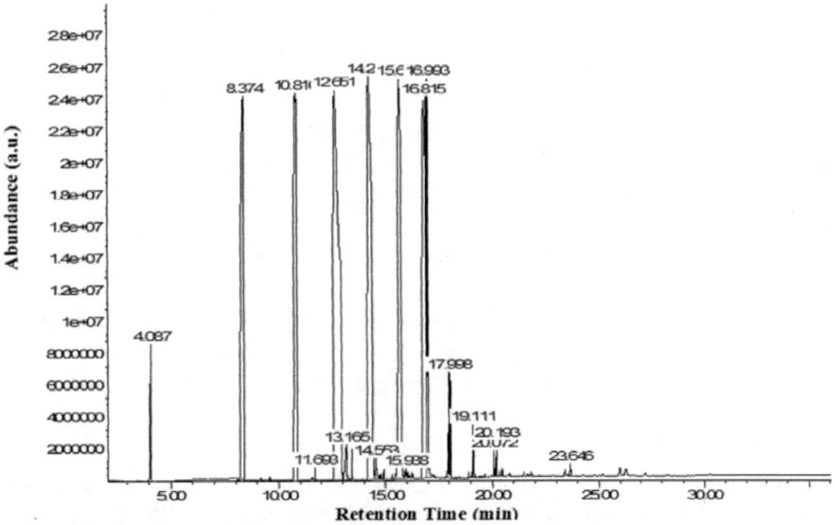

Figure 9. Chromatogram of the conversion of coconut oil into biodiesel using cellulose-Fe_3O_4/ZnO nanocomposite catalysts.

Table 1. Biodiesel from coconut oil transesterification results.

Biodiesel (Methyl Esther type)	Retention time	Area (%)
Hexanoic acid, methyl ester	4.089	1.29
Octanoic acid, methyl ester	8.374	11.40
Decanoic acid, methyl ester	10.811	9.81
Undecanoic acid, methyl ester	11.693	0.08
Dodecanoic acid, methyl ester	12.653	23.89
Myristic acid, methyl ester	14.249	17.12
Hexadecanoic acid, methyl ester	15.671	13.28
11-Octadecenoic acid, methyl ester	16.819	13.76
Octadecanoic acid, methyl ester	16.994	6.22
Tetracosanoic acid, methyl ester	20.194	0.20

4. Conclusion

Cellulose-Fe_3O_4/ZnO nanocomposites were successfully synthesized from the combination of Fe_3O_4/ZnO composites with Cellulose biopolymers, where Fe_3O_4/ZnO was dispersed on the surface of cellulose. This result was supported by the characterization of FTIR, XRD, SEM, and SEM-Mapping. The best conversion biodiesel product was obtained using a catalyst amount of 0.6% wt and the reaction time was 120 minutes with a conversion of 90.6%. Test results using GC-MS instruments show that the most formed methyl esters are dodecanoic acid methyl esters (lauric acid methyl esters) with a retention time of 12,653 minutes and with an area percentage of 23.89%.

References

[1] Zik N A F A, Sulaiman S and Jamal P 2020 *Renew. Energy* **155** 267
[2] Borah M J, Devi A, Borah R and Deka D 2019 *Renew. Energy* **133** 512
[3] Srivastava N and Gaurav K 2019 *J. Biochem Tech* **10** 1
[4] Helmiyati G H, Budiman A Y and Ramadhani S 2020 *Rasayan J. Chem.* **13** 298
[5] Roman F F, Ribeiro A E, Queiroz A, Lenzi G G, Chaves E S and Brito P 2019 *Fuel* **239** 1231
[6] Mahmood K H, Iqbal T, Haider A C, Javaid A and Iqbal C I 2020 *Fuel* **277**
[7] Tulashie S K and Kotoka F 2020 *Therm. Sci. Eng. Prog.* **16**
[8] Rasyid R, Sabara Z, Ainun P. H, Juradin R and Malik R 2018 *IOP Conf. Ser. Earth Environ. Sci.* **175** 2
[9] Acosta P I, Campedelli R R, Correa E L, Bazani H A G, Nishida E N, Souza B S and Mora J R 2020 *Fuel* **271**
[10] Correia L M, Saboya R M A, de Sousa Campelo N, Cecilia J A, Rodríguez-Castellón E, Cavalcante C L and Vieira R S 2014 *Bioresour. Technol.* **151** 207
[11] Helmiyati H and Masriah I 2019 *AIP Conf. Proc.* **2168**
[12] Helmiyati and Anggraini Y 2019 *Int. J. Technol.* **10** 798
[13] Zhang J, Yan X, Hu M, Hu X and Zhou M 2018 *J. Mol. Liq.* **249** 772
[14] Yang C, Wang J, Fan H, Hu Y, Shen J, Shangguan J and Wang B 2018 *Energy and Fuels* **32** 6064
[15] Dehghan S, Kakavandi B and Kalantary R R 2018 *J. Mol. Liq.* **264** 98
[16] Saxena V, Sharma S and Pandey L M 2019 *Mater. Lett.* **237** 232
[17] Baskar G and Soumiya S 2016 *Renew. Energy* **98** 101
[18] Ali M A, Al-Hydary I A and Al-Hattab T A 2017 *Bull. Chem. React. Eng. Catal.* **12** 460
[19] Lestari I and Helmiyati H 2020 *IOP Conf. Ser. Mater. Sci. Eng.* **763**
[20] Helmiyati H and Suci R P 2019 *AIP Conf. Proc.* **2168**
[21] Shekofteh-gohari M and Habibi-yangjeh A 2016 *J. Ind. Eng. Chem.* **44** 174

[22] Maleki A, Haji R F, Ghassemi M and Ghafuri H 2017 *J. Chem. Sci.* **129** 457
[23] Hassan N, Ismail K N, Ku Hamid K H and Hadi A 2018 *IOP Conf. Ser. Mater. Sci. Eng.* **358**
[24] Wang J, Lou Y, Xu C, Song S and Liu W 2016 *Sep. Purif. Technol.* **159** 57
[25] Jayalakshmi G, Saravanan K, Pradhan J, Magudapathy P and Panigrahi B K 2018 *J. Lumin.* **203** 1
[26] Widayat, Wibowo A D K and Hadiyanto 2013 *Energy Procedia* **32** 64

IOP Conf. Series: Materials Science and Engineering **959** (2020) 012017 doi:10.1088/1757-899X/959/1/012017

Adsorption capability for Congo Red on exfoliated graphene-decorated $CoFe_2O_4$ nanocomposite: Kinetic, isotherm, thermodynamic and recyclability studies

L V Tan[1,2,*], N T Hong-Tham[3,4] and P V Thinh[3,4]
[1] NTT Hi–Tech Institute, Nguyen Tat Thanh University, Ho Chi Minh City 700000, Vietnam
[2] Ben Tre's Department of Science and Technology, Ben Tre Province 86000, Vietnam
[3] Center of Excellence for Green Energy and Environmental Nanomaterials, Nguyen Tat Thanh University, Ho Chi Minh City 700000, Vietnam
[4] Center of Excellence for Functional Polymers and NanoEngineering, Nguyen Tat Thanh University, Ho Chi Minh City 700000, Vietnam

* Email: lvtan@ntt.edu.vn

Abstract. In the present work, we investigated kinetics and isotherms of the adsorption process of EG@$CoFe_2O_4$ adsorbent toward Congo Red dye in aqueous solution. Four kinetic models (pseudo first-order, pseudo second-order, Elovich, and Bangham) and four isotherm models (Langmuir, Freundlich, Temkin, and Dubinin–Radushkevich) were estimated elucidate the adsorption mechanism. Current results show that the pseudo second-order model is the most appropriate kinetics for the adsorption and that the monolayer behavior (Langmuir model), rather than multilayer behavior (Freundlich equation), is more suitable to describe the adsorption. The adsorption process was confirmed to be endothermic and spontaneous via thermodynamic study. On the other hand, via Boehm's titration we confirmed that EG@$CoFe_2O_4$ can contain functional groups (carboxylic, phenolic, laconic groups and basic groups) with the respective amount of 0.020, 0.044, 0.032, and 0.156 mmol/g. The EG@$CoFe_2O_4$ showed high maximum adsorption capacity (98.60mg/g), and good recyclability.

1. Introduction

Great industrial developments of mankind have increasingly elevated people's living needs and, at the same time, introduced a wide range of pollution issues. One of which is the discharge of organic dyes to the environment [1,2]. In printing and dyeing industries, dyes that are derived from coal-tar based hydrocarbons are usually prevalent [3,4] and thus their unabated release have raised numerous problems due to harmful effect on human health and ecosystem around the water [5-7].

Congo Red (CR) is a direct diazo dye that is toxic to many organisms and is a suspected carcinogen and mutagen [8,9]. In addition, its low degradability, high toxicity, and high stability to photodegradation have urged for development of many CR treatment methods such as catalysts, biogas, biological, advanced oxidation processes (AOPs) and physicochemical treatments [10-12]. Among them, the adsorption method has been seen an optimal method because it is cost-effective and potentially gives high-performance dye removal. This prompts for attempts aiming at manufacturing low-cost adsorbents and investigating their efficiencies for the removal of dyes from aqueous solution. Against

CR dye , notable adsorbents might include pine bark [13] or $ZnFe_2O_4/SiO_2$/Tragacanthin gum magnetic nanocomposite, both of which shown excellent adsorption capacity toward the dye [14-15]. However, these materials were sophisticatedly synthesized, and the manufacturing underwent numerous stages. With the advent of nanotechnology, researchers have explored intensively the use of carbon nanomaterials for water purification with the hope that new and successful pathways can be opened to solve current water shortage [16-18]. Among natural sources that could be utilized to manufacture adsorbents, natural graphite is a promising candidate due to its abundance and inexpensiveness and the existence of a routine that could facilely transform it into exfoliated graphite (EG) with many surface functional groups. However, EG material is difficult to recover from the aqueous solution after wastewater treatment. To resolve this, recent studies have attempted to attach ferrites nanoparticles (NPs), such as Fe_3O_4 [19], Co_3O_4 [20], $CoFe_2O_4$ [21] and $NiFe_2O_4$ [22] into graphene structure, producing composites that could be magnetically recovered.

In this study, we synthesized EG@$CoFe_2O_4$ by sol-gel method. Kinetic, isotherm and thermodynamic adsorption was investigated. Besides, Boehm' titration method was also used to determine the content of functional groups onto the surface of EG@$CoFe_2O_4$.

2. Materials and methods

2.1. Chemicals and instruments
The precursor used in this study was natural graphite flake (GF) with the particle size of 60 mesh, gathered and selected from Yen Bai province, Vietnam. H_2SO_4 (98%), H_2O_2 (30%) and Congo red dye were purchased from Merck. The adsorbent was characterized via XRD using the D8 Advance Bruker powder diffractometer with Cu-Kα beams used as excitation sources. S4800 instrument (Japan) operating at an accelerating voltage source (15 kV) was used to capture SEM images of the sample at the magnification of 7000x. Chemical bonds and functional groups of the as-synthesized adsorbent were elucidated via infrared FT-IR spectra obtained by the Nicolet 6700 spectrophotometer. To measure CR concentration, UV-vis spectrophotometer was used at wavelength of 500 nm.

2.2. Synthesis of EG
Microwave irradiation method was used to produce EG porous material from the natural flaky graphite [22]. First, natural flaky graphite was introduced into a mixture containing H_2SO_4 (98%) and H_2O_2 (30%) at a volume ratio of 100: 7and allowed to stand at room temperature for 2 hours. Afterwards, the solid was subjected to washing and then neutralization using H_2O and diluted NaOH solution respectively. The powder was exfoliated by microwave irradiation (750 W, 10 sec), affording the EG that would be used for subsequent studies.

2.3. Synthesis of $CoFe_2O_4$
Conventional polymerized complex routine was adopted to fabricate manganese-based magnetic nanoparticle of $CoFe_2O_4$ [26]. First, a mixture containing 93 g of citric acid, 140 mL of ethylene glycol and distilled water (2: 5 by volume) was heated to 80 °C under air atmosphere. Afterwards, the mixture was added with 0.303 g of $Co(NO_3)_2·6H_2O$, followed by heating at 130°C for 2 hours. Then, the obtained polymeric resin precursor was subjected to heating in a furnace at 1000°C for 2 h and allowed to cool down at room temperature, affording the sample for further studies.

2.4. Synthesis of EG@$CoFe_2O_4$
The synthesis procedure of EG@$CoFe_2O_4$ followed a previously reported routine [26]. First, 0.7 g of $Fe(NO_3)_3·9H_2O$ and 0.25 g of $Co(NO_3)_2·6H_2O$ were introduced into 50 mL of H_2O under stirring, followed by heating to 90 °C. Afterwards, dropwise addition of 50 mL of citric acid solution (0.02 M) was carried out, followed by stirring for 60 min. Then, the solution was added with 0.8 g of EG and had pH adjusted to weakly basic (pH 8-9) with NH_3 solution. After 30 min, a slow addition of NH_3 solution

for the second time into the beaker (pH 10) was carried out. The mixture was dried at 80 °C and calcined at 700 °C during 120 min to obtain as-received sample.

2.5. Experimental batch

The batch experiments were carried out to evaluate adsorption capacity of the synthesized EG toward CR dye. First, the adsorbent was added to 100 mL of CR solution (20–60 mg/L), followed by agitation on a shaking platform. Preliminary results suggested the period after which the adsorption attained the equilibrium was 210 min. After the reaction has finished, a sample of the solution was withdrawn using a syringe and had its dye concentration determined by using UV-vis spectrophotometer at 500 nm. Following formula was used to determine the removal efficiency (*H*%) and adsorption capacity (*Q*):

$$\mathrm{H}\,(\%) = \left(1 - \frac{\mathrm{C}}{\mathrm{C_o}}\right) \times 100 \tag{1}$$

$$Q_t = \left(\frac{C_o - \mathrm{C}}{\mathrm{m}}\right) \times \mathrm{V} \tag{2}$$

where C_o and *C* are, respectively, the initial and equilibrium dye concentrations, *V* is the volume of solution, and *m* represents the weight of adsorbent.

3. Results and discussion

3.1. Adsorption kinetics

Herein, the behavior of adsorption process was examined by estimating four kinetics including Pseudo-first-order, Pseudo-second order, Elovich model and Bangham model [27,28]. The results were given in table 1 and table 2. The coefficient of determination, R^2, was determined to assess the model compatibility to the data. Thus, high R^2 values show a good agreement between experimental and data predicted from the corresponding model.

From table 1, R^2 values of the pseudo second-order model (0.99983–0.99990) were so far higher than those of pseudo first-order model. The figure 1 (a, b) also confirmed the best linearity of pseudo second-order model. Similar trends could also be deduced with the $CoFe_2O_4$ adsorbent. Specifically, table 2 and figure 2(a, b) presents R^2 (0.98934-0.99709) what were better than the other values (0.79212-0.93388). Moreover, the error between Q_e value and Q_2 value is minimal.

Thus, the pseudo second-order model is recommended as the appropriate model to describe the adsorption of CR toward EG@$CoFe_2O_4$ and $CoFe_2O_4$, affirming the chemisorption as the effective mechanism that controls the adsorption process. This result also accept the role of electrostatic attraction between adsorbate and adsorbent through rate-controlling steps.

On the other hand, Elovich and Bangham models are two equations can be used to determine the adsorption kinetic of CR onto EG@$CoFe_2O_4$ and $CoFe_2O_4$. According to the results from table 1, Bangham model seem to fit well with the experimental data since their goodness of fit coefficients (R^2=0.8469-0.96706) are much higher than those by Elovich model (R^2=0.78181-0.94339). From the analysis results, it was revealed that the adsorption data of CR over $CoFe_2O_4$ suggested that Bangham model (R^2=0.94306-0.97319) is more fitted than Elovich (0.92916-0.97017), which was only better fitted with the adsorption data in the CR concentration range at 30-60 mg/L. Otherwise, the CR adsorption rates (α, mg/g.min) far exceeded CR desorption rates (β, g/mg) over EG@$CoFe_2O_4$ and $CoFe_2O_4$. Moreover, the CR adsorption rates (α, mg/g.min) onto EG@$CoFe_2O_4$ were higher than onto $CoFe_2O_4$. This indicates that the adsorption of CR towards EG@$CoFe_2O_4$ was more favourable than over $CoFe_2O_4$.

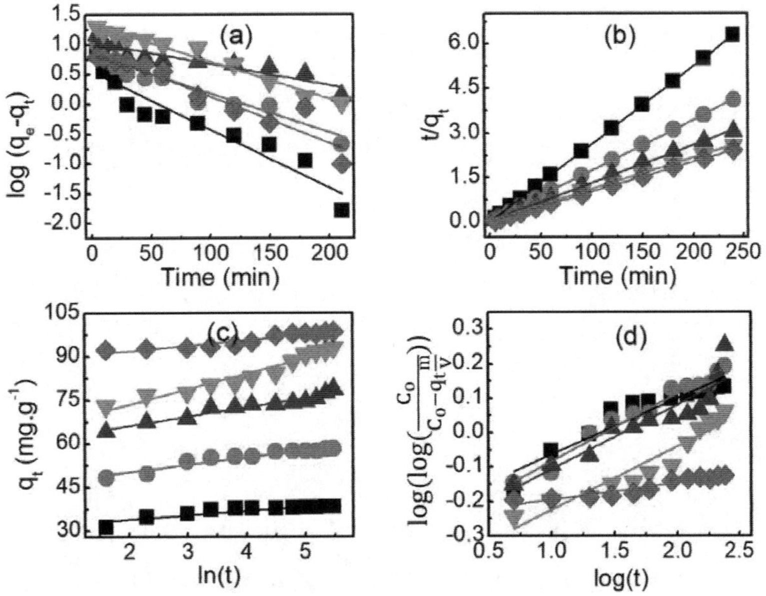

Figure 1. Kinetic models for adsorption of CR using EG@$CoFe_2O_4$ (a) Pseudo-first-order, (b) Pseudo-second-order, (c) Elovich model, and (d) Bangham model.

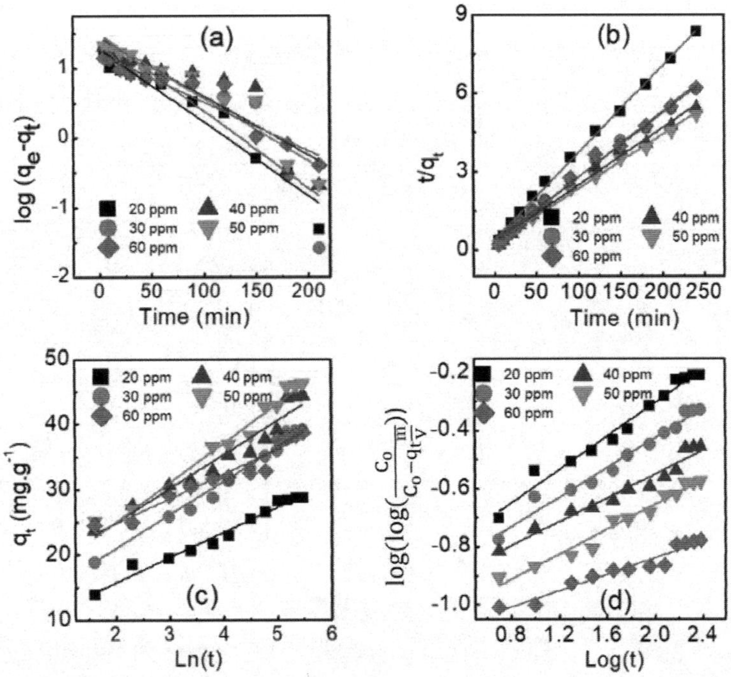

Figure 2. Kinetic models for adsorption of CR using $CoFe_2O_4$ (a) Pseudo-first-order, (b) Pseudo-second-order, (c) Elovich model, and (d) Bangham model.

IOP Conf. Series: Materials Science and Engineering **959** (2020) 012017 doi:10.1088/1757-899X/959/1/012017

Table 1. Kinetic model parameters for adsorption of CR using EG@$CoFe_2O_4$

Models	Parameters	Concentrations [mg/L]				
		20	30	40	50	60
Pseudo-first-order (3):	k_1 [min^{-1}/[mg/L]$^{1/n}$]	0.0225	0.0150	0.0083	0.0143	0.0179
	Q_1 [mg/g]	3.56	6.89	11.01	21.95	8.19
$\log(Q_1 - Q_t) = \log Q_1 - \frac{k_1.t}{2.303}$	R^2	0.89184	0.88621	0.83756	0.97057	0.89939
Pseudo-second-order (4):	k_2 [g/[mg.min]]	0.0219	0.0083	0.0039	0.0020	0.0069
	Q_2 [mg/g]	38.18	58.55	78.00	93.89	99.01
$\frac{t}{Q_t} = \frac{1}{k_2.Q_2^2} + \frac{t}{Q_2}$	$H = k_2Q_2^2$	31.95	28.42	23.76	17.75	67.93
	R^2	0.99999	0.9999	0.99887	0.99883	0.99992
Elovich (5)	β [g/mg]	0.675	0.400	0.314	0.191	0.529
$Q_t = \frac{1}{\beta}.\ln(\alpha.\beta) + \frac{1}{\beta}.\ln(t)$	α [mg/[g.min]]	2318.55	8434.41	1619.66	426.77	26566.09
	R^2	0.78181	0.91989	0.93366	0.94339	0.8879
Bangham (6)	k_B [mL/[g/L]]	0.13635	0.12042	0.11687	0.08991	0.13139
$\log\log\left(\frac{VC_o}{V.C_o - Q_t.m}\right) =$	α_B	0.16366	0.19781	0.18775	0.18398	0.04864
$\log\left(\frac{k_B}{2.303.V}\right) + \alpha_B.\log(t)$	R^2	0.86822	0.96706	0.8469	0.90856	0.88444

Table 2. Kinetics model parameters for adsorption of CR using $CoFe_2O_4$

Models	Parameters	Concentrations [mg/L]				
		20	30	40	50	60
Pseudo-first-order (3):	k_1 [min^{-1}/[mg/L]$^{1/n}$]	0.0243	0.0255	0.0195	0.0207	0.0162
	Q_1 [mg/g]	19.42	31.49	28.13	29.46	16.59
$\log(Q_1 - Q_t) = \log Q_1 - \frac{k_1.t}{2.303}$	R^2	0.93388	0.80165	0.79212	0.90297	0.89697
Pseudo-second-order (4):	k_2 [g/[mg.min]]	0.0027	0.0018	0.0016	0.0015	0.0025
	Q_2 [mg/g]	30.08	40.70	45.70	48.22	39.51
$\frac{t}{Q_t} = \frac{1}{k_2.Q_2^2} + \frac{t}{Q_2}$	$H = k_2Q_2^2$	2.45	2.98	3.26	3.52	3.89
	R^2	0.99709	0.99503	0.98934	0.99527	0.99356
Elovich (5)	β [g/mg]	0.2607	0.1953	0.1930	0.1659	0.2645
$Q_t = \frac{1}{\beta}.\ln(\alpha.\beta) + \frac{1}{\beta}.\ln(t)$	α [mg/[g.min]]	31.4832	41.6771	86.3147	50.9793	360.5440
	R^2	0.97579	0.97017	0.93842	0.95447	0.92916
Bangham (6)	k_B [mL/[g/L]]	0.03035	0.02663	0.02517	0.01893	0.01742
$\log\log\left(\frac{VC_o}{V.C_o - Q_t.m}\right) =$	α_B	0.28722	0.25533	0.20561	0.21529	0.14132
$\log\left(\frac{k_B}{2.303.V}\right) + \alpha_B.\log(t)$	R^2	0.975	0.97319	0.95719	0.96749	0.94306

IOP Conf. Series: Materials Science and Engineering **959** (2020) 012017 doi:10.1088/1757-899X/959/1/012017

3.2. Adsorption isotherm

The fitting of the adsorption equilibrium data into various isotherm models allows for identification of the suitable isotherm model and optimal adsorbent dosage and elaborates the mechanism by which adsorbents interact with solute [29,30]. Herein, four isotherm models: Langmuir, Freundlich, Temkin and Dubinin–Radushkevich (D–R) isotherm models were adopted and the R^2 values were used as the indicator to evaluate the well fitness.

The Langmuir, Freundlich, Temkin and D-R model parameters are listed in table 3. It is shown that CR adsorption onto EG@$CoFe_2O_4$ and $CoFe_2O_4$ seems to follow the Langmuir model, as evidenced by slightly higher R^2 compared to those of other models. This suggests that the solute might be adsorbed onto a surface with finite identical sites via monolayer adsorption. Figure 3 presents isothermal models of CR adsorption onto EG@$CoFe_2O_4$ and $CoFe_2O_4$ at varying initial CR Concentration (20, 30, 40, 50, 60 mg/L). Langmuir isotherm fittings were the most appropriate isotherm to describe the data, indicated by high linear correlation coefficient (R^2) of 0.99714 and 0.967 with EG@$CoFe_2O_4$ and $CoFe_2O_4$, respectively. It is visually indicated that the isotherms data points were linear over the whole CR concentration range, demonstrating that sorption data fit well with the Langmuir model. Moreover, drawing from table 3, respective coefficients of 1/n were 0.2206, 0.1604 onto EG@$CoFe_2O_4$ and $CoFe_2O_4$, respectively, indicating a favourable adsorption.

Table 3. Parameters of Langmuir, Freundlich, Temkin and Dubinin–Radushkevich isotherms for adsorption of dyes onto EG@$CoFe_2O_4$ and $CoFe_2O_4$, respectively.

Model	Parameters	Value	
		EG@$CoFe_2O_4$	$CoFe_2O_4$
Langmuir (7):	k_L [L/mg]	2.2651	7.0786
$\frac{1}{Q_e}=\left(\frac{1}{Q_mK_L}\right)\cdot\frac{1}{C_e}+\frac{1}{Q_m}$	Q_m [mg/g]	102.669	41.067
	$R_L = 1/(1+K_L.C_o)$	7.3×10^{-3}	2.4×10^{-3}
	R^2	0.99714	0.967
Freundlich (8):	k_F [mg/g]/[mg/L]$^{1/n}$	63.36	25.01
$\ln Q_e=\ln K_F+\frac{1}{n}\ln C_e$	1/n	0.2206	0.1604
	R^2	0.93715	0.40498
Temkin (9):	k_T [L/mg]	0.117	0.064
$Q_e=B_T\ln K_T+B_T\ln C_e$	B_T	14.1751	5.7023
	R^2	0.93958	0.36522
D-R (10):	B [kJ^2/mol^2]	2.56	0.172
$\ln Q_e=\ln Q_m-B\varepsilon^2$	Q_m [mg/g]	83.867	43.460
	E [J/mol]	441.94	170.49
	R^2	0.74324	0.80043

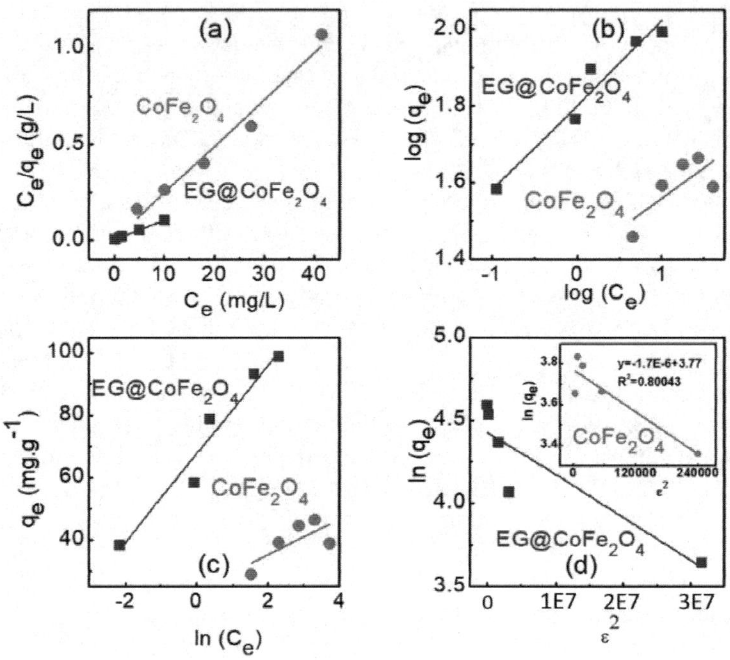

Figure 3. Isotherm models of EG@$CoFe_2O_4$ and $CoFe_2O_4$ (a) Langmuir isotherm model, (b) Freundlich isotherm model, (c) Temkin isotherm model, and (d) Dubinin–Radushkevich isotherm model

3.3. Thermodynamic parameters

The intercept and the slope derived from plotting ln(K_C) against 1/T might be used to calculate values of enthalpy change (ΔH°) and entropy change (ΔS°). Gibb's free energy ΔG° is given by

$$\Delta G^o = -RT \ln(K_C) \tag{3}$$

where R is the gas constant, T is the absolute temperature and K_C is thermodynamic equilibrium constant was calculated by $K_C = \frac{q_e}{C_e}$, where K_c was calculated by

$$\ln K_C = -\left(\frac{\Delta H^o}{R}\frac{1}{T} + \frac{\Delta S^o}{R}\right). \tag{4}$$

Table 4 summarized free energy change and thermodynamic parameters. First, negativity of the enthalpy change indicates the exothermic nature of the CR adsorption onto EG@$CoFe_2O_4$. On the other hand, the adsorption process is spontaneous, reflected by the negative value of Gibb's free energy, which is more negative with elevating temperature. Apparently, at higher temperatures, at around 40, higher adsorption has actually occurred.

Table 4. Thermodynamic parameters of CR sorption process by EG@$CoFe_2O_4$.

Samples	Temp [K]	ΔG° [kJmol^{-1}]	ΔH° [kJmol^{-1}]	ΔS° [Jmol^{-1}K^{-1}]
1	283	–482.5	57250	205.01
2	293	–821.8		
3	303	−1161.1		
4	313	−1500.4		

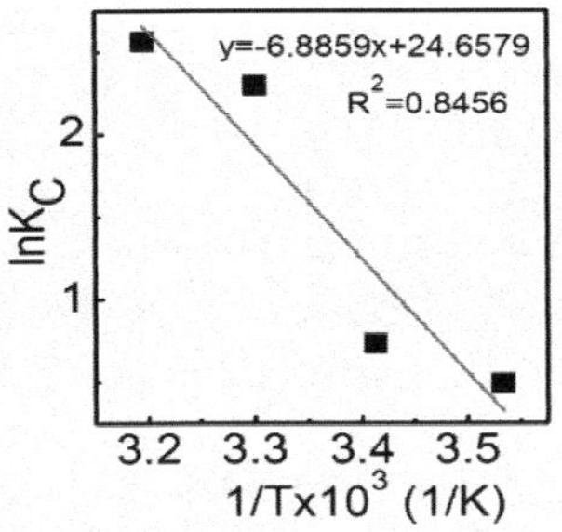

Figure 4. Thermodynamic plot for CR removal by EG@$CoFe_2O_4$.

3.4. Boehm' titration method

The Boehm assay was used to differentiate and quantify functional groups, in which different bases including $NaHCO_3$, Na_2CO_3 and NaOH were used. To be specific, $NaHCO_3$, (pK_{NaHCO3} = 6.37) neutralizes only carboxy1 groups on the carbon surface, Na_2CO_3 ($pK_{Na_2CO_3}$ = 10.25) titrates carboxyls and lactones, and NaOH (pK_{NaOH} = 15.74) neutralizes carboxyl, lactone and phenolic groups [17]. In this report, quantity of functional groups was expressed as mmol of Phenolic, laconic, carboxylic and basic per gram of EG@$CoFe_2O_4$. The results are shown in table 5. It is shown that EG@$CoFe_2O_4$ contained functional groups (carboxylic, phenolic, laconic groups and basic groups) with the amount of 0.020, 0.044, 0.032, and 0.156 mmol/g, respectively. Unlike EG@$CoFe_2O_4$, these functional groups were not found for $CoFe_2O_4$. Thus, it is suggested that EG@$CoFe_2O_4$ might exhibit improved adsorption of CR compared with $CoFe_2O_4$ where no surface functional groups were found.

Table 5. Number of functional groups on surface of EG@$CoFe_2O_4$.

	Number of functional groups [mmol/g]				
Sample	Carboxylic	Phenolic	Laconic	Total oxygenated	Total basic
$CoFe_2O_4$	0	0	0	0	0
EG@$CoFe_2O_4$	0.020	0.044	0.032	0.096	0.156

3.5. Recyclability evaluation of EG@CoFe2O4

Reutilization is a major advantage when using an adsorbent in water treatment processes. Recyclability assessment also provides useful information about adsorption stability during the adsorbent cycles. The recycling capability of the EG@$CoFe_2O_4$ was quickly tested and the results were shown in figure 6. After every run, appreciable reduction in adsorption capability was observed. In detail, the degradation ratio of CR by EG@$CoFe_2O_4$ adsorbent reached almost 100% after the first cycle, then quickly decreased in following runs and ultimately only attained 12.2 % after the fifth run. This is possibly explained to repeated exposure of active sites to CR molecules that resulted in reduced adsorption capacity.

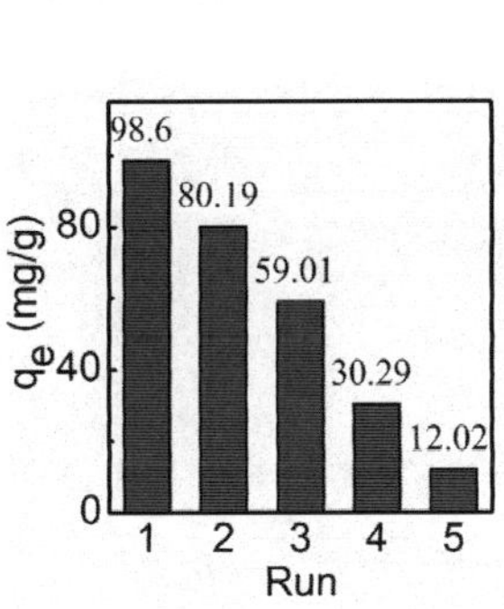

Figure 5. Recycling utilization of EG@$CoFe_2O_4$.

4. Conclusion

In this study, we demonstrated that the adsorption of CR at different CR concentrations over EG@$CoFe_2O_4$ followed chemisorption mechanism (pseudo-second order), and monolayer adsorption behavior (Langmuir equation). The adsorption process was confirmed to be endothermic and spontaneous in nature. Compared with $CoFe_2O_4$, EG@$CoFe_2O_4$ showed remarkably higher maximum adsorption capacity, which was confirmed by results obtained from Langmuir model. These results are consistent with the results of the Boehm's titration and suggest the promising potential of EG@$CoFe_2O_4$ in adsorption of CR dye from aqueous solution.

References

[1] Katheresan V, Kansedo J and Lau S Y 2018 *J. Environ. Chem. Eng.* **6** 4676
[2] Bharali D and Deka R C 2017 *J. Environ. Chem. Eng.* **5** 2056
[3] Anantharaj S, Karthick K and Kundu S 2017 *Mater. Today Energy* **6** 1
[4] Nhung N T H, Quynh B T P, Thao P T T, Bich H N and Giang B L 2018 *IOP Conf. Ser. Earth Environ. Sci.* **159** 012015
[5] Starukh G 2017 *Nanoscale Res Lett* **12** 391
[6] Litefti K, Freire M S, Stitou M and González-Álvarez J 2019 *Sci. Rep.* **9**
[7] Etemadinia T, Barikbin B and Allahresani A 2019 *Surf. Interfaces* **14** 117
[8] Liu P, Huang Y, Wang L, Zong M and Zhang W 2013 *Materials Letters* **107** 166
[9] Zong M, Huang Y, Wu H, Zhao Y, Wang Q and Sun X 2014 *Materials Letters* **114** 52
[10] Liu P, Huang Y and Sun X 2013 *Mater. Lett.* **112** 117
[11] Sun X, He J, Li G, Tang J, Wang T, Guo Y and Xue H 2013 *J. Mater. Chem. C* **1** 765
[12] Contescu A, Contescu C, Putyera K and Schwarz J A 1997 *Carbon* **35** 83
[13] Nguyen T T, Nguyen T N T, Bach L G, Nguyen D T and Bui T P Q 2019 *IIUM Eng. J.* **20** 202
[14] Udhaya P A and Meena M 2019 *Mater. Today Proc.* **9** 528
[15] Livani M J, Ghorbani M and Mehdipour H 2018 *New Carbon Mater.* **33** 578
[16] Nguyen T N T, Nguyen-Tran D H, Bach L G, Du Truong T H, Le N T T and Nguyen D H 2019 *Prog. Nat. Sci.: Mater. Int.* **29** 612
[17] Niu J, Chi H, Gao P, Liu Z, Yin X and Gu Z 2019 *Prog. Nat. Sci.: Mater. Int.* **29** 655
[18] Thuong H T T, Kim C T T, Quang L N and Kosslick H 2019 *Prog. Nat. Sci. Mater. Int.* **29** 641
[19] Zhang T, Li K and Pan J H 2019 *Prog. Nat. Sci. Mater. Int.* **29** 679
[20] Zhang X, Kawai H, Yang J and Goh K E J 2019 *Prog. Nat. Sci. Mater. Int.* **29** 667
[21] Xiao Y, Li X, Zai J, Wang K, Gong Y, Li B, Han Q and Qian X 2014 *Nano-Micro Lett.* **6** 307
[22] Thi H T N *et al. Key Eng. Mater.* **814** 463
[23] Fu M, Jiao Q and Zhao Y 2013 *J. Mater. Chem. A* **1** 5577
[24] Pham V T, Nguyen H-T T, Tran T V, Nguyen D T C, Le H T N, Nguyen T T, Vo D-V N, Le T H N and Nguyen D C 2019 *J. Chem.* **2019** 1
[25] Xianhua H, Xiaoqin T, Shejun H, Xinyu W, Yumei G and Xiang L 2017 *Rare Metal Mat. Eng.* **46** 1169
[26] Pham T V *et al.* 2020 *Mater. Today Proc.* **22** 2341
[27] Eum Y J, Hwang S-O, Ryu J, Kim J-W, Koo C Y, Lee J-Y and Lee H Y 2014 *Ferroelectrics* **465** 76
[28] Bach L G, Van Tran T, Nguyen T D, Pham T V and Do S T 2018 *Res. Chem. Intermed.* **44** 1661
[29] Van Tran T, Dai Cao V, Nguyen V H, Hoang B N, Vo D V N, Nguyen T D and Bach L G 2020 *J. Environ. Chem. Eng.* **8** 102902
[30] Nguyen H D T, Nguyen H T, Nguyen T T, Thi A K L, Nguyen T D, Bui Q T P and Bach L G 2019 *Materials* **12** 1913

IOP Conf. Series: Materials Science and Engineering **959** (2020) 012018 doi:10.1088/1757-899X/959/1/012018

Silver nanoparticles (AgNPs) as photocatalyst in the photodegradation of rhemazol brilliant blue

I E Suprihatin[1,*], G A D Lestari[2], R Mardhani[1] and V Edoway[1]
[1] Department of Chemistry, Faculty of Mathematics and Natural Sciences, Udayana University, Jimbaran, Bali, Indonesia.
[2] Study Program Pharmacy, School of Pharmacy Mahaganesha, Jl. Tukad Barito Timur No. 57 Denpasar, Bali, Indonesia.

[*] E-mail: eka_suprihatin@unud.ac.id

Abstract. The *phytosynthesis* of silver nano particles (AgNPs) and its use as a photocatalyst to photodegrade rhemazol brilliant blue are discussed. The AgNPs was prepared by reducing Ag^+ from 0.01 M $AgNO_3$ using frangipani (*Plumeria rubra*) water extract. The photodegradation was conducted by irradiating the suspension of the coloring agent and the AgNPs with UV rays at 259 nm. The optimum pH, AgNPs volume, and irradiation time were determined by measuring the degradation percentages at various pH, AgNPs volumes, and irradiation times. The study found that the optimum photodegradation occurred at pH 4, using 10 mL AgNPs, and irradiated for 60 minutes. Under the optimum conditions, AgNPs is capable of effectively photodegrading (97.84 ± 1.09) % of 600 ppm rhemazol brilliant blue

1. Introduction

Among the known nanomaterials, silver nanoparticle (AgNPs) is getting a lot of attention because of their non-toxic nature and various benefits, including as antibacterial agents for various types of microorganisms that cause various types of diseases [1]. Apart from its use in biology, AgNPs has the potential to be applied in various fields, including as an optical sensor and as a photocatalyst [2]. Its photo catalytical activity in the photodegradation of dyes has been reported in several studies. Vanaja *et al.* [3] reported that 95% of methylene blue dye can be degraded within 72 hours. Saeed *et al.* [4] successfully degraded 90% of methylene orange dye within 6 hours, at pH 2, using a photocatalyst from polyacrylonitrile nanofiber modified AgNPs. AgNPs can also degrade 98.3% of the dye methylene purple 6B for 4 hours and the methylene orange dye in 1 hour [5]. Research conducted by Kulkarni & Bhanage [6] showed that Ag nanoparticles (AgNPs) from sugar cane juice were able to degrade all samples of methyl orange dye in 21 minutes and degrade almost all samples of methylene blue dye within 50 minutes. Jyoti & Singh [7] synthesized AgNPs from *Zanthoxylum armatum* leaves which can degrade several dyes very quickly and efficiently.

Silver nanoparticles can be made by physical or chemical methods, which have their respective weaknesses. The former is expensive and difficult [8], while the latter causes toxic effects due to the hazardous materials used. A relatively new method that can be used to synthesize a catalyst into nanoparticle size is a reduction of metal ions using metabolites from plants, referred to as biosynthesis [9] or phytosynthesis [6]. This method can produce nanoparticles that are safe and environmentally friendly (Green synthesis) because it can minimize the use of harmful inorganic materials. This synthesis method is very effective because it is fast, non-toxic and environmentally friendly technology [10]. In principle, AgNPs biosynthesis is reducing Ag^+ to Ag in an autocatalytic reactions by secondary metabolites such as terpenoids, flavonoids, alkaloids, quaternary isoquinoline alkaloids, aporphyrine alkaloids etc. In several studies, these compounds are able to produce or form nanoparticles from a metal [10, 11].

In this study, AgNPs was synthesized using a water extract of frangipani flower (*Plumeria rubra*) which is very abundant in Bali. This flower water extract contains bioactive compounds such as tannins, phenols and vitamin C [12]. The AgNPs produced is used as a photocatalyst to degrade the brilliant blue Rhemazol dye which is widely used in the textile dyeing industry in Bali.

2. Experimental section

2.1. Materials

The materials used in this study included *Plumeria rubra* powder, Rhemazol brilliant blue dye, $AgNO_3$ (Sigma-Aldricht), demineralized water, and distilled water.

2.2. Apparatus

Equipment used apart from the usual laboratory glass wares and equipment, were black plastic sheets, irradiation boxes with a UV lamp (Philips TUV 15 W/G15 T8). The instruments were UV-Vis spectrophotometer (Shimadzu 2600) and Particle size analyzer (PSA from Malvern).

2.3. Procedure

Twenty grams of dry *Plumeria rubra* powder was added into 100 mL demineralized water and heated at 60°C for 15 minutes. The mixture was then left to stand for 24 hours, and filtered. The filtrate was then mixed with 0.01 M $AgNO_3$ solution (1:10 v/v) and heated at 60°C for 20 minutes. The AgNPs produced was characterized by measuring its absorbance in the range of 400-500 nm [13] and by measuring its particle size using PSA.

The optimum volume of the photocatalyst was determined by varying the volume of the catalyst by 5, 10, 15, 20, and 25 mL in 25 mL of 50 ppm aqueous dye solution. The optimum pH was determined by irradiating the25 mL of 50 ppm of dye solution at pH 2, 4, 6, 8, and 10. The optimum irradiation time was determined by irradiating 25 mL of 50 ppm of dye solution with UV light at the optimum pH, the optimum amount of catalyst for 30, 60, 90, 120, 150, and 180 minutes.

Photodegradation capacity was studied by irradiating 25 mL of 50, 100, 200, 300, 400, 500, and 600 ppm rhemazol brilliant blue under the optimum conditions. The percentage of degradation was obtained using the equation:

$$D\ (\%) = \frac{C_0 - C_t}{C_0} \times 100\%. \tag{1}$$

3. Results and discussion

3.1. The characteristics of the AgNPs

The growth of AgNPs can easily be observed using UV-Vis absorption. Qualitative analysis of AgNPs growth can be carried out based on the specific characteristics of the Surface Plasmon Resonance (SPR) of the nanoparticles. SPR is a collection of oscillations against electron conduction that occurs on the surface of the material. The presence of an electromagnetic field generated by electromagnetic waves causes the excitation of the oscillatory collection of electrons in metal nanoparticles, which is called Localized Surface Plasmon Resonance [14]. The change in color from yellow to reddish brown indicates the formation of silver nanoparticles as a result of the SPR phenomenon. The UV-Vis spectra of the silver particles produced is shown in figure 1. The spectra show that the particles, having maximum absorbance between 400 and 500 nm, can be categorized as nanoparticles [13]. The spectra also indicate that the AgNPs is stable for at least 29 hours to stand at room temperature. This stability is an advantage for users in case there is a delay in the application of the AgNPs.

Figure 2 shows the result of the particle size distribution analysis. It shows that the silver particles produced in this research have a very good size uniformity at 18.98 nm. This confirms that the silver particles produced fulfil the nanoparticle category. More importantly, the AgNPs can be used as photocatalyst for the degradation of pollutant molecules such as Rhemazol brilliant blue.

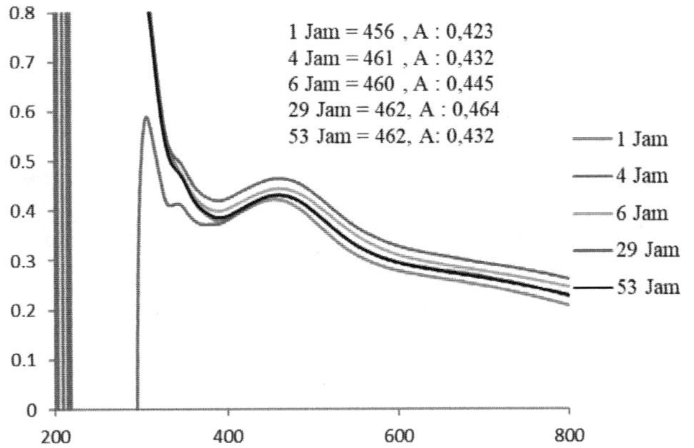

Figure 1. UV-Vis spectra of AgNPs after 1, 4, 6, 29, and 53 hours aging.

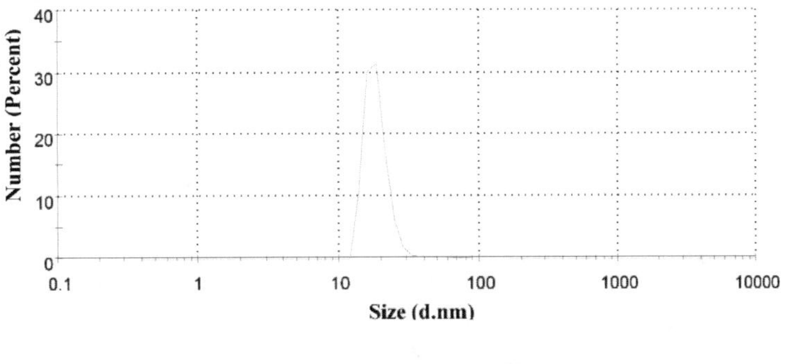

Figure 2. The size distribution of the AgNPs.

3.2. The Optimum volume of photocatalyst

Since the AgNPs present as a suspension, the amount of the photocatalyst used in the photodegradation is measured as volume. This study reveals that to photodegrade 25 mL of 50 mg/L Rhemazol brilliant blue we only need 10 mL of the AgNPs suspension, which is about 0.1 mmol $AgNO_3$. Figure 3 shows the degradation percentages change with the volume of the AgNPs suspension. It is clearly shown that using more than 10 mL AgNPs decrease the percentage of the degradation. It is apparently due to reducing clarity of the mixture which prohibits light reaching the dye molecules. As a consequence, further photodegradation cannot be achieved.

3.3. Optimum pH for the photodegradation

Figure 4 shows that the optimum photodegradation occurs at pH 4. Upon UV irradiation, the silver atoms excite their 4d electrons to 5sp leaving holes in the valence band [15]. The excitation is enhanced under the acidic condition, making its surface positively charged. Since Rhemazol brilliant blue is an anionic molecule, it easily gets in contact with the photocatalyst. In the same instance, the electron holes react with H_2O to form hydroxyl radicals (OH•) which can degrade the dye. The electrons excited readily react with oxygen to produce more radicals (O_2•) which also degrade the dye molecules. Degradation percentage tends to decrease with increasing pH. The percentage decrease in alkaline degradation may be due to the alkaline OH^-, which, although can form hydroxyl radicals, the excessive number of the OH^- react with hydroxyl radicals (OH •) or the hydroxyl radical recombine into new compounds [16, 17]. Moreover, the optimum pH is also beneficial for the photodegradation procedure of Rhemazol brilliant blue, since the pH is actually the original pH of an aqueous solution of the dye agent; thus no additional acid or base would be required for adjusting the system to achieve the optimum pH.

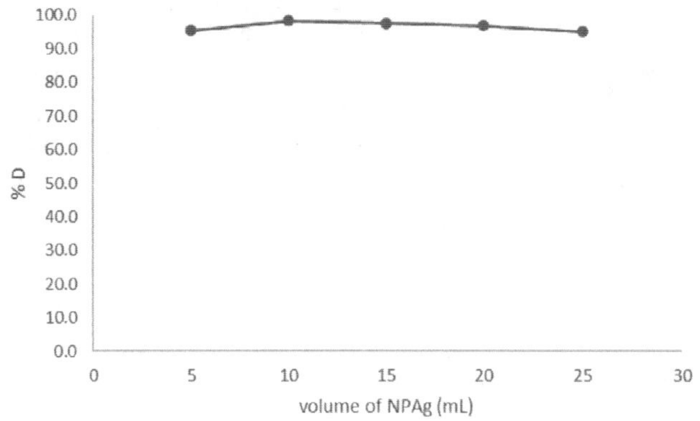

Figure 3. The photodegradation percentages at various photocatalyst volumes.

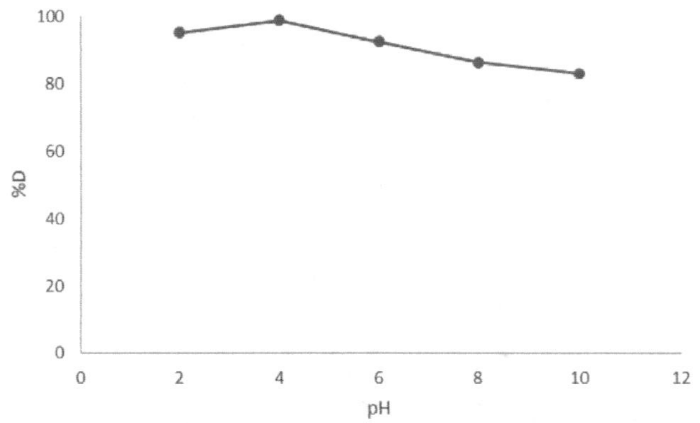

Figure 4. The change of photodegradation percentage with pH.

3.4. Optimum irradiation time

From figure 5 it can be seen the effect of irradiation time on the percentage of degradation, where the percentage of degradation increases to up to 99% within 60 minutes irradiation. The percentage decreases with longer irradiation time than 60 minutes, although the figure is still higher than 90 %, which is very good. This means that the photocatalytic activity has reached its maximum at 60 minutes, which is very effective, much faster than that using other catalysts such as metal oxide-clay composites [18].

3.5. The effectiveness of photodegradation of rhemazol brilliant blue

The photodegradation capacity was investigated by irradiating aqueous dye solution with a varied. This is confirmed by the fact that irradiation process without AgNPs can only reduce the concentration of the dye by less than 50%. If this procedure is applied to liquid waste containing dye agents, it will leave less than 10 ppm dye if the initial concentration is as high as 600 ppm which is unlikely. Therefore, this

AgNPs is an effective photocatalyst for the photodegradation of rhemazol brilliant blue. In our previous work this AgNPs was also effective as a photocatalyst for degrading indigosol blue [19,20] and methylene blue. Indigosol blue and Rhemazol brilliant blue are anionic dye, while methylene blue is cationic, thus it may suggest that photodegradation using AgNPs catalyst is an effective method to remove dye pollutant from liquid waste.

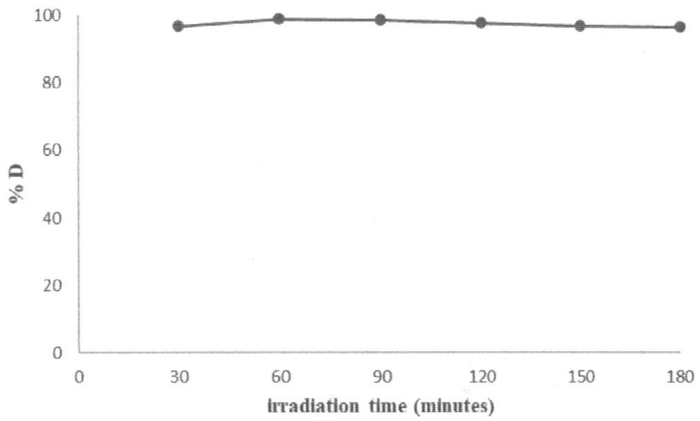

Figure 5. The change in degradation percentage with irradiation time.

Table 1. The degradation percentages of rhemazol brilliant blue by UV light with and without AgNPs.

Initial dye concentration (ppm)	Degradation percentage (%)
50*	44.68 ± 1.05
100	99.77 ± 1.21
200	99.69 ± 1.18
300	99.45 ± 1.19
400	99.22 ± 1.16
500	98.67 ± 1.11
600	97.84 ± 1.09

*without AgNPs

4. Conclusion
The AgNPs is proven to have the photocatalytic capability in degrading rhemazol brilliant blue molecules. Under pH 4, 10 mL of the AgNPs suspension effectively removes 97.84 ± 1.09 % of 25 mL of 600 ppm Rhemazol brilliant blue during 60-minute UV irradiation. This result proves that AgNPs can be used as a photocatyst with high capacity to photodegrade rhemazol brilliant blue, and potentially capable of treating waste containing dye agents.

5. Suggestion for future work
AgNPs theoretically works with visible light, which is preferable to UV. Future research should study the photocatalytic activity of the AgNPs using visible light or sunlight.

Acknowledgements
We wish to thank Wiji Rahayu S.Si. of PT. DKSH Jakarta for providing access to PSA.

References

[1] Kim D, Jeong S and Moon J 2006 *Nanotechnology* **17** 4019
[2] Haryono A, Dewi S, Harmami S B and Randy M 2008 *J. Ris. Ind.* **2** 156
[3] Vanaja M, Paulkumar K, Baburaja M, Rajeshkumar S, Gnanajobitha G, Malarkodi C, Sivakavinesan M and Annadurai G 2014 *Bioinorg. Chem. Appl.* **2014** 1
[4] Saeed K, Khan I, Shah T and Park SY 2015 *Fibers Polym.* **16** 1870
[5] Sinha T, Ahmaruzzaman M and Bhattacharjee A 2016 *Indian J. Chem. Technol.* **23** 462
[6] Kulkarni A A and Bhanage B M 2014 *ACS Sust. Chem. Eng.* **2** 1007
[7] Jyoti K and Singh A 2016 *J. Gen. Eng. Biotech.* **14** 311
[8] Gopinath V, Mubarakali D, Priyadarshini S, Priyadharshini N M, Thajuddin N and Velusamy P 2012 *Colloids Surfaces. B* **96** 69
[9] Kumar V and Yadav S K 2009 *J. Chem. Tech. Biotech.* **84** 15
[10] Akhtar M S, Panwar J and Yun YS 2013 *ACS Sust. Chem. Eng.* **1** 591
[11] Yalkhin M, Sulfikar and Rasyid M 2015 *J. Sains* **4** 28
[12] Wrasiati LP, Hartati A and Yuarini DAA 2011 *J. Biol.* **15** 39
[13] Yousefzadi M, Rahimi Z and Ghafori V 2014 *J. Agardh. Mater. Lett.* **137** 1
[14] Willets K A, Hall W P, Sherry L J, Zhang X, Zhao J and Duyne R P V 2007 *Nanobiotech.* **II** 159
[15] Chen X, Zheng Z, Ke X 2010 *Green Chem.* **12** 414
[16] Sun Z, Chen Y, Ke Q, Yang Y and Yuan J 2002 *J. Photochem. Photobiol. A* **149** 16974
[17] Ghozali A I, Sugiyo W and Latifah 2012 *Indo. J. Chem. Sci.* **1** 79
[18] Suprihatin I E, Sibarani J, Ariati N K and Sulihingtyas W D 2019 *J. Phys. Conf. Ser.* **1341** 032028
[19] Lestari G A D, Suprihatin I E and Sibarani J 2019 *J. Kim. Sains Aplikasi* **22** 200
[20] Lestari G A D, Suprihatin I E and Sibarani J 2020 *Cakra Kimia* **8** 35

IOP Conf. Series: Materials Science and Engineering **959** (2020) 012019 doi:10.1088/1757-899X/959/1/012019

Synthesis of Carbon Nanotubes (CNT) by Chemical Vapor Deposition (CVD) using a biogas-based carbon precursor: A review

D D Saputri, A M Jan'ah, and T E Saraswati*
Department of Chemistry, Faculty of Mathematics and Natural Sciences,
Sebelas Maret University, Jl. Ir. Sutami No.36A Kentingan Surakarta 57126 Indonesia

* Email: teguh@mipa.uns.ac.id

Abstract. Carbon is the fourth most abundant element in the universe. Among all carbon allotropes, carbon nanotube (CNT) with a cylindrical tube structure becomes an attractive carbon nanomaterial due to its unique physical and chemical properties. The typical gas sources that are usually used in CNT synthesis are mostly hydrocarbons, e.g., alcohol, methanol, methane (CH_4), ethylene (C_2H_4), acetylene (C_2H_2), propane (C_3H_8), and cyclohexane (C_6H_{12}). Unfortunately, those carbon precursors are not environmentally friendly due to their origin from fossil fuels. Considering its continuity production, utilizing "green" alternatives and inexpensive materials will be more promising for industrial-scale production. Biogas is one of the alternative ideas as a carbon precursor in the process of CNT synthesis. Biogas consists of 40–60% methane (CH_4), 40–60% carbon dioxide (CO_2), and traces of nitrogen (N_2), oxygen (O_2), hydrogen sulfide (H_2S), hydrogen (H_2), and ammonia (NH_3). The most commonly used to produce CNT is CH_4 through chemical vapor deposition (CVD). CVD is the most expectant and scalable method for future power and electronic devices with its economical and straightforward friendly design.

1. Introduction

Carbon is one of the chemical elements that is the fourth most abundant element in the universe after oxygen. Carbon was found a long time ago in the form of charcoal or soot, then its admission as an element emerged in eighteenth later [1]. Allotropy or allotropes form if one element exists in two or more different structures, due to its different bonding way [2], resulting in a new form of materials with the same content. Diamond, graphite, carbine, fullerenes, nanotubes, graphene, and amorphous carbon are well-known as carbon allotropes, which have various unique structures with each remarkable properties [3]. Due to their remarkable properties, carbon allotropes have various applications such as catalyst supports, adsorbents, gas-storage, separation technology, battery electrodes, porous template materials, fuel cells, and biological cells. Carbon shows many outstanding chemical and mechanical properties, e.g., high melting and sublimation points, and easy to form diverse compounds due to its ability to form long carbon-to-carbon chains [4].

Carbon nanotube (CNT) is one of the most attractive nanomaterials among the other carbon allotropes, which has a cylindrical tube structure. CNT is a furled graphene sheet in which every layer consists of a hexagonal carbon ring arranged by sp^2 hybridized atom carbons [5]. CNT was discovered by Sumio Iijima, a scientist from Japan, in 1991. Firstly, it was made from the arc discharge method when synthesizing fullerenes. CNT seemed like extended fullerenes with a diameter of around 0.7 nanometers and a length up to several micrometers [6, 7]. Iijima and other scientists found this material as a remarkable substance. Many studies have been done to improve and develop CNT, creating CNT with high strength, stiffness, and electrical conductivity for amplifying the material composites and nano-electronic applications [8].

CNT is categorized into two types depending on the amount of graphite layer rolled up, i.e., single-walled carbon nanotubes (SWCNT) and multi-walled carbon nanotubes (MWCNT) [9]. As is shown in Figure 1, SWCNT consists of one layer of graphite, similar to ordinary straw with a diameter of 1.4 nm or less. In comparison, MWCNT has more than one layer of graphite with a diameter of around 2−25 nanometers [10]. Carbon nanotubes have a type of sp^2 carbon hybridization [11]. SWCNT is classified in different structure forms, i.e., zigzag, armchair, and chiral structures. The chiral vector in CNT is marked with (n, m) indices. Integers n and m directions influence the number of unit vectors in graphene honeycomb crystal lattice. Zigzag structure means if m is zero; armchair structure means if the m = n; then chiral structure means the other state of zigzag and armchair. In addition, MWCNT has two structural models, namely the Russian doll model (smaller diameter) and the parchment model (a sheet of graphene rolled up around itself) [12].

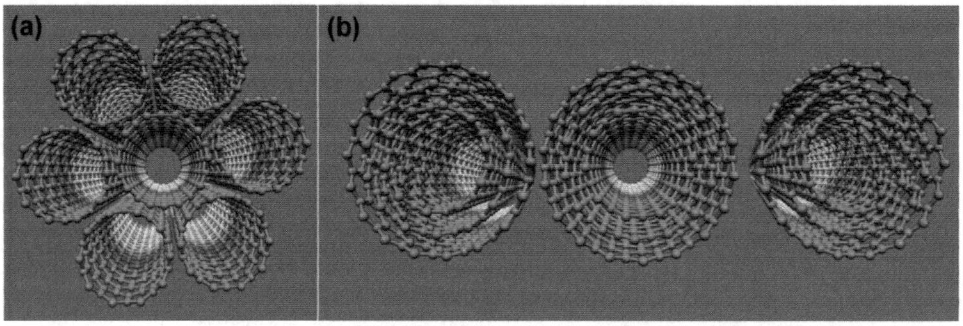

Figure 1. A bundle of (a) SWCNT and (b) MWCNT.

In the CNTs synthesis, there are several parameters to determine the success of CNT growth, such as a catalyst, temperature of substrate, gas mixture, and gas pressure [13]. Carbon precursor gas decides a high impact on the growth of CNT. The typical gases used in CNT synthesis are mostly hydrocarbons, such as alcohol, methanol, methane (CH_4), ethylene (C_2H_4), acetylene (C_2H_2), propane (C_3H_8), and cyclohexane (C_6H_{12}). Unfortunately, most gas sources come from fossil fuels; thus, they have several disadvantages, such as their unrenewable material and the high cost [14]. Therefore, this review will discuss an alternative gas precursor in the synthesis of carbon nanotubes, i.e., biogas from waste material. The utilization of biogas can minimize the use of unrenewable gas as a carbon precursor in the synthesis of CNT.

2. Synthesis carbon nanotubes

CNT is one of the carbons allotropes, which has recently been developed. Because of its chemical, electrical, mechanical, and thermal properties [6], CNT are potentially transformed into many applications like biotechnology [15], nanoelectronics [16], and thermal management. The synthesis methods to make CNT are also continuously developed and discovered.

Various methods are studied to gain CNT, such as arc discharge, chemical vapor deposition well-known as CVD [17], arc discharge, laser ablation [18], and electrolytic [19]. The main parameters in CNT synthesis are substrate temperature, gas pressure, gas mixture, and catalyst [20]. The arc discharge method can be applied by resulting energy from DC between graphitic anode and cathode with a chamber filled with helium gas. The arc occurred creates a high-temperature plasma (>4000 K), which sublimes the carbon precursors in the anode. The arc discharge method brings advantages, such as fast production of CNT with few defects; however, the method is not easy-controlled and has low productivity. Next, laser ablation is a method that prepares CNT at high temperature (±1200 °C) using a laser, in which the process was generally conducted in an inert atmosphere. This method transforms carbon species to be graphitic materials with a metal catalyst, using a laser applied to vaporize the carbon. Laser ablation offers CNTs synthesized for higher purity than those generated by arc discharge, but less productivity of CNT. Among the various methods to synthesize CNTs, CVD is currently the most widely used due to its low cost, high productivity, scalability, and easy controllability [21].

Compared with other methods, chemical vapor deposition (CVD) is a low cost, high productivity, scalability, easy controllability, perchance, and scalable method for future power and electronic devices [22]. Impurities that come out from the synthesis of CNT affect the quality impact of CNT. However, sometimes the metal impurities give additive strength properties to CNT [23]. Liu *et al.* [24] reported that the most effective catalyst metals for carbon nanotube production using CVD are iron (Fe), nickel (Ni), and cobalt (Co) because of their high carbon solubility, excellent chemical and mechanical properties, and diffusion rate of carbon.

The CVD process typically uses a furnace chamber. Initially, all gases inside the chamber are removed to avoid the oxidation process which might be caused by oxygen-containing gases in the atmospheric air. The carbon precursor gases (particularly hydrocarbon) are then escorted into the reaction chamber together with an inert gas like helium or argon. The furnace is then heated to vaporize and decompose the gases resulting in the reaction between the reactive species coming from gases with catalyst, producing the deposition carbon on the substrate. Hydrogen as the residue gas will be exhausted from the chamber by the elimination system [25]. The schematic process of CVD is shown in Figure 2. The vapor decomposition occurs in the range of temperature 600-1200°C. The vapor hydrocarbons will interact with metal catalyst until they break down into hydrogen and carbon. Carbon will dissolve into the metal, while hydrogen will evaporate [26]. The economically-friendly and straightforward experimental setup becomes one reason to choose the CVD to be a method for synthesizing CNT. Many hydrocarbons can be used as precursors in various states of compounds, i.e., gas, liquid, or solid. As a result, it makes some possibilities to build many kinds of CNT growth, such as thick and thin films, coiled or straight nanotubes, powder, entangled or aligned, or desired architecture of nanotubes on a patterned substrate [6, 7].

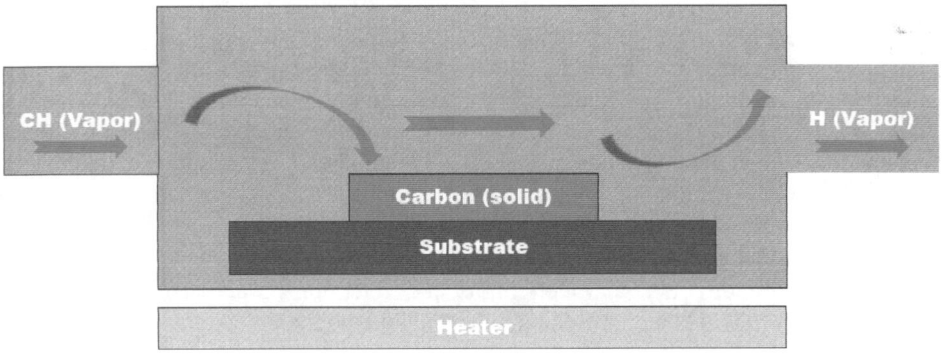

Figure 2. Schematic of the CVD process.

Based on substrate-catalyst interaction, the CNT growth is classified into two types, tip-growth and base growth. Tip-growth means the catalyst presence above the CNT, while in base-growth, the catalyst presence at the base of CNT. Tip-growth is a weak substrate-catalyst interaction. CNT starts growing from the metal base and keeps growing up to have enough space to decompose hydrocarbons depending on the metal concentration gradient. At this mechanism, CNT keeps pushing apart the metal from the substrate. In contrast, in base growth, CNT is not strong enough to push metal because of the strong substrate-catalyst interaction, resulting in the growth mostly happens above the metal [24, 26].

3. Biogas from waste materials as carbon precursor of synthesis CNT

Successful CNT growth depends on the effective decomposition of the carbon precursor with the formation of stable catalysts that impact the arrangement of carbon lattice [27]. Common carbon precursors usually used are acetylene, xylene, benzene, toluene, and methane. However, those carbon precursors usually come from fossil fuels. The unrenewable resources have disadvantages for a sustainable environment, such as high cost and low availability. Considering its continuity production, utilizing a "green" carbon precursor will be an alternative way to synthesis CNT. The benefit of using natural precursors is having less cost, long-term material, pathogen reduction, and saving energy consumption [28]. One of the alternative carbon precursors to synthesis CNT is biogas.

Biogas is a mixture gas that has components about 40%–60% methane (CH_4), 40%–60% carbon dioxide (CO_2), along with traces of oxygen (O_2), hydrogen (H_2), nitrogen (N_2), hydrogen sulfide (H_2S), then ammonia (NH_3) [29]. Biogas is the result of anaerobic digestion or fermentation, which exploits kinds of waste like animal manure [30], agricultural waste [31], and organic waste [32]. Waste generation increases every year, coming from agricultural production and household consumption. Around the world, it represents about 1.3 billion tons of waste per year [33].

The source of biogas can be obtained from cattle dung, animal manure, food waste or digested slur, straw, human and animal feces, leaves, weeds, leaf, poultry waste, domestic rubbish, and industrial wastes [28, 34]. Indonesia, as a country with a big population, carries out a high strength to implement from waste to energy, including the generation of biogas by anaerobic digestion. Many studies have reported specific types of utilization waste materials as potential economic sources in bioconversion for making biogas [35].

Cattle manure or cattle dung is categorized as biomass, which means alternative source energy from living materials such as plants and animals. There are few ways to convert biomass into energy by burning material (woody things or grasses): decomposing biomass with various conditions like temperature, pressure, and catalyst to acquire combustible yield, or applying an anaerobic digestion process in which microorganisms split up biomass without oxygen to produce biogas (particularly CO_2 and CH_4) [36]. Biomass can be sustainable and renewable feedstock that comprises carbon-rich material, chemicals, and biofuels [37].

Biomass can be converted into biogas by four steps, as shown in Figure 3. They consist of hydrolysis, acidogenesis, acetogenesis, and methanogenesis, resulting in methane gas. In the first stage of hydrolysis, this process makes a complex proteins, carbohydrates, and fats which were contained in waste, into monomers by bacterial cellulosome and exoenzymes. Following the first stage, in the second step of acidogenesis, the degradation continually proceeds into short-chain acid. Then, for acetogenesis, that short-chain will be converted to be CH_3COOH, CO_2, and H_2. The last step is methanogenesis, which the intermediates are transformed into CO_2 and methane [38]. Biogas technology works as a complete system to put forward the conversion energy, factors of waste material, and environmental and renewable sources [39]. Other factors that make the bioconversion of organic waste successfully produce biogas are temperature, environmental conditions, pH, retention time, organic mixture material, and reactor configuration [33].

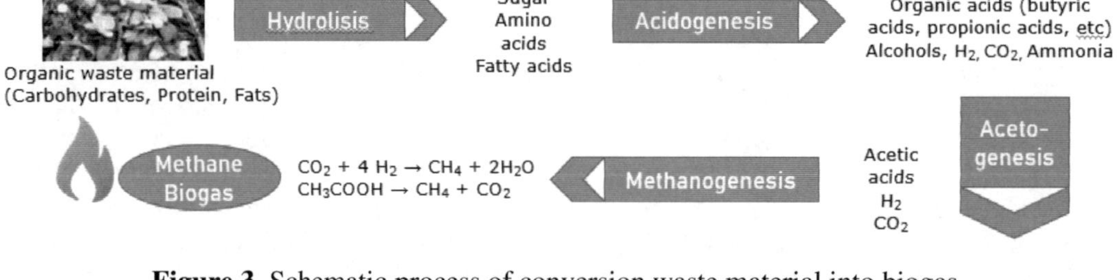

Figure 3. Schematic process of conversion waste material into biogas.

Budiyono *et al.* studied making biogas from solid waste (cattle manure) and liquid waste (rumen fluid) [40]. Solid waste is filtered from the slurry form but keeping it full of bacteria. All the waste collected needed to be homogenized. Both of various wastes are heated to 105 and 600 °C. The biodigester system used a batch system. The biodigester chambers were created from polyethylene bottle plugs, then connected with biogas measurement. The heater temperature of the biodigester was controlled by a water bath. As a result, the biogas formed was measured by the liquid displacement method. Rumen fluid made a positive impact on increasing the production rate of biogas. Rumen fluids gave influence into cumulative biogas production and kinetic rate of biogas production. The optimum temperature for bacteria was at 38,5 °C to degrade the organic substrate. By the adding of rumen fluid inoculums, the process produced more biogas (A: 194,4 (ml/gVS); meanwhile, the process without adding rumen fluids resulted in lower yield (A: 58,16 ml/gVS).

Godi *et al.* had other experiments in making bio-digester, which consisted of a digestion chamber, slurry inlet and outlet, gas exit, cap, and scrambler [41]. The methane gas was not only the gas resulted; the other gas (carbon (IV) oxide and hydrogen sulfide) was also detached from the digestion process chamber. Consequently, it required something that can get rid of those gases. In their experiment, $Ca(OH)_2$ was used as a gas remover for undesired gases. The steps of making the biodigester were initially by the cow dung. First, waste prepared from cow dung was blended with water in the chamber to make a slurry. The mixture was moved down into the digester through the inlet. Second, they made a pipe connection between the gas source pipe and the digester chamber and also with the collecting container. The pipe also was passed by the solution of prepared calcium hydroxide to remove another gas. Next, another outlet pipe was linked up with the conical flask. By giving candle wax on the connected pipe, the gas leak was reduced. The calibrated drum was let to have fermentation; then, the produced gas was collected through the upward delivery method. The process took time for almost 20 days for fermentation from cow dung into biogas because cow feed mainly contains fibrous substances so that it took longer to decay by microorganisms. As a result, the total methane yielded is 1570 ml, and the total biogas yield is 2300 ml. Therefore, the production of methane compounds is 68% of the total composition of biogas generated from cow dung.

4. Synthesis of carbon nanotubes using biogas-based carbon precursor

Biogas converted from waste material is potentially used as the carbon precursor in the synthesis of CNT, especially for the methane gas produced from biogas bioconversion of waste. Khavarian and Mohammed [42] studied the biogas reforming over MWCNT. They used a nanocomposite catalyst such as Co-Mo/MgO catalyst and methane (CH_4) gas in CVD. The CVD process was conducted in a continuous rotary reactor at temperature 825 °C and was supplied with N_2 as a carrier gas. The average diameter of the MWCNT ranges between 10 and 30 nanometers. The increasing of the CO_2/CH_4 feed ratio influenced higher methane conversion and a lower H_2/CO syngas ratio as CO formation reactions.

Baddour *et al.* [43] also explained the process of synthesis CNT using the CVD furnace. The CVD system is composed of a quartz tube, gas inlet, thermocouple, and pump. The quartz tube was placed in the furnace. Then, the substrates are located inside the quartz tube in the middle of the furnace cap. In this experiment, acetylene (C_2H_2) was used as a carbon precursor and nitrogen (N_2) as the carrier gas, which entered the quartz tube from the gas inlet. The furnace system temperature was set at a temperature of 650, 700, and 800 °C with substrate pre-heating at 850°C for 30 minutes and at 650, 700, 800, and 850°C without pre-heating. The nanotubes resulting in are kind of MWNT with a range of diameter around 20 to 70 nanometers.

In another study, Vamathevan developed the process of synthesizing carbon nanotubes from biogas [44]. The study made the chamber consisted of argon and biogas cylinders that were connected to the reactor. The sample of catalyst material was kept in the sample holder. After the holder was inserted to the gas cylinder, the synthesis of carbon nanotubes using biogas is ready to run. The reactor was heated up to 550°C when the biogas was delivered into the reactor for one hour. In the end, after an hour of gas supply, the reactor got cooled down to room temperature. The product of the reaction was then taken out and was characterized. This work successfully developed a process to synthesize carbon nanotubes using biogas and the catalysts of nickel particles deposited on the support surface. However, the results of a product characterization suggested that the synthesis process of the CNT produced using biogas needs to be more developed by optimizing the synthesis parameters.

5. Conclusion

Carbon precursor gas takes a high impact on the growth of CNT. The commonly used hydrocarbon gases have several disadvantages, such as limited availability, unrenewable, and high cost. The synthesis of CNT can be conducted by CVD method, which is the most profitable method to synthesis CNT that brings many conveniences. Utilization of biogas converted from waste material as a carbon precursor of synthesis CNT confidently provides an alternative sustainable gas resource. The biogas can be produced from anaerobic digestion or fermentation, which exploits kinds of waste like animal manure, agricultural waste, and organic waste. The biogases produced can be used as alternative gas to synthesis CNT as the carbon precursor via CVD method.

Acknowledgment

The author would like to thank the Ministry of Research and Technology/National Agency for Research and Innovation, the Republic of Indonesia, under the Research Grant No. 452/UN27.21/PN/2020, provided by LPPM Sebelas Maret University (PNBP 2020).

References

[1] Weeks M E 2018 *Discovery of The Element* (Sagwan Press)
[2] McNaught A D and Wilkinson A 1997 *Compendium of chemical terminology* vol 1669 (Blackwell Science Oxford)
[3] Falcao E H and Wudl F 2007 *J. Chem. Technol. Biotechnol.* **82** 524
[4] Rahman T, Fadhlulloh M A, Nandiyanto A B D and Mudzakir A 2015 *Jurnal Integrasi Proses* **5** 120-31
[5] Holban A M, Grumezescu A M and Andronescu E 2016 *Surface Chemistry of Nanobiomaterials,* ed A M Grumezescu (William Andrew Publishing) pp 301−327
[6] Sengupta J 2018 *Handbook of Nanomaterials for Industrial Applications* ed C Mustansar Hussain (Elsevier) pp 172−194
[7] Meyyappan M, Delzeit L, Cassell A and Hash D 2003 *Plasma Sources Sci. Technol.* **12** 205
[8] Öncel Ç and Yürüm Y 2006 *Fuller. Nanotub. Car. N.* **14** 17
[9] Shashikumar N S, Gireesha B J, Mahanthesh B and Prasannakumara B C 2018 *Multidiscip. Model. Mater. Struct.* **14** 769
[10] Caglar B 2010 *Production of carbon nanotubes by PECVD and their applications to supercapacitors* Master's Thesis (Universitat de Barcelona)
[11] Manawi Y M, Samara A, Al-Ansari T and Atieh M A 2018 *Materials* **11** 822
[12] Eatemadi A, Daraee H, Karimkhanloo H, Kouhi M, Zarghami N, Akbarzadeh A, Abasi M, Hanifehpour Y and Joo S W 2014 *Nanoscale Res. Lett.* **9** 393
[13] Barnard J S, Paukner C and Koziol K K 2016 *Nanoscale* **8** 17262
[14] Alfarisa S, Abu Bakar S, Mohamed A, Hashim N, Kamari A, Md Isa I, Mamat M H, Rahman Mohamed A and Rusop Mahmood M 2015 *Adv. Mat. Res.* **1109** 50
[15] Saito Y, Hamaguchi K, Hata K, Uchida K, Tasaka Y, Ikazaki F, Yumura M, Kasuya A and Nishina Y 1997 *Nature* **389** 554
[16] Martin C R and Kohli P 2003 *Nat. Rev. Drug Discov.* **2** 29
[17] Gattia D M, Vittori Antisari M, Marazzi R, Pilloni L, Contini V and Montone A 2006 *Mater. Sci. Forum* **518** 23
[18] Batani D, Vinci T and Bleiner D 2014 *Laser Part. Beams* **32** 1
[19] Novoselova I A, Oliinyk N F, Volkov S V, Konchits A A, Yanchuk I B, Yefanov V S, Kolesnik S P and Karpets M V 2008 *Physica E Low Dimens. Syst. Nanostruct.* **40** 2231

[20] Gohier A, Minea T M, Point S, Mevellec J Y, Jimenez J, Djouadi M A and Granier A 2009 *Diam. Relat. Mater.* **18** 61
[21] Kwon S-J, Seo H-K, Ahn S and Lee T-W 2019 *Adv. Sustain. Syst.* **3** 1800016
[22] Quinton B T, Barnes P N, Varanasi C V, Burke J, Tsao B-H, Yost K J and Mukhopadhyay S M 2013 *J. Nanomater.* **2013** 356259
[23] Williams P T 2020 *Waste and Biomass Valorization* (Springer)
[24] Liu X, Shen B, Wu Z, Parlett C M A, Han Z, George A, Yuan P, Patel D and Wu C 2018 *Chem. Eng. Sci.* **192** 882-
[25] Kumar M and Ando Y 2010 *J. Nanosci. Nanotechnol.* **10** 3739
[26] Venkataraman A, Amadi E V, Chen Y and Papadopoulos C 2019 *Nanoscale Res. Lett.* **14** 220
[27] Nessim G D, Seita M, Plata D L, O'Brien K P, John Hart A, Meshot E R, Reddy C M, Gschwend P M and Thompson C V 2011 *Carbon* **49** 804
[28] Kumar R, Singh R K and Singh D P 2016 *Renew. Sustain. Energ. Rev.* **58** 976
[29] Köhring M, Böttger S, Willer U and Schade W 2015 *Sensors* **15**
[30] Bidart C, Fröhling M and Schultmann F 2014 *Renew. Sustain. Energ. Rev.* **38** 537
[31] Karellas S, Boukis I and Kontopoulos G 2010 *Renew. Sustain. Energ. Rev.* **14** 1273
[32] Zhang C, Su H, Baeyens J and Tan T 2014 *Renew. Sustain. Energ. Rev.* **38** 383
[33] Khanal S K, Nindhia T G T and Nitayavardhana S 2019 *Sustainable Resource Recovery and Zero Waste Approaches* (Elsevier) pp 165–74
[34] Adeniran K A, Ahaneku I E, Itodo I N and Rohjy H A 2014 *Agric. Eng. Int.: CIGR Journal* **16** 126
[35] Khalil M, Berawi M A, Heryanto R and Rizalie A 2019 *Renew. Sustain. Energ. Rev.* **105** 323
[36] Hoa L Q, Vestergaard M d C and Tamiya E 2017 *Sensors* **17**
[37] Wang Z, Shen D, Wu C and Gu S 2018 *Green Chem.* **20** 5031
[38] Faisal S, Yusuf Hafeez F, Zafar Y, Majeed S, Leng X, Zhao S, Saif I, Malik K and Li X J A S 2019 *Appl. Sci.* **9** 59
[39] Arvanitoyannis I S and Tserkezou P 2008 *Int. J. Food Sci. Technol.* **43** 958
[40] Budiyono I, Nyoman I, Widiasa, Sunarso S and Seno J 2010 *Int. J. Chem. Biol. Eng.* **3** 39
[41] Godi N Y, Zhengwuvi L B, Adulkadir S and Kamtu P M 2013 *J. Mech. Eng. Res.* **5** 1
[42] Khavarian M, Mohamed A R, Ibrahim S M and Noorsal K 2017 *AIP Conference Proceedings* **1901** 030008
[43] Baddour C E, Fadlallah F, Nasuhoglu D, Mitra R, Vandsburger L and Meunier J-L 2009 *Carbon* **47** 313
[44] Vamathevan K 2011 *The Development of the Process to Synthesize Carbon Nanotubes from Biogas* Master's Thesis (University of Moratuwa)

IOP Conf. Series: Materials Science and Engineering **959** (2020) 012020 doi:10.1088/1757-899X/959/1/012020

Sol-gel synthesis and photocatalytic activity of biomimetic calcium manganese oxide catalysts

W Setyarini[1], Nurhayati[1], N Dewi, R Anggraini[1], S S Siregar[1], E Amiruddin[2] and A Awaluddin[1,*]
[1] Department of Chemistry, Faculty of Mathematics and Natural Sciences, Universitas Riau, Kampus Bina Widya, Km 12.5 Simpang Baru, Pekanbaru, Riau, Indonesia
[2] Department of Physics, Faculty of Mathematics and Natural Sciences, Universitas Riau, Kampus Bina Widya, Km 12.5 Simpang Baru, Pekanbaru, Riau, Indonesia

[*] E-mail: amirawaluddin01@gmail.com

Abstract. The water contaminations by dye residues are increasingly becoming serious concerns worldwide due to dye toxicity and persistent characteristics. Manganese oxide-based catalysts having similar structures with manganese compounds found in nature (biomimetic) have been considered as a very promising and effective photocatalyst for the degradation of organic pollutions in wastewater. This paper focuses on the synthesis of calcium manganese oxide catalysts from the octahedral layered birnessite-type manganese oxide and calcium carbonate via sol-gel method with citric acid as the complexing agent. The as-synthesized oxide was then tested as a photocatalyst for the degradation of methylene blue (MB) dye. The calcium manganese oxides prepared by the two different mole ratios of $CaCO_3/MnO_2$ resulted in the similar crystallinity and crystal phases but difference in the crystal sizes. The photocatalytic activities of the prepared calcium manganese oxides for the degradation of methylene blue are compared to that of the calcium manganese oxide prepared from tunnel cryptomelane-type manganese oxide. Both the calcium manganese oxides prepared from the different manganese oxide phases show the remarkable performance for the photocatalytic degradation of methylene blue after 10 minutes of reaction times. The birnessite-prepared calcium manganese oxide displays slightly higher photocatalytic performance than cryptomelane-prepared calcium manganese oxide.

1. Introduction

Water pollution due to organic contaminants has become a serious concern all over the world. The organic pollutants containing residual dyes have an inverse effect to environments and human health because they are toxic, mutagenic, and carcinogenic. They are also highly visible, persistent, and non-biodegradable since they have a polymeric structure. The different strategies have been applied to degrade this refractory material such as biological, physical, and chemical methods. The biological methods have been considered to be the most widely applied and economical to remove the contaminants but suffers from the length of time used for the process. In addition, the method cannot be applied if the pollutants are toxic to the microbes used in the process. Although the physical method such as adsorption is also widely used to treat organic wastewater containing dyes and is simple and effective, this method creates additional problems due to the transfer of the pollutant to the adsorbent.

The advanced oxidation processes (AOPs) are perceived as highly promising pathways to treat the recalcitrant materials such as dyes by using the strong powerful hydroxyl radical •OH as the main oxidant. The hydroxyl radical, whose standard redox potential is 2.8 V, immediately attack and non-selectively degrade most organic contaminants. Hydroxyl radicals react via hydrogen abstraction, electrophilic addition to double bond, and electron transfer reactions.

The wide ranges of technique are applied to generate hydroxyl radical in the AOP method such as Fenton and related process, electrochemical oxidation, wet air oxidation, photocatalytic reaction, and so forth. Photocatalytic reactions are widely used for the treatment of dyes and use photocatalysts, which usually semiconductor materials using light of energy greater than the semiconductor band-gap to promote an electron from valence band to the conduction band. This leaves an electron vacancy or hole in the valence band. The hole is highly oxidative and immediately reacts with organic molecules such as dyes adsorbed in the photocatalyst surface. The hole also reacts with adsorbed water molecules, generating hydroxyl radicals that in turn degrade organic molecules.

Manganese oxides-based catalysts have been reported as an active photocatalyst for the abatement of organic wastewater. Among a wide range of manganese oxides photocatalysts, calcium manganese oxides have been regarded as biomimetic redox catalysts because they are ubiquitous in the earth and have structural similarities to the μ-oxo-Mn_4Ca cluster in the Photosystem-II protein used for the water splitting in natural photosynthesis. The calcium manganese oxide-based perovskite structures are reported as an excellent candidate for photocatalysts [1]. Several studies have succeeded in synthesizing $CaMnO_3$ using the solid reaction method with $CaCO_3$ and $MnCO_3$ as starting material [2]. The previous synthesis of perovskite based-manganese oxides has been conducted by sol-gel and ceramic methods with different starting materials. The cryptomelane-type manganese oxide has been used to synthesize calcium manganese oxide with distinctive properties. It is believed that the different crystal structures of manganese oxides may have the profound effect on the resulting calcium manganese oxides. The manganese oxides with different crystal structures and properties have been successfully synthesized by sol-gel method [3–6], ceramic method [7], and precipitation method [8–10].

In the present study, calcium manganese oxide catalysts were synthesized by sol-gel method via the reaction between birnessite-type manganese oxide with calcium carbonate. The manganese oxide was first synthesized by ceramic method using potassium permanganate and citric acid, which is then reacted with calcium carbonate with two different mole ratios. The use of the birnessite-type manganese oxide as the starting material for the synthesis of calcium manganese oxide-based perovskite has never been reported previously and would play a crucial role in the production of calcium manganese oxides with distinctive structure and properties. The as-synthesized calcium manganese oxide materials are applied as a photocatalyst for the degradation of methylene blue dye.

2. Experimental

2.1. Materials

$KMnO_4$, glucose, HCl, $CaCO_3$, citric acid, and HNO_3 were all purchased from Merck (Germany), without any further purification.

2.2. Synthesis of MnO_2

The octahedral layered manganese oxide was prepared via a solvent-free method [6]. The mole ratios of $KMnO_4$ and glucose are set to 3:1 and grounded into a fine powder. The resulting fine-powder of $KMnO_4$ and glucose were then mixed and ground in one mortar. The mixture was calcined at 700°C for 7 hours. The final product was cooled up to room temperature, washed using HCl 0.1 M and distilled water. The black product was dried at 110°C. As compared, the octahedral tunnel cryptomelane was synthesized using the method reported in the literature [3].

2.3 Synthesis of calcium manganese oxide
The oxide was prepared via sol-gel method. The appropriate amount of $CaCO_3$ and MnO_2 are shown in Table 1. The fine-powder was dissolved in 200 mL HNO_3 and stirred for ± 3 minutes. The mixture was added 25.2168 g citric acid and stirred until 7 hours. The solution is heated at 100°C to gel, dried at 110°C until 12 h and calcined at 270°C (2h). The product was calcined at 700°C.

Table 1. Amount of $CaCO_3$ and MnO_2.

	Mole ratio	Amount (g)
$CaCO_3$	3:4	1.50
MnO_2	3:4	1.74
$CaCO_3$	1:1	2.50
MnO_2	1:1	2.17

2.4 Characterization
X-ray powder diffraction (XRD) was performed on a Shimadzu 700 Maximax X-ray diffractometer operating at 40 kV and a current of 30 mA with Cu Kα radiation (X = 1.54060 A°). The data were collected with scan speed 2° per minute in 2θ, from 2θ = 10° to 90°.

2.5 Photocatalytic activities
The photocatalytic activities of the sample were evaluated via the photocatalytic oxidation of 30 ppm MB under UV light irradiation. A 250 watt (Philips) UV lamp (350-400 nm) was used as a light source. Prior to illumination, 50 mg catalyst was added to the MB solution (100 mL, pH 2). The degradation efficiency of MB was analyzed using a UV-Vis spectrometer. Peaks were observed to be present at 660 nm and were assigned as the absorption, which was indicative of the degradation of MB.

3. Result and discussion
The synthesis of calcium manganese oxide in this study used nitric acid as the solvent and citric acid as the chelating agent. X-ray diffraction was used to characterize the phase, purity, crystallinity, and the size crystal of the as-synthesized manganese oxide and the corresponding calcium manganese oxides. Figure 1 (a) shows the XRD patterns of the as-synthesized manganese oxide, which indicates the presence of peaks at 2Θ: 12.53; 25.2; 35.2 ; 37.3; 39.88; 42.4; 40; 65.5 corresponding of diffraction planes of (001), (002), (201), (111), (201), (112), (113) and (321), respectively. The oxide belongs to layered manganese oxide, birnessite (JCPDS 80-1098). The products of the reaction between calcium carbonate and birnessite with the two different mole ratios (3:4 and 1:1) are the calcium manganese oxides (JCPDS 076-1133). The XRD patterns of the calcium manganese oxides are shown in figure 1 (b) with a strong peak at 2Θ of 30° due to the presence of un-reacted $CaCO_3$. The XRD results suggested that the calcination temperature at 700°C was not high enough to decompose and convert fully $CaCO_3$ into $CaMnO_3$, which is typical stoichiometric calcium manganese oxide perovskite. The different mole ratios (3:4 and 1:1) used in this experiment merely resulted in the difference in peak intensity due to $CaCO_3$ indicating the difference in calcium content in the as-synthesized calcium manganese oxides. Kim et al. (2014) synthesized $CaMnO_3$ and oxygen-deficient perovskite $Ca_2Mn_2O_5$ from $CaCO_3$ and manganese nitrate using citric acid as the complexing agent via sol-gel method [11]. The product $CaMnO_3$ was then calcined at 900°C to generate crystalline $CaMnO_3$, which was further reductive annealing to generate $Ca_2Mn_2O_5$. The $Ca_2Mn_2O_5$ perovskite displayed higher oxygen evolution reaction than that $CaMnO_3$ perovskite due to the presence of oxygen vacancies. Zhang et al. (2016) also indicated that the presence of an oxygen vacancy in the $Ca_2Mn_2O_5$ induced enhancement of photochemical water oxidation compared to $CaMnO_3$ [12]. In this study, the

perovskite $CaMnO_3$ was first synthesized by using the birnessite-type manganese oxide prepared from sol-gel method. Farag et al. (2017) prepared $CaMnO_3$ with distinctive properties from the reaction of $CaCl_2$ and $MnCl_2$ using sol-gel method with the increased electrical conductivity in higher temperatures [2]. Gagrani et al. (2020) studied the different calcium manganese oxide nanoparticles ($Ca_3Mn_3O_8$, $CaMn_2O_4$, and $CaMnO_3$) used as photocatalysts for the degradation of the dye [1]. The photocatalytic activities of as-prepared calcium manganese oxides increase in the order $Ca_3Mn_3O_8 < CaMnO_3 < CaMn_2O_4$. It is believed that the presence of oxygen vacancy in certain perovskite is responsible for the enhancement of their catalytic activity.

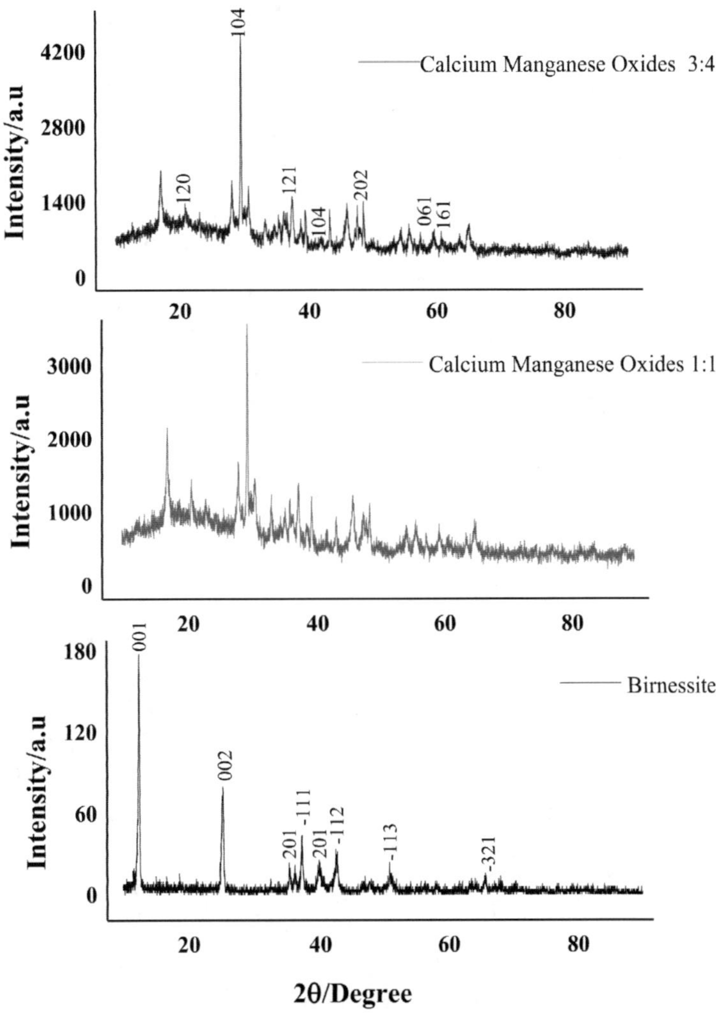

Figure 1. XRD patterns of birnessite MnO_2 (black-line), calcium manganese oxides $CaMnO_3$ 3:4 (blue-line), and calcium manganese oxides $CaMnO_3$ 1:1 (green-line).

The crystal sizes of the birnessite-type manganese oxide and the corresponding calcium manganese oxides with the different mole ratios are shown in table 2. The birnessite displayed much lower crystal size compared to the calcium manganese oxides. The two calcium manganese oxide synthesized with the different mole ratios indicated the marked difference in the crystal sizes. The one synthesized with the 3: 4 mole ratio resulted in the perovskite with the much higher crystal size than that of 1:1 mole ratio calculated by Scherrer equation shown in table 2. Thus, the difference in mole ratios of starting materials plays a crucial role in the production of the difference crystal sizes of calcium manganese oxide.

Figure 2 displayed the catalytic activities of the calcium manganese oxide as a photocatalyst for the degradation of MB at the different pH ranges. The results indicated that the working pH plays a crucial role in the degradation of MB using the calcium manganese oxide as the photocatalyst. The moderate acidic condition (pH 5) and basic pH (pH 8) displayed less than 60% of MB degradation after 120 minutes of degradation reaction. On the other hand, the highly acidic condition (pH 2) resulted in considerable degradation of MB even after 10 minutes of degradation reaction.

Figure 3 displayed the MB degradation using three different photocatalysts, the birnessite-type manganese oxide, the calcium manganese oxide prepared from birnessite with the ratio of 3:4 ($CaCO_3$/birnessite) and the calcium manganese oxide prepared from cryptomelane-type manganese oxide with the ratio of 3:4 ($CaCO_3$/cryptomelane). It is obvious (inset in figure 3) that the as-synthesized calcium manganese oxides prepared from two different crystal structures of manganese oxides show the much higher catalytic activities than that of the MnO_2 (birnessite) at the same experimental condition. The calcium manganese oxide catalysts prepared from different crystal structures also resulted in the difference in catalytic performance. The calcium manganese oxide prepared from layered birnessite-type manganese oxide shows slightly higher catalytic activity than that prepared from tunnel cryptomelane-type manganese oxide.

Table 2. Average crystal sizes of birnessite and calcium manganese oxide.

Samples	Average crystal sizes (nm)
Birnessite	30.273
Calcium manganese oxides 3:4	149.035 x 10^3
Calcium manganese oxides 1:1	100.422

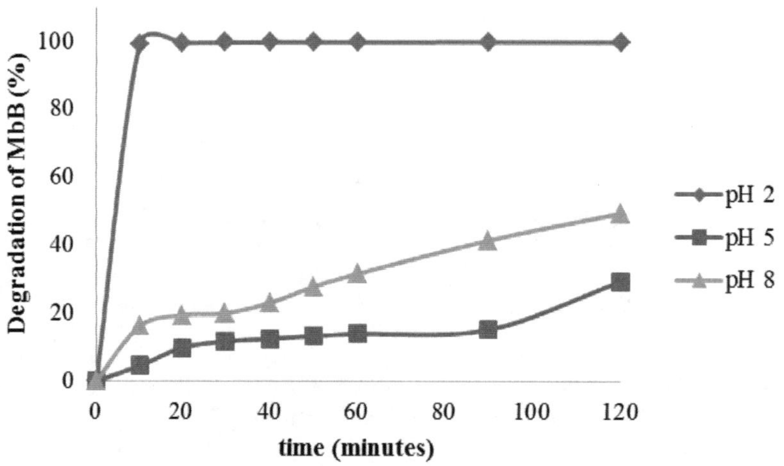

Figure 2. Degradation of methylene blue (MB) by the calcium manganese oxide at pH 2 (red line), pH 5 (blue line), and pH 8 (green line).

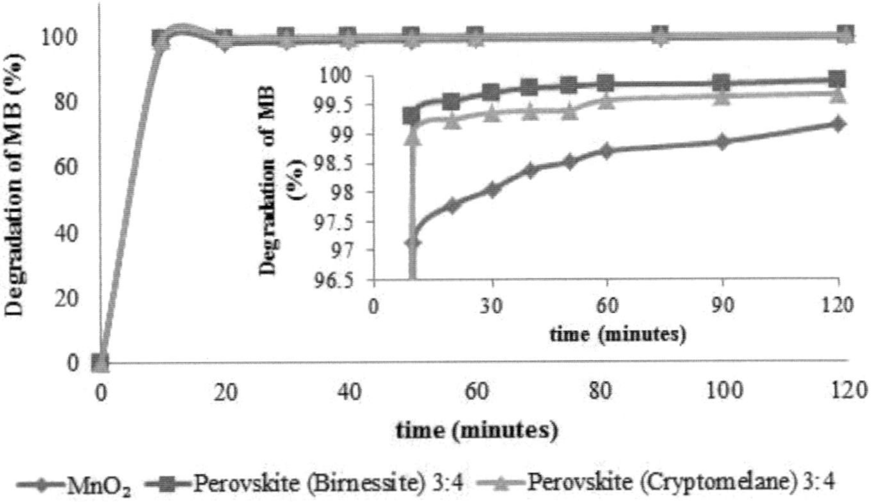

Figure 3. Degradation of MB by MnO_2 (blue line), calcium manganese oxide prepared via 3: 4 mole ratio of $CaCO_3$/birnessite (red line), and calcium manganese oxide prepared via 3: 4 mole ratio of $CaCO_3$/cryptomelane (green line). The inset shows the difference in the percentage of degradation (magnification).

4. Conclusion

The calcium manganese oxides with perovskite structure have been successfully synthesized from calcium carbonate and birnessite-type manganese oxide with two different mole ratios. The oxides have a similar crystal structure and crystallinity but difference in the crystal size. The as-synthesized calcium manganese oxide displayed much higher catalytic activity as photocatalysts compared to birnessite-type manganese oxide for the degradation of methylene blue. The working pH wastewater seems to play a key role for the degradation of MB dye.

Acknowledgments

This study was funded by the Ministry of Education and Culture, under the scheme of "Thesis Magister" with the contract no. 063/SP2H/LT/DRPM/2020 on behalf of Prof. Dr. Amir Awaluddin, M.Sc.

References

[1] Gagrani A, Sousa S, Monteiro O C and Tsuzuki T 2020 *J. Solid State Chem.* **288** 121390
[2] Farag H K and Youssef A M 2017 *Int. J. Chemtech. Res.* **10** 371
[3] Awaluddin A, Astuti L, Linggawati A, Siregar S S, Prasetya and Saputra L 2018 *AIP Conf. Proc.* **2026** 020075
[4] Awaluddin A, Amiruddin E, Anggraini R, Indriyani N and Siregar S S 2020 *IOP Conf. Ser. Mater. Sci. Eng.* **845** 012014
[5] Awaluddin A, Anggraini R, Siregar S S, Muhdarina and Prasetya 2019 *MATEC Web Conf.* **276** 06005
[6] Ching S, Roark L J, Duan N and Suib S L 1997 *Chem. Mater.* **9** 750
[7] Siregar S S and Awaluddin A 2018 *IOP Conf. Ser. Mater. Sci. Eng.* **345** 012005
[8] Absus S, Zulfa R, Awaluddin A, Anita S and Siregar S S 2018 *AIP Conf. Proc.* **2049** 020009
[9] Zulfa R, Absus S, Awaluddin A, Anita S, Siregar S S and Zulfikri 2018 *J. Phys. Conf. Ser.* **1097** 012051
[10] Subramanian N, Viswanathan B and Varadarajan T K 2014 *RSC Adv.* **4** 33911
[11] Kim J, Yin X, Tsao K C, Fang S and Yang H 2014. *J. Am. Chem. Soc.* **136** 14646
[12] Zhang G, Dong W, Huang X and Zou J 2016 *Catal. Commun.* **89** 117

IOP Conf. Series: Materials Science and Engineering **959** (2020) 012021 doi:10.1088/1757-899X/959/1/012021

Synthesis of indium sulfide photoanode, its characterization and application for degradation of methylene blue and methyl orange

Gunawan[1,*], A Haris[1] and W Septina[2]
[1] Department of Chemistry, Faculty of Science and Mathematics, Diponegoro University, Jl. Prof. H. Soedarto, S.H.,Tembalang, Semarang 50275, Indonesia
[2] Hawaii Natural Energy Institute, University of Hawai'i at Mānoa (UHM), 1680 East West Road, POST 109 Honolulu, HI 96822, United States

[*] E-mail: gunawan@live.undip.ac.id

Abstract. The researches on methylene blue (MB) and methyl orange (MO) destructions were carried out using indium sulfide (In_2S_3) photoanode in photoelectrocatalytic system. It consisted of In_2S_3 thin film as an anode synthesized by chemical bath deposition (CBD) on FTO and platinum as a cathode. Characterization with EDX showed the presence of In_2S_3 with a composition close to In_2S_3 for 10 min deposition. The In_2S_3 band gap measurement was obtained 2.49 eV. Photocurrent measurement revealed that In_2S_3 had *n*-type semiconductor property. The application of In_2S_3 thin film as photoanode in a photoelectrocatalytic cell for MB and MO treatments gave degradation's yields *ca.* 71 and 55 %, respectively, under illumination with a 20 watt UV lamp, a bias voltage of 0.5 V for 150 min and with sample concentration of 25 mg/L.

1. Introduction

Photoelectrocatalytic (PEC) cell is a very useful way to treat wastewater organic pollutants that can convert them into non toxic product. Many researchers have focussed this method both for producing electicity and hydrogen generation [1]. The active site of the photoelectrocatalytic cell is anode electrode that made from *n*-type semiconductor, it is called as photoanode, while the cathode electrode is made from platinum. Then, they are immersed in innert solution such as Na_2SO_4, and the electrodes are connected by wire with and without bias potential. Under illumination, photoexitation will be occured at photoanode producing electrons and holes. The electrons generated from this illumination will go from valence band (V_b) to conduction band (V_b), and will flow to external circuit and retured back to catode to reduce water to generate hydrogen. Whereas, the holes producing from photoanode will oxide organic wastes, or react with water or OH^- to form OH^*. The radical of hydroxy (OH^*) is very strong oxidant (2.8V) that can be able to destruct organic wastes resulting carbondioxide, water and other by-products [2].

Now days, several researchers have concerned to series of TiO_2 nanotubes [3–7]. They found photoanode structural properties such as surface area and thickness that affected the ability of the photoanode to utilize light and would hinder the transport of electrons generate by the light [6–8]. Photoanode of ZnO/Zn had been studied with a closed or open circuit system and without using

IOP Conf. Series: Materials Science and Engineering **959** (2020) 012021 doi:10.1088/1757-899X/959/1/012021

electrolytes [9]. However, the use of electrolytes can increase the performance of photoelectrocatalytic cell [1,10–12] and sodium sulfate is ussually utilized as it is innert electrolyte and it is remained in the solution when photoelectrocatalytic process occurs, since there is no reduction-oxidation reaction taken place for sodium and sulfate ions. The presence of electrolyte in photoelectrocatalytic cell can improve the conductivity of the solution and decrease the internal resistance of the cell. A decrease in internal resistance can cause a larger potential difference at the two electrodes and stimulate the flow of electrons at the cathode, which will leave a hole in the photoanode that is able to oxidize strongly.

Hitherto, there are not much applications of In_2S_3 thin film as a photoanode in a photoelectrocatalytic system. Thin film of In_2S_3 has been widely used as a buffer layer in solar cells [13] and photoelectrochemical water splitting [14–16]. In this research, synthesis of In_2S_3 photoanode for photoelectrocatalytic cell and its application for degradation of methylene blue and methyl orange was conducted. Characterization of the In_2S_3 and its application for destruction of MB and MO with varied electrolytes were studied.

2. Experimental

2.1. (Flourine doped-tin oxide) FTO preparation
Before used for the In_2S_3 deposition, FTO glass was cut into 1 cm x 2 cm and immersed in acetone and sonificated for 10 min. Then soaked into distilled water for 10 min until homogenized with a sonificator. Finally, soaked again in acetone and sonificated for 10 min and it was ready to be used as a glass substrate.

2.2. In_2S_3 Deposition by Chemical Bath Deposition (CBD)
Synthesis of In_2S_3 using CBD method was carried out in an acid cupboard. In 100 mL beaker, 25 mM $InCl_3$, 100 mM thioacetamide and 100 mM acetic acid were added to the beaker, then the FTO was inserted into the beaker vertically using a clamp holder. Furthermore, the magnetic bars were inserted in the beaker and in a water bath on the magnetic strirrer equipped with an electric heater. While stirring at a speed of 100 rpm the bath was heated until the solution in the beaker reached a temperature of 65 °C. After about 60 min a yellow deposit of In_2S_3 appeared, and varied deposition times started from this point. Finally, the beaker was removed from the water bath, then the FTO coated with In_2S_3 was taken out and washed with distilled water and dried by spraying with nitrogen gas.

2.3. Characterization
The obtained In_2S_3 thin film was characterized using EDX (Phenom Pro-X desktop SEM with EDX), SEM (JSM-6510LA), XRD (PANalytical X Pert^3 Powder X-ray diffractometer (Cu Kα, Ni filter)), Raman (Jasco NRC 3100 Laser), and determination of the band gap used a UV-Vis spectrophotometer (PG Instruments T60 UV Vis Spectrophotometer), as well as the photocurrent test (*J-V*) with artificial light (1.5 AM solar simulator) in 1 M Na_2SO_3 solution was measured by a potentiostat (Corr Test-150) under chopped irradiation.

2.4. Application of In_2S_3 photoanode for photoelectrocatalysis cell
Dye waste treatment using photoelectrocatalytic cell is schematically showed in figure 1. Photoanode and platinum were assembled and put into the beaker filled with dye to be destructed. Furthermore, the irradiation was carried out using a 20 watt UV lamp (366 nm) and a bias voltage of 0.5 V. The parameters studied in these dye waste treatment processes consisted of the treatment times, the uses of no and varied electrolytes of NaOH, HCl, Na_2SO_4. The dye concentration analysis after the treatment was carried out using a UV-Vis spectrophotometer.

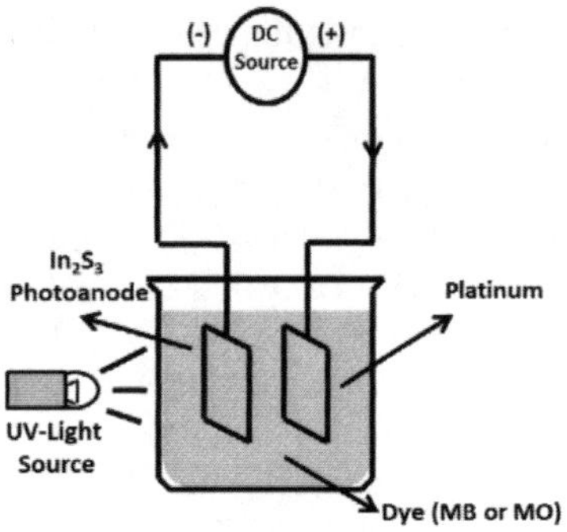

Figure 1. Schematic diagram of dye destrustion using photoelectrocatalytic cell.

3. Results and discussion

The deposition of In_2S_3 on FTO glass is shown in figure 2. It is seen that the thin film was formed evenly on the surface of FTO. A fine film was obtained with 10 min deposition time, and if it was continued, larger In_2S_3 crystals grew to form bulk crystals. However, the presence of bulk crystals will inhibit the semiconductor properties of the photoanode as there was very small photocurrent as it was observed from J-V curve measured by potentiostat.

Figure 2. In_2S_3 thin film deposited on the FTO surface.

3.1. Composition analysis of In_2S_3 thin films with EDX

The composition analysis of In_2S_3 was performed using EDX and shown in table 1 as stated with weight percentages.

Table 1. Percentage of In_2S_3 on FTO glass.

	Weight percentage prepared in		
Element	10 min	20 min	30 min
O	58.00	77.01	50.79
S	12.05	1.26	16.73
In	6.90	0.39	9.88
Sn	23.06	21.34	22.61

From table 1 can be seen that there are 4 elements from the EDX observation, namely oxygen, sulfur, indium and tin. The dominant elements are tin and oxygen. Tin and oxygen come from the constituents of FTO, namely fluorine doped-tin oxide. These observations show that the composition is close to In_2S_3 for the sample prepared at 10 min. This is also supported by the existence of a large photocurrent of the sample (in the next discussion). Although the 10 min sample does not exactly match the percentage of In_2S_3.

3.2. Characterization of the In_2S_3 thin film surface by SEM

Figure 3 shows SEM images of In_2S_3 on FTO with varied times of deposition, it can be seen that the crystals of In_2S_3 have diameters less than 500 nm. Nevertheless. the FTO substrate is not well covered with In_2S_3 thin film precipitate. Even, the SEM is depicted that the film formed does not grow well for

20 min deposition. After 30 min deposition, In_2S_3 crystals form agglomeration not thin film anymore such as in 10 min deposition.

Figure 3. SEM images of In_2S_3 on FTO for deposition times of: (a) 10, (b) 20, and (c) 30 min.

3.3. Analysis In_2S_3 thin film using XRD

Figure 4 shows the X-ray diffraction of In_2S_3 deposited on FTO with a glass support. Because In_2S_3 is amorphous the diffractograms of In_2S_3 samples deposition at varied times do not show any differences. The XRD peaks are dominated by the FTO peaks. The peaks at $2\theta = 34.29^o$, 38.32^o and 62.15^o are the peaks from SnO_2 of the substrate used (FTO).

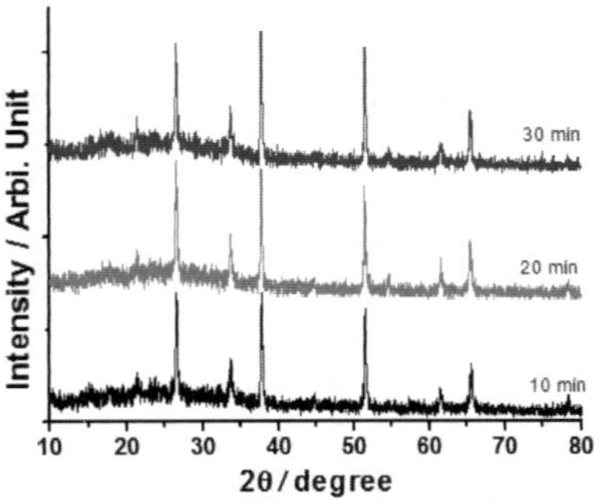

Figure 4. XRD of In_2S_3 thin film on FTO deposited with time variations of 10, 20, 30 min.

3.4. Analysis In_2S_3 with a Raman Spectrophotometer

Raman analysis shows the presence of spectra in the 200 and 300 cm^{-1} shoulder areas which are the typical wave number regions for In_2S_3 as shown in figure 5, namely spectra of In_2S_3 powder and In_2S_3 thin film on glass deposited for 10 min.

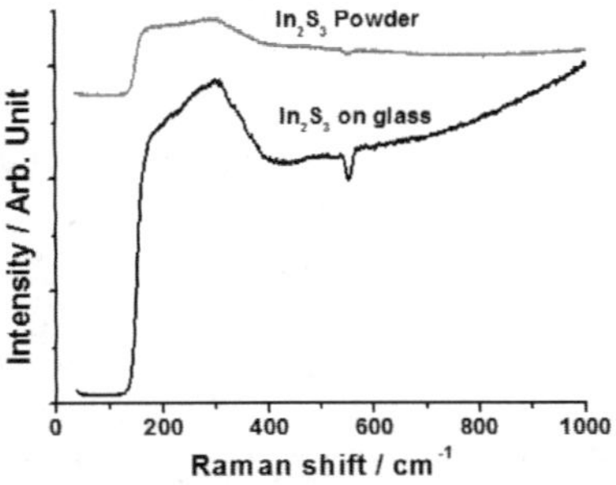

Figure 5. Raman spectra of In_2S_3 powder and In_2S_3 on glass deposited for 10 min.

3.5. Measurement of the absorbance of the In_2S_3 thin film

In_2S_3 thin film deposited for 10 min was then passed through the UV-Vis spectrophotometer pathlight to see the absorbance of the In_2S_3 thin film at various wavelengths. Figure 6 shows that the resulting thin film shows transparency of more than 60% in the visible and near infrared regions at wavelengths greater than 500 nm. At shorter wavelengths (below 500 nm) there is a decrease in transparency due to

IOP Conf. Series: Materials Science and Engineering **959** (2020) 012021 doi:10.1088/1757-899X/959/1/012021

the photoexitation of the band gap (E_g). Furthermore, by calculation band gap (hυ) using the relationship hυ = 1239/λ and insertion the wavelength value from figure 6 (496.7 nm) the In_2S_3 band gap can be obtained equals 2.49 eV.

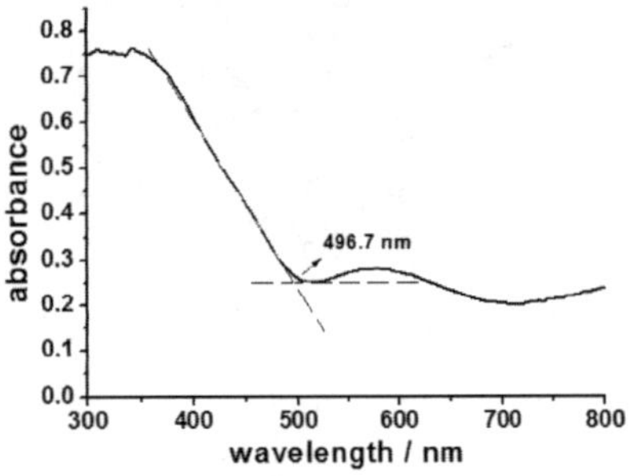

Figure 6. Absorbance *vs* wavelength curve of In_2S_3 thin film at 10 min deposition time.

3.6. Measurement of photocurrent of In_2S_3 thin film

Photocurrent test was done by positioning In_2S_3 photoanode as the working electrode, Ag/ AgCl as the reference electrode and Pt as the counter electrode. Furthermore, to see the presence of photocurrent, the three electrodes were inserted into the 1M Na_2SO_3 solution as a hole scavenger and during the measurement process with the Potentiostat and Galvanostat CS-150 the light was turned off and on to see the difference in photocurrent that occured. Measurements were made from potentials of 0.8 to -0.8 volts. Figures 7a, b, c show the resulting In_2S_3 thin film has a *n*-type semiconductor property. For 10 min deposition, In_2S_3 gives a better photocurrent profile compared to the 20 and 30 min depositions which is indicated by a higher and flatter dark pattern and light current. At a potential of 0.5 V gives a photocurrent ca. 0.01 mA/cm^2 for In_2S_3 deposited for 10 min. Further, application of In_2S_3 thin film deposited for 10 min as a photoanode for dye destruction was used, since it has a better photocurrent.

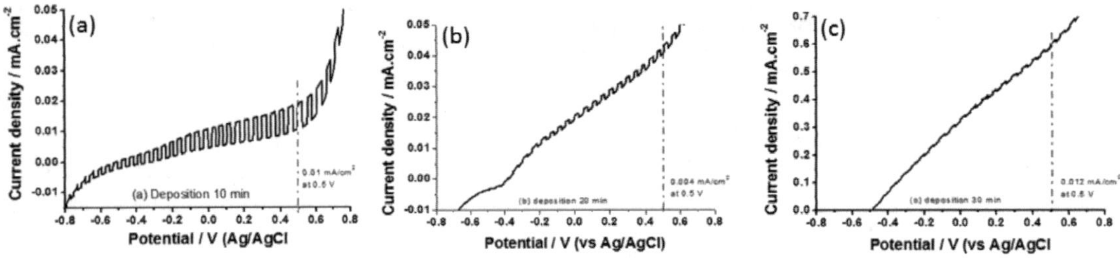

Figure 7. J-V curves for In_2S_3 thin films deposited for (a) 10, (b) 20, (c) 30 min.

3.7. Application of the In_2S_3 thin film as a photoanode for the destruction of MB and MO

The application of the In_2S_3 thin film photoanode deposited for 10 min in a photoelectrocatalytic cell to degrade MB and MO was carried out by irradiation with a 20 watt UV lamp (366 nm) at a distance of 20 cm and a bias voltage of 0.5 V. Figure 8 shows degradation of MB without and with the addition of 0.1M electrolytes of sodium sulfate, NaOH and HCl, respectively with In_2S_3 photoanode for 2.5 h with a bias voltage of 0.5 V. From this figure can be seen MB without using electrolytes provides the smallest decrease in treatment compared to using electrolytes. Whereas the use of HCl provides the smallest reduction compared to sodium sulfate and NaOH electrolytes. When NaOH electrolyte used gave the greatest treatment close to 71% for 150 min. The presence of hydroxide ions in solution increases a further possible methylene blue destruction to occur.

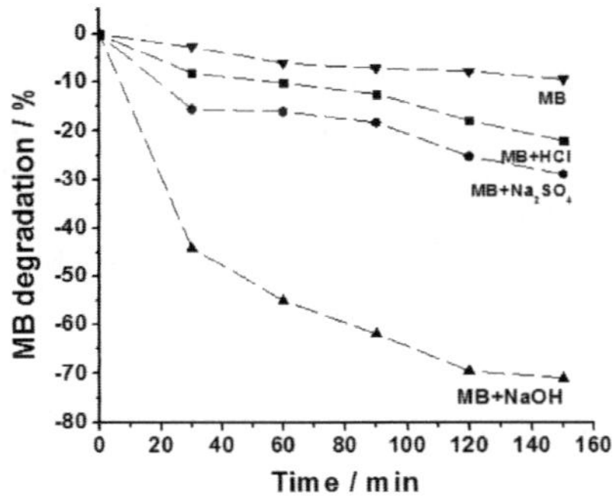

Figure 8. Degradations of MB without and with electrolytes (Na_2SO_4, NaOH and HCl) treated using In_2S_3 photoanode with a bias voltage of 0.5 V.

Methyl orange (MO) was also treated in the similar ways as with MB and the results with varied electrolytes as shown in figure 9.

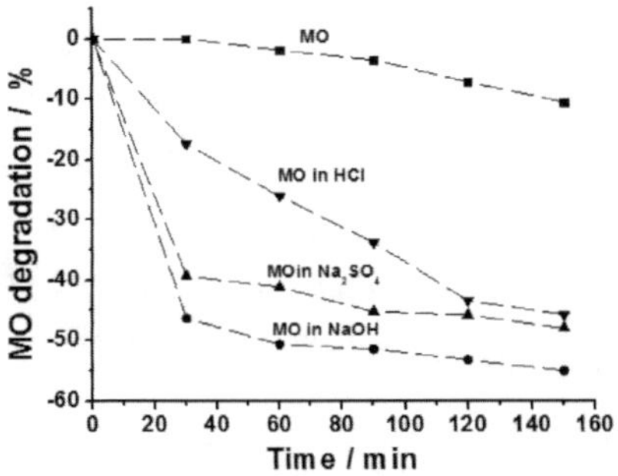

Figure 9. Degradations of MO without and with electrolytes (Na_2SO_4, NaOH and HCl) treated using In_2S_3 photoanode with a bias voltage of 0.5 V.

Figure 9 shows that the treatment patterns are similar when using MO compared to MB although the greatest destruction occurred only up to 55%. This can be the structure of MO is more complicated than that of MB, so it takes a longer process to break down into a smaller molecules.

3.8. UV-Vis measurements of MB and MO at varied electrolytes

UV-Vis measurements of MB and MO were adjusted according to the electrolyte used, because the maximum wavelength is greatly influenced by the type of electrolyte used, as shown in figure 10. Therefore, the reduced MB and MO absorbances after photoelectrocatalytic processes were measured based on the related wavelength of the sample's electrolyes applied. Figure 10a shows that MB gives the same maximum peaks for MB, MB with HCl and Na_2SO_4 and gives a different maximum peak for MB with NaOH. However, for MO with HCl (figure 10b) gives a different maximum peak compared to other treatments that give similar maximum peaks namely for MO with NaOH, Na_2SO_4 electrolytes and without electrolyte.

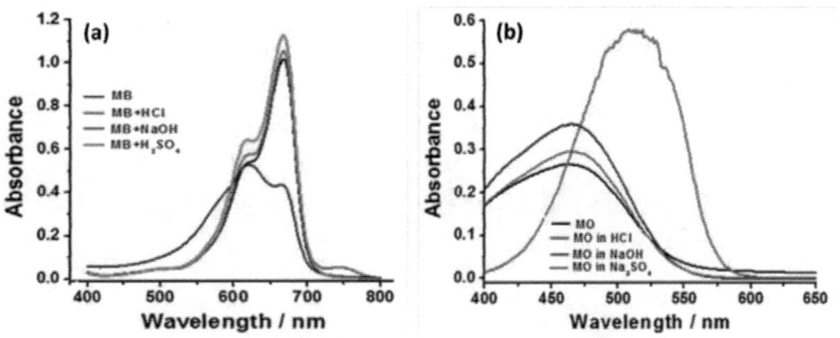

Figure 10. Curve profiles for MB (a) and MO (b) measured without and by using electrolytes (HCl, Na_2SO_4 and NaOH).

4. Conclusion

The synthesis of In_2S_3 thin film by chemical bath deposition (CBD) had been successfully carried out. Characterization with EDX obtained the In_2S_3 composition close to 10 min deposition. SEM image revealed that it did not provide much information regarding In_2S_3 deposition, but for 30 min deposition gave agglomeration deposition. XRD analysis showed the presence of amorphous In_2S_3 which had sharp peaks. From absorption measurement of In_2S_3 thin film deposited for 10 min could be derive a band gap of 2.49 eV. A *n*-type semiconductor property was showed by photocurrent test of In_2S_3 thin film. Application of In_2S_3 deposited 10 min as a photoanode in a photoelectrocatalyst cell for methylene blue and methyl orange treatments resulted degradations up to 71 and 55 %, respectively, for 25 mg /L samples for 2.5 h.

Aknowledgement

The authors were thankful to the Faculty of Science and Mathematics at Diponegoro University for supporting this work (Non APBN FSM UNDIP 2019).

References

[1] Antoniadou M, Kondarides DI, Labou D, Neophytides S and Lianos P 2010 *Sol. Energy. Mater. Sol. Cells* **94** 592
[2] Konstantinou IK and Albanis TA 2004 *Appl. Catal. B Environ.* **49** 1
[3] Liu Y, Zhou B, Li J, Gan X, Bai J and Cai W 2009 *Appl. Catal. B Environ.* **92** 326
[4] Liu Y, Li J, Zhou B, Chen H, Wang Z and Cai W 2011 *Chem. Commun. (Camb)* **47** 10314
[5] Liu Y, Li J, Zhou B, Li X, Chen H, Chen Q, Wang Z, Li L, Wang J and Cai W 2011 *Water Res.* **45** 3991
[6] Li J, Chen Q, Bai J and Zhou B 2013 *J. Hazard. Mater.* **262** 304
[7] Li K, Xu Y, He Y, Yang C, Wang Y and Jia J 2013 *Environ. Sci. Technol.* **47** 3490
[8] Liu Y, Zhou B, Bai J, Li J, Zhang J, Zheng Q, Zhu X and Cai W 2009 *Appl. Catal. B Environ.* **89** 142
[9] Lee SL, Ho IN, Ong SA, Wong YS, Voon CH, Khalik WF, Yusoff NA and Nordin N 2017 *Chemosphere* **166** 118
[10] Wang B, Zhang H, Hu XY, Xuan J and Leung MKH 2014 *Chem. Eng. J* **253** 174
[11] Yang J, Liao W, Liu Y, Muruganathan M and Zhang Y 2014 *Electrochim. Acta* **144** 7
[12] Li L, Xue S, Chen R, Liao Q, Zhu X, Wang Z, He X, Feng H and Cheng X 2015 *Electrochim. Acta* **182** 280
[13] Jiang F, Ozaki C, Gunawan, Harada T, Tang Z, Minemoto T, Nose Y and Ikeda S *Chem. Matter.* **28** 3283
[14] Gunawan G, Septina W, Ikeda S, Harada T, Minegishi T, Domen K and Matsumura M 2014 *Chem. Commun.* **50** 8941
[15] Gunawan, Septina W, Harada T, Nose Y and Ikeda S 2015 *ACS App. Mater. Interfaces* **7** 16086
[16] Jiang F, Gunawan, Harada T, Kuang Y, Minegishi T Domen K and Ikeda S 2015 *J. Am. Chem. Soc.* **137** 13691

IOP Conf. Series: Materials Science and Engineering **959** (2020) 012022 doi:10.1088/1757-899X/959/1/012022

The potency of *Hibiscus tiliaceus* leaves as antioxidant and anticancer agents via induction of apoptosis against MCF-7 cells

Y Andriani[1,*], M Sababathy[1], H Amir[2], PR Sarjono[3], DF Syamsumir[1], S Sugiwati[4] and MNI Kassim[1]

[1] Institute of Marine Biotechnology, Universiti Malaysia Terengganu, Mengabang Telipot 21030, Kuala Nerus, Terengganu, Malaysia

[2] Educational Chemistry Program, Faculty of Teacher Training and Education, Bengkulu University, Bengkulu, 38371, Indonesia.

[3] Department of Chemistry, University of Diponegoro, Semarang, Indonesia

[4] Research Centre for Chemistry, Indonesian Institute of Sciences, Kawasan Puspitek Serpong, Tangerang, 15314, Indonesia

[*] E-mail: yosie.hs@umt.edu.my

Abstract. *Hibiscus tiliaceus* is one of the herbal medicines that have been used as traditional medicine for a long time. Antioxidant and anticancer potencies of this plant have reported by some researchers. However, there are no studies reported on antioxidant and anticancer potencies of *H. tiliaceus* leaves collected from Terengganu-Malaysia, especially against breast cancer cells (MCF-7). Different solvents in the extraction process and different sampling areas were chosen in this study compared to other studies. These could affect the chemicals content of *H. tiliaeus* leaves as well as on their bioactivities. Hence, the objectives of this study were to investigate the antioxidant and anticancer potencies of *H. tiliaceus* leaves from Terengganu, Malaysia, against MCF-7. The sample was extracted by solvent-solvent partitioning using hexane and ethyl acetate. Antioxidant and cytotoxicity properties were carried out by DPPH free radical scavenging activity and MTT, respectively. The morphological features were stained by Annexin- V/PI and DAPI. Results revealed that *H. tiliaceus* leaves exhibited strong DPPH free radical scavenging and cytotoxic activities against MCF-7. Morphological features showed the cells were put to death by both early and late apoptosis. Our results found that *H. tiliaceus* leaves have potency as antioxidant and anticancer agents against MCF-7 cells.

1. Introduction

Breast cancer is one kind of cancer diseases that becomes a leading cause of death among women worldwide. An estimated of 2.09 million new cases of breast cancer and deaths were estimated to occur among women in 2018. Nearly 70% of deaths from cancer occur in low and middle-income countries [1]. Cancer is one of the situations where too little apoptosis occurs, subsequent in malignant cells that will not die [2]. The mechanism of apoptosis is complex and involves many pathways. Regardless of being the cause of problem, apoptosis shows a significant role in the treatment of cancer as it is a common goal of many treatment strategies [2]. Current cancer treatments, chemotherapy and chemo therapeutic drugs have adverse side effects to the patients, which may be divided into temporary and permanent effects. According to Komen, temporary effects include languishment, soreness in mouth, throat and skin, hair loss, frequent urination and signs of anorexia On the other hand, poor salivary function, menopause, sterility and nerve damage have been recognized as permanent effects which are also equally rarely [3].

Some herbs are known to have anticancer properties [4–5] that can be easily obtained or grown at one's own localities, inexpensive and gives minimal side effects [6] to patients treated with the herbs as therapeutic agents. Besides, herbs are also recommended to be given to cancer patients who develop side effects during or after mainstream therapies like chemotherapy and immunotherapy. According to Lobo *et al.* [7], a high amount of antioxidant found in vegetables and fruits have the capability to reduce oxidative damage caused by free radicals and reactive oxygen molecules. Antioxidant inhibits oxidation process by preventing or slowing down damage to cells due to production of free radicals that exist as unstable molecules with unpaired electron and are highly reactive. Free radicals have been reported to naturally produce and build up in the body but hasten with exposure to ultraviolet light, tobacco smoke, alcohol, pesticides and pollutions that leads to oxidative stress [8]. Oxidative stress happens when free radicals are produced in excess and may establish a deleterious process by which it may alter or damage cellular structures and functions [9–11]. Antioxidants also naturally exist as vitamin E and phenolic compounds such as flavonoids and alkaloids that contributes to its renowned ability to fight cancer. Antioxidant prevents carcinogenesis by inhibiting the initiation of the cancer cell formation by reactive radicals [12].

One of potential plants is Waru Laut (*H. tiliaceus*). It has long been used as a traditional medicine for treatment of coughs and fevers, bronchitis and skin diseases (figure 1) [13]. Its methanol leaves extract is reported to have good antioxidant potential, and some chemical constituents such as phenolics, flavonoids, tannins, steroids, and terpenoids [14–15]. Those chemical constituents have been correlated to the antibacterial property of the *H. tiliaceus* [14,16–17], and they could be responsible to its anti-biofilm potency [14].

Figure 1. *Hibiscus tiliaceus* leave and flower.

Hibiscus can be commonly found in Asia and they are extensively diversified in species, for instance *H. rosa-sinesis* and *H. tiliaceus* are the two of the many favored species among botanists and flower lovers. In the current study *H. tiliaceus* is selected to investigate antioxidant and anticancer potential due to its medicinal properties that mostly the bark and leaves of the plant are traditionally used to treat many diseases for years long [18]. It has been evidenced that the aqueous extract from the wood and flowers of *H. tiliaceus* has also been used to treat skin diseases [19]. Previous studies have also reported that this species has high antioxidant properties determined in AEAC (ascorbic acid equivalent antioxidant capacity) [20].

Furthermore, other researchers used ethanol extract and proved that it has antioxidant property [21]. Samsudin et al. [12] using 3 different solvents in extraction process also reported that *H. tiliaceus* exhibited very good antioxidant and antibacterial activities. From the outcome of these two researchers, it is proven that the *H. tiliaceus* is a good candidate for the antioxidant studies. Due to the fact that there is a high radical scavenging activity presence in this plant [20], *H. tiliaceus* was also used for cytotoxicity study, such as colon cancer and gastric cancer. The chemical constituents that present in these plants contributed to the various activities of the samples. Rahman and Khan [22] had carried out a research to identify the anti-cancer potential of South Asian plants. One of the species that was listed is *H. tiliaceus*. Phytochemicals presented in this plant are quarcetin, kaempferol and B-sitosterol. Even some studies have reported on cytotoxicity activity of *Hibiscus* species against some cancer cells [18, 23], however, there is no study reported on the cytotoxicity of *H. tiliaceus* leaves (collected from Terengganu, Malaysia) and the morphological features of MCF-7 cells death after treated by H. *tiliaceus* leaves. Thus, it is eagerly needed to discover the potency of *H. tiliaceus* as anticancer and antioxidant potential agents.

2. Methodology

2.1 Sample collection and extraction process
Sample of *H. tiliaceus* leaves was collected around coastal area of UMT, Terengganu, Malaysia in April, 2019. The voucher specimen of sample has been deposited at Institute of Marine Biotechnology, Universiti Malaysia Terengganu with the voucher number of TER0619001. The part of leaves was cut into a small pieces, and then put into the -80 ºC for further drying using freeze dryer for 3 days. Dried sample was then ground to be fine powder and used for extraction process. Extraction by cold maceration using methanol solvent was done first to yield methanol extract. Solvent-solvent partitioning was continued on methanol extract using hexane and ethyl acetate, subsequently. Total samples obtained from the extraction process were two, namely hexane (HF) and ethyl acetate (EF) fractions.

2.2 Antioxidant activity: DPPH free radical scavenging assay
Antioxidant capability was achieved by DPPH free radical scavenging assay based on Kumaran and Karunakaran [24]. Quercetin and DMSO were used as a positive control and negative control, respectively. Two-fold serial dilution of EF fraction in DMSO (10, 5, 2.5, 1.25, 0.625, 0.313, 0.156 mg/ml) was prepared in 96 well plates. Two hundred microliter of methanolic DPPH solution (6 x 10^5 M) was loaded into all wells and left in a dark room under room temperature for 30 minutes. The absorbance of the sample was measured with ELISA reader (Multiskan ascent, Thermo Electron Corporation) at 517 nm. Free radical scavenging activity in the sample was then calculated.

2.3 Cytotoxicity activity: MTT assay
MCF-7 cells were obtained from the American Type Culture Collection (Rockville, MD, USA), All organic solvents, media and other reagents were of analytical grade and purchased from Sigma or Merck. Cytotoxicity properties of *H. tiliaceus* leaves samples against MCF-7 cells were determined by MTT assay (Adapted from Andriani *et al.*) [15]. The cells were incubated at 37 °C in an atmosphere of 5% CO_2 until cells confluency reached at least 80% to proceed with MTT assay. The cells were trypsinised and counted using hemocytometer. 100 µl of $1\text{x}10^5$ cells were seeded into each wells of 96-well plate and the plate was placed in the incubator for 24 hours. After the incubation period, culture medium in each wells are discarded completely and added a 100 µl mixture of EF diluted with culture medium by two-fold serial dilution of six different concentrations (60; 30; 15; 7.5; 3.75; 1.875 µg/ml). The concentration of positive control was prepared same as sample. Furthermore, 96- well plate was incubated for 72 h then 20 µl of MTT (5 mg/ml dissolved in PBS) and was incubated for 72 h followed by addition of 20 µl of MTT (5 mg/ml dissolved in PBS) into each well and incubated for another 4 hours. After the incubation, all wells are completely emptied and replaced with 100 µl of DMSO and incubated for 10 more minutes. Enzyme linked immunosorbent assay (ELISA) reader is used to measure the absorbance at 570 nm. IC_{50} value is determined from the calculation of percentage of cell viability using the following formula:

$$(\text{Absorbance sample}/\text{Absorbance blank}) \text{ X } 100 \quad (1)$$

2.3.1 Morphological features of cells death (Adopted from Andriani et al.) [25]
The concentration of samples used in this part was determined based on the IC_{50} value obtained in

cytotoxicity analysis. Cells were seeded at 5 × 10^3 cells/well in 96-well plate and incubated for 24 h. Subsequently, medium was discarded and replaced with new medium containing EF fraction of *H. tiliaceus* leaves at the concentration of 10 µg/ml and incubated for 24 h. Vincristine sulfate was used as positive control and vehicle as negative control. After 24 h incubation, medium was discarded and Annexin-V/PI/DAPI reagent was added to each well. Cells were then incubated for 10 - 15 min at 15-25 °C. The morphological feature of cells death was analyzed by ImageXpress Micro XLS Widefield High-Content Analysis System (HCS) (Sunnyvale, USA).

3. Results and discussion

3.1. Antioxidant property
Based on figure 2, EF of *H. tiliaceus* leaves exhibited strong DPPH free radical scavenging activity (IC_{50} < 2 µg/ml) compared to the hexane fraction (HF) with low DPPH free radical scavenging activity (IC_{50}=10 µg/ml). The DPPH radical scavenging assay was used in this study since it has been long used to determine and measure compounds of free radical scavenging activities to evaluate antioxidant properties of natural products mostly plant samples [26]. The DPPH is a stable free radical that is able to react with compounds to donate a hydrogen atom [27].

A high antioxidant property of EF could be correlated the chemicals constituent content in this fraction. The EF fraction was obtained from extraction process using ethyl acetate solvent. Bata *et al.* [28] suggested that ethyl acetate is an effective solvent for the extraction of phenols and flavonoids compounds group. Even different solvent has used in the extraction process (hexane, dichloromethane, ethyl acetate, and methanol), ethyl acetate fraction of *H. tiliaceus* leaves still showed strong antioxidant property [15]. They also reported that this fraction was rich by phenolic compounds. Phenolic compounds like flavonoids and phenols have been reported on their effective ability to exhibit potent anti-cancer activities as well as combat various diseases associated with oxidative stress [29]. Since EF has strong antioxidant activity (figure 2), it was selected for cytotoxicity test against MCF-7 cells (figure 3).

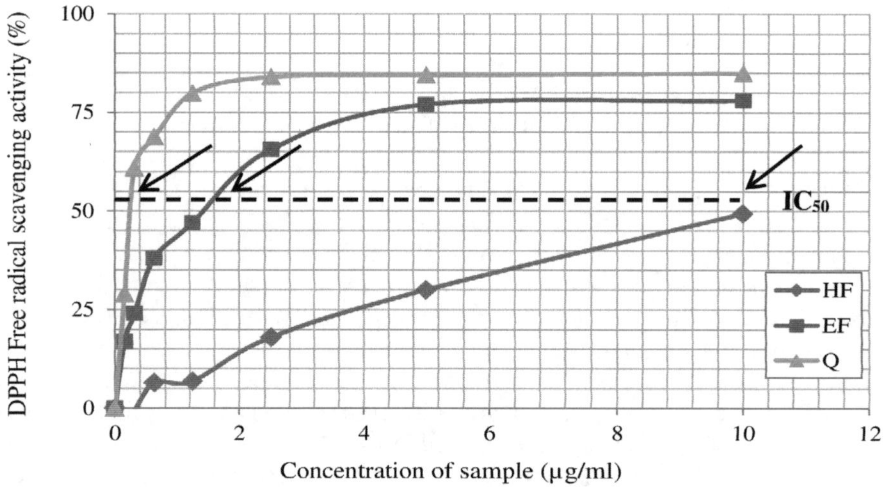

Figure 2. DPPH free radical scavenging activity of HF and EF of *H. tiliaceus* leaves as compared to the Quercetin (Q), respectively. All data were calculated in 3 replicates and black arrow indicates the IC_{50} value.

3.2. Cytotoxicity
Cytotoxic activity of the *H. tiliaceus* was tested using MTT assay that is interpreted by the expression

of IC_{50} value measuring the inhibition of cell proliferation by 50% obtained by linear regression of concentration against percentage of cytotoxicity activity [30]. Cytotoxicity effect was exhibited by the decrease in the number of viable cells and their morphology changes after being treated with the sample. MTT assay is colorimetric test that measures the intensity of the colour produced due to the metabolic activity of living cell that appears as coloured product. The MTT is reduced to formazan catalysed by mitochondrial succinate dehydrogenase of living cells. The colour intensity of the formazan that is measured using the ELISA reader gives a proportional value of the number of viable cells after treated with the sample of different concentrations by which in this experiment was 60, 30, 15, 7.5, 3.75 , 1.875 µg/ml. Cytotoxicity study was carried out using ethyl acetate extract of *H. tiliaceus* on human breast cancer cell line, MCF-7. Our result showed that there was a dose-dependent inhibition as the cells were treated with higher concentrations (figure 3).

According to Andriani *et al.* [31], sample extracts that possess less than 30 µg/ml of IC_{50} is considered to be a potential anticancer agent. Our result also showed that the sample has moderate cytotoxic property against MCF-7 cells with the IC_{50} value=10 µg/ml (figure 3). Cytotoxic property of EF in this study (figure 3) could be correlated to its antioxidant activity (figure 2). Sample with antioxidant activity can inhibit the propagation of the oxidative chain reaction. Propagation stage is mainly caused by the presence of free radicals [32], a component that causing cancer [12]. Thus, samples with antioxidant properties can be used to prevent the form of cancer cells by free radicals.

Several studies on the cytotoxic effect of *H. tiliaceus* against cancer cell line have been conducted in the past years. A study was carried out by Artanti *et al.* [30] indicated that there is a correlation between antioxidant and cytotoxicity activity based on the IC_{50} value obtained from the study when the ethanolic extract of *Hibiscus rosa-sinensis L.*, *Hibiscus tiliaceus L.*, and *Malvaviscus arboreus Cav.* was introduced to T47d cell line, which is another type of breast cancer cell line. Other study that demonstrates the cytotoxicity effect of *H. tiliaceus* includes the crude extract of leaves and bark that showed the presence of cytotoxic compounds and suggested that both the extracts of different parts of *H. tiliaceus* (leaves and bark) have the ability to kill cancer cells and potential to be tested for anti-tumour and anti-cancer properties. Uddin *et al.* [33] demonstrated cytotoxicity properties of 32 Bangladeshi plant extracts including methanolic extract of *H. tiliaceus* that was tested against healthy mouse fibroblast and three human cancer cell lines (gastric, colon and breast cancer cells), however the sample only showed selective cytotoxicity effect on breast cancer cells (IC_{50}=1.1– 1.6 mg ml^{-1}) and proposed that it could be used as a source or lead to designing anti-cancer drug in the future.

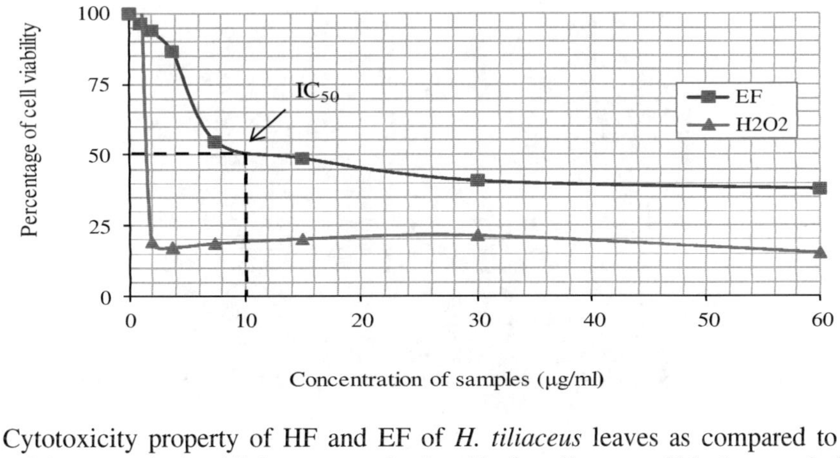

Figure 3. Cytotoxicity property of HF and EF of *H. tiliaceus* leaves as compared to the hydrogen peroxide (H_2O_2), respectively. All data were calculated in 3 replicates and black arrow indicates the IC_{50} value.

3.2.1. Morphological features of cells' death.

IOP Conf. Series: Materials Science and Engineering **959** (2020) 012022 doi:10.1088/1757-899X/959/1/012022

Our findings showed that the cells died by both early and late apoptosis after the cells treated by EF (figure 4c) compared to the positive control (figure 4b). Furthermore, DAPI stain positive (blue colour) was presented for both life and dead cells. The result with Annexin V/PI negative and DAPI positive specify the viable cells (Figure 4a). In general, Annexin V/PI method was successfully applied in the current study to exhibit induction of apoptosis in MCF-7 cells after treated with EF. MCF-7 cells treated with EF showed clear apoptotic features after treatment period (24 hours incubation). Most cells remained intact but they still didn't lose their shape. Similar to the Anexin/PI staining, DAPI also indicates cells that experience apoptotic cell death from colour changes, which is in the current study upon treated with EF. Bright blue colouration with intact nuclear structure showed viable cells for control which is in contrast to the cells that were stained bright red or pink in colour in the nucleus that indicates the presence of dead cells or those of that has undergone apoptosis. Interestingly, a significant phenomenon was observed after 24 hours of incubation with EF where chromatins appeared shorter or has undergone chromatin condensation (nuclear pyknosis) which is one of the known hallmarks of apoptosis and blabbing of the MCF- 7 cell membrane. It showed similar morphology features compared to the control drug, vincristine sulfate. Hence, our results demonstrate that EF is proven to have potencies in inducing apoptosis against MCF-7 cells (figure 4c).

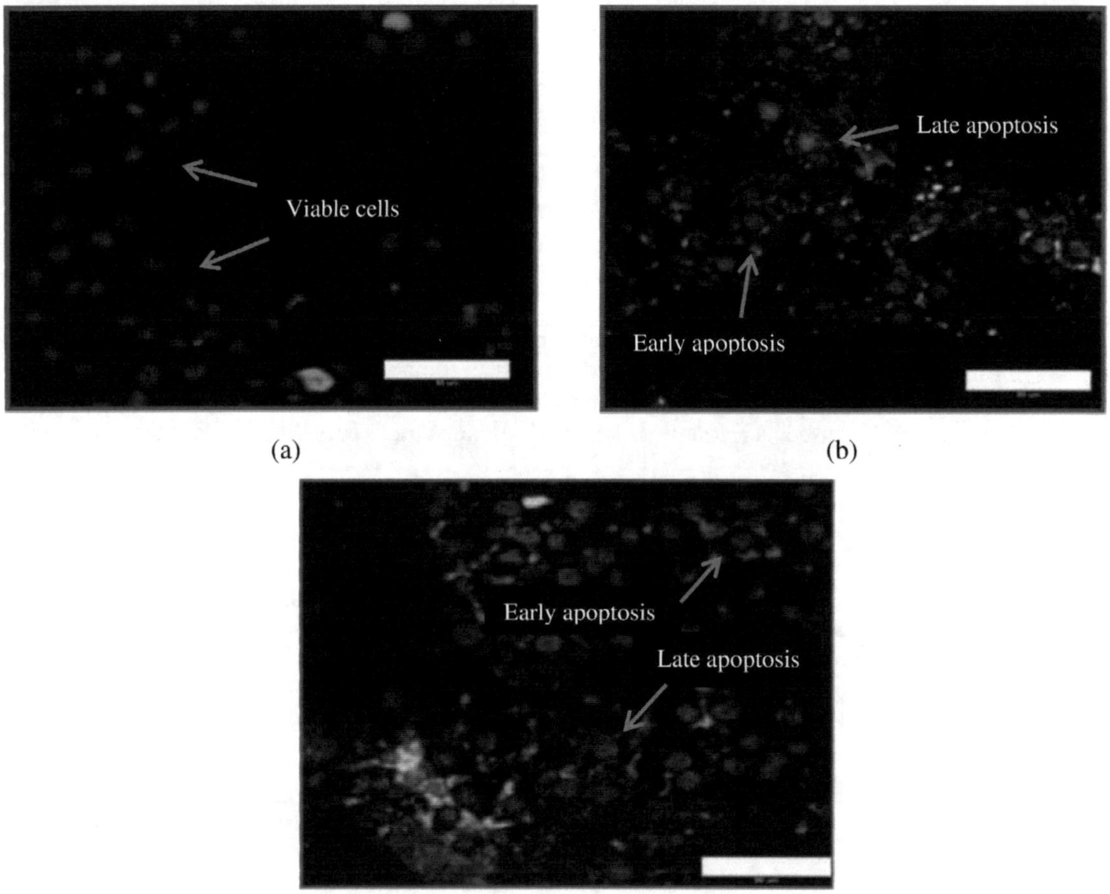

(a) (b)

(c)

Figure 4. Apoptotic cell death of MCF-7 cells among group of negative control (a), vincristine sulfate (b), and upon treatment with EF of *H. tiliaceus* leaves (c) (Magnification = 10x).

According to Hingorani *et al.* [34], reported that the determination of apoptotic cell death with

Annexin V method is indicated from the staining of the cells whereby cells that undergo early apoptosis upon treatment appears green (PI negative) meanwhile red (PI positive) indicates that the cells undergo late apoptosis. In addition, Vermes *et al.* [35] stated that PI is widely used to determine if cells are viable, apoptotic, or necrotic in conjunction with Annexin V through differences in plasma membrane integrity and permeability. Usually, combination of the Annexin V/ PI is used for studying apoptotic cells as described by Cornelissen *et al* [36]. The integrity of the nuclear membranes and plasma will be decreased in late apoptotic and necrotic cells [37–38]. Degreasing of membrane integrity allowing PI to pass through the membranes, into nucleic acids, and shows as red fluorescence [35,39].

4. Conclusion

Our results provide evidence that *H. tiliaceus* leaves have potency as antioxidant and anticancer potential agents against MCF-7 cells. However, further study at molecular levels using various incubation time, and using different staining such as tunnel, annexin V-FITC, caspase, and could be explored deeply.

Acknowledgement

The authors wish to thank the Institute of Marine Biotechnology for its good environment and research facilities during experiment.

References

[1] WHO 2019 Cancer- early diagnosis and screening. https://www.who.int/activities/ promoting-cancer-early-diagnosis [Accessed on 4 June 2020]
[2] Wong R S Y 2011 *J. Ex.Clin. Canc. Res.* **1** 87
[3] Komen and Susan G 2019 Chemo Side-Effects Info at Susan G Komen ww5.komen.org/BreastCancer/LongTermSideEffectsofChemotherapy.html [Accessed on 2 July 2020]
[4] Kaefer C M and Milner J A 2011 Herbs and spices in cancer prevention and treatment *Herbal Medicine: Biomolecular and Clinical Aspects* (2nd Ed) eds I F F Benzie and S Wachtel-Galor (Boca Raton: CRC Press) chapter 17
[5] Luo H, Vong C T, Chen H, Gao Y, Lyu P, Qiu L, Zhao M, Liu Q, Cheng Z, Zou J, Yao P, Gao C, Wei J, Ung C O L, Wang S, Zhong Z and Wang Y 2019 *Chinese Med.* **48** 1
[6] Aziz S M A, Aeron A and Kahil T 2016 *Microbes in food and health: Health benefits and possible risks of herbal medicine* (Berlin: Springer)
[7] Lobo V, Patil A, Phatak A and Chandra, N 2010 *Pharmacogn Rev.* **8** 118
[8] Phaniendra A, Jestadi D B and Periyasamy L 2015 *Indian J. Clin. Biochem.* **1** 11
[9] Halliwell B and Gutteridge J M C 2007 *Free radicals in biology and medicine* (Oxford: Clarendon Press)
[10] Droge W 2002 *Physiol. Rev.* **82** 47
[11] Willcox J K, Ash S L and Catignani G L 2004 *Crit. Rev. Food Sci. Nutr.* **44** 275
[12] Saladin K S 2016 *Anatomy and Physiology: The Unity of Form and Function* (New York: McGraw-Hill Education)
[13] Vanzella C, Bianchetti P, baraini S, Vanzin S I, Melecchi M I, Caramão E B and Siqueira R. 2012 *BMC Complem. Altern. M.* **12** 41
[14] Andriani Y, Mohamad H, Kesaven B, Abdullah M I and Amir H 2017 *J. Sustain. Sci. Manag.* **12** 45
[15] Andriani Y, Ramli N M, Syamsumir D F, Kassim M N I, Jaafar J, Azis N A, Marlina L, Musa N S and Mohamad H 2019 *Arab. J. Chem.* **12** 3555
[16] Samsudin M S, Andriani Y, Sarjono P R and Syamsumir D F 2019 *Alotrop* **3** 175
[17] Riihinen K R, Ou Z M, Godecke T, Lankin D C, Pauli G F and Wu C D 2014 *Fitoterapia* **97** 78
[18] Chen J J, Huang S Y, Duh C Y, Chen I S, Wang T C and Fang H Y 2006 *Planta Med.* **72** 935
[19] Hemaiswarya S, Poonkothai M, Raja R and Anbazhagan C 2009 *Egyp. J. Biol.* **11** 52
[20] Wong S K, Lim Y Y and Chan E W C 2010 *Ethnob. Leaflets* **14** 781

[21] Salem M Z M, Oliver-Perez and Salem A Z M 2014 *Life Sci.* **2014** 1
[22] Rahman M M and Khan M A 2013 *Nat. Prod. Bioprospec.* **3** 74.
[23] Abdul-Awal S, Nazmir S, Nasrin S, Nurunnabi T and Uddin S 2016 *SpringerPlus* **5** 10.1186/s40064-016-2891-0
[24] Kumaran A and Karunakaran R J 2006 *Food Chem.* **97** 109
[25] Andriani Y, Mohamad H, Kassim M N I, Rosnan N D, Syamsumir D S, Saidin J, Sifzizul T M T and Amir H 2017b *J. Appl. Phram. Sci.* **09** 030
[26] Hu F L, Lu R L, Huang B, and Ming L 2004 *Fitoterapia.* **1** 14
[27] Aqil F, Munagala R, Jeyabalan J, Joshi T, Gupta R C and Singh I P 2014 The Indian blackberry (jamun), antioxidant capacity, and cancer protection *Cancer: Oxidative Stress and Dietary Antioxidants* ed V R Preedy (London: Academic Press) chapter 10 pp 101–113
[28] Bata M and Rahayu S 2017 *Curr. Bioac. Compd.* **13** 157
[29] Arts I C and Hollman P C 2005 *Am. J. Clin. Nutr.* **81** 317S
[30] Artanti A N, Rahmadanny N and Prihapsara F 2018 *Mater. Sci. Eng.* **349** 012006
[31] Andriani Y, Effendy E W M, Sifzizul T M T and Mohamad H. 2011 *Int. J. Pharm. Scie Res.* **2** 1700
[32] Tachakittirungrod S, Okonogi S and Chowwanapoonpohn S 2007*Food Chem.* **2** 381
[33] Uddin S J, Grice I D and Tiralongo E 2011 *Evid-based compl. Alter. Med.* **2011** 1
[34] Hingorani R, Deng J, Elia J, McIntyre C and Mittar D 2011 *BD Biosci.* **2011** 1
[35] Vermes I, Haanen C and Reutelingsperger C 2000 *J. Immunol. Methods* **243** 167
[36] Cornelissen M, Philippe J, De Sitter S and De Ridder L 2002 *Apoptosis* **7** 41
[37] Kroemer G, Dallaporta B and Resche-Rigon M 1998 *Annu. Rev. Physiol.* **60** 619
[38] Denecker G, Vercammen D, Declercq W and Vandenabeele P 2001 *Cell. Mol. Life Sci.* **58** 356
[39] Boutonnat J, Barbier M, Ronot X and Seigneurin D 2000 *Morphologie* **84** 11

IOP Conf. Series: Materials Science and Engineering **959** (2020) 012023 doi:10.1088/1757-899X/959/1/012023

The effect of leaching agent on molecularly imprinted membrane urea transport process based on polyeugenoxy acetic acid

M C Djunaidi[1,*], T Wahyuni[1], R A Lusiana[1], D S Widodo[1] and Pardoyo[1]
[1] Chemistry Department, Faculty of Mathematics & Natural Sciences, University of Diponegoro

[*] E-mail: choliddjunaidi@live.undip.ac.id

Abstract. Research on membrane synthesis as a hemodialysis membrane candidate has been carried out with molecular imprinting technique using MIM (Molecularly Imprinted Membrane) based on polyeugenoxy acetic acid. Polieugenoxy acetic acid was obtained from the synthesis of eugenol derivatives. This research aims to produce a selective urea membrane based on polyeugenoxy acetic acid for hemodialysis process and to know the effect of different leaching agents in the template release in the transport test. Polyeugenoxy acetic acid that has been bound with urea was mixed with *Poly Vinyl Alcohol* (PVA), a crosslinking agent for *Poly Ethylene Glycol Diglycidyl Ether* (PEGDE), and *N-Methyl 2 Pyrrolidone solvent* (NMP) to form the hydrogel. The hydrogel was then dissolved by NMP to make a membrane. The process of releasing the urea template used ethanol and HCl to obtain urea MIM (*Molecularly Imprinted Membrane*) and transports was done to find out the difference in membrane transportability. The polymer base and urea MIM obtained were characterized using FTIR, SEM-EDX, and TGA / DTA, while urea MIM was used to transport urea, creatinine, and vitamin B_{12} through the transport process and was analyzed by UV-VIS spectrophotometer. The results obtained are that MIM has better transports results than NIM. The percentage of transport of urea, creatinine, and vitamin B_{12} by MIM membranes is 45.02%; 33.63%; 14.28%, and the percentage of transport of urea, creatinine, and vitamin B_{12} by NIM membranes was 43.36%; 27.42%; 9.9%. In addition, there is an influence of eluent variations (leaching agent) on the transport of urea, creatinine, and vitamin B_{12}, where the optimum transport results are shown in washing using HCl eluents. Percentage of urea transport by membrane washed with eluent variations of HCl, Ethanol, and Aquadest was 45.02%; 40.12%; 39.17%, the percentage of creatinine transport in HCl, Ethanol, and Aquadest variations was 47.03%; 44.2%; 36.6%, and the percentage of vitamin B_{12} transport in the HCl, Ethanol and Aquadest variations respectively was 8.87%; 18.27%; 3.65%.

1. Introduction

Chronic kidney failure (CKD) is a global public health problem with an increased prevalence and poor prognosis. According to the Global Burden of Disease 2010 results, CKD was the 27th leading cause of death in the world in 1990 and increased to 18th in 2010 [1]. The data showed that the prevalence of CKD is going to increase by about 1,8 % over a year [2]. Currently, hemodialysis is a biomedical method that is used to replace kidney function in removing urea and creatinine from the body that was found in the blood [3]. In general, the high levels of urea and creatinine indicate the kidney inability in disposing these toxic compounds.

Hemodialysis use a semipermeable membrane to selectively dispose the harmful component from the kidney. A good semipermeable membrane should not pass proteins or substances that are still beneficial for the body, instead it should be able to pass urea molecules in the blood, have good strength and mechanical resistance, and should also be biocompatible. Hemodialysis membrane based on cellulose, chitosan, and polysulfone have some disadvantages such as causing an inflammatory response in patients, only the small molecules can pass through, lack of active sites, low mechanical strength, and adsorption of blood proteins to the surface. [4-6]. Therefore, in this study, the latest clean material used was eugenol, where eugenol has three functional groups which capable of being functional monomers that are potential for selective adsorption processes with molecular imprinting technology, and hemocompatibility properties, to minimize the weaknesses of the basic ingredients of hemodialysis membrane, and also improve the properties of hemodialysis membrane.

MIM (Molecularly Imprinted Membrane) is a membrane that contains introduction sites to molecules made with molecular printing techniques. This printing technique can give the membrane the ability selectively recognizing target molecules in solution during the transport process [7]. An important step in the imprinted technology concept is the release of target molecule templates, which will produce areas that capable of recognizing the target molecule or target analyte. Djunaidi (2016) has conducted research on the adsorption of selective phenol molecules with different eluent solvents which are methanol, water, ethanol, and chloroform [8]. Another research by Lee et al (2008), described the molecularly imprinted creatinine membrane (MIM) using poly (ethylene-vinyl alcohol) material with eluent used to release (leaching) target molecules from the template such as ethanol and water deionization [9]. However, the above research does not explain its effect to the transports applications results. Therefore, in this research, variations of leaching agents such as aquadest, HCl, and ethanol were used to find the optimal eluent to release the template molecules, to know its effect on the transport of urea, creatinine, and vitamin B_{12}, and hopefully, to increase the selectivity of MIM (molecularly imprinted membrane) when compared to NIM (Non-Imprinted Membrane).

2. Materials and method

2.1. Materials

The materials purchased from SIGMA-Aldrich were Eugenol and BF_3-diethylether while other reagents were purchased from Merck, NaOH, chloroacetic acid , Chloroform, Methanol, Poly vinyl alcohol (PVA), urea, creatinine, vitamin B_{12}, PEGDE (*Polyethylene glycol diglycidyl ether*), 1-Methyl 2-pyrrolidone and Diethyl ether. Demineralized water was purchased from Bratachem.

2.2. Instrumentation

The instruments used were FTIR Spectrophotometer (Shimadzu Prestige 21), analytical balance (Ohauss), UV-Vis Spectrophotometry (LW-V-200-RS), pH meter (Trans Instrument), SEM-EDX (Phenom Pro X Desktop with EDX), and TGA (DTA/TG Exstar SII 7300).

2.3. Method

2.3.1. Synthesis of polyeugenol

5 grams of eugenol were added to a three-neck flask and then added with 1 mL of BF_3-diethyl ether. The mixture was stirred using a stirrer for 4 hours and once every 1 hour the addition of 0.25 mL BF_3-diethyl ether was done. After the reaction lasts for 4 hours, the polymerization was stopped by adding 1 mL of methanol. The formed gel was then dissolved with diethyl ether and washed with distilled water until it reached a neutral pH. The solution was then dried by adding Na_2SO_4 anhydrous. After being completely free of water, the solution was evaporated at room temperature. The formed

IOP Conf. Series: Materials Science and Engineering **959** (2020) 012023 doi:10.1088/1757-899X/959/1/012023

precipitate was then dissolved with distilled water, dried and weighed. The obtained result was analyzed by FTIR.

2.3.2. Synthesis of polyeugenoxy acetic acid
A total of 5 g of polyeugenol was put into a boiling flask, then 33% NaOH solution (33 grams NaOH in 100 mL) as much as 17.5 mL was added. Subsequently, the mixture was stirred for approximately 30 minutes and added 12.5 mL of 50% chloroacetic acid solution (50 g in 100 mL water) little by little with a dropper pipette while continuing to stir. The mixture was heated in a water bath with a temperature of 80–90°C. Heating was carried out for 2 hours, then let cooled and acidified with 6 M HCl to pH = 1. Then extracted with diethyl ether as much as 3 times, 50 mL in each extraction. The ether extract was combined and extracted with 5% w/v sodium bicarbonate 3 times 30 mL each, then the water layer was acidified with 6 M HCl to pH = 1. Then filtering, drying and weighing were carried out. The result obtained was analyzed by FTIR.

2.3.3. Synthesis of MIM in-situ
A total of 1 g polyeugenoxy acetic (PA) acid was contacted with 1000 ppm urea as much as 20 mL. The results are then dried. PA which has been contacted with urea coupled with 0.5 g PVA dissolved in 2,5 mL of NMP added with 0.1 M sodium hydroxide as a catalyst and PEGDE and heated at 120–140°C for 20–30 minutes and allowed to stand overnight to form a gel. Gel with a weight of 0.35 grams for membrane thickness was then added with NMP solvent and heated to obtain the right viscosity to do the casting process. The casting membrane was then heated at 80°C until the membrane is dry and molded. It was then soaked with 2 M KCl until the membrane is detached from the petridish. The membrane sheet was then cut to the size of the diffusion cell. The removal of urea with HCl and ethanol for 24 hours produced the MIM urea membrane. Then the obtained membrane is mounted on the diffusion cell tool.

2.3.4. Synthesis of NIM (Non-imprinted Membrane)
NIM was synthesized in the same way as MIM urea but without the binding stage of urea first. The PA was used without contacting the urea first.

2.3.5. Urea transport
Transport urea with membrane produced using a diffusion cell apparatus, in the feed phase (feed, urea) is 300 mg/L of urea and in the receiving phase is phosphate buffer pH 7.4, each 50 mL and then stirred for 24 hours. Sampling was done every one hour until 8 hours and followed by sampling at the 24 hours.

2.3.6. Creatinine transport
Creatinine transport with the membrane produced using a diffusion cell device, in the feed phase (feed, creatinine) was 25 mg/L creatinine and in the receiving phase was a phosphate buffer of pH 7.4 each 50 mL and then stirred for 24 hours. Sampling was done every one hour until 8 hours and followed by sampling at the 24 hours.

2.3.7. Vit B12 transport
Transport of vitamin B_{12} with a membrane produced using a diffusion cell apparatus, in the of feed phase (feed, urea) is 20 mg/L of vitamin B_{12} and in the receiving phase is a phosphate buffer at pH 7.4, each 50 mL and then stirred for 24 hours. Sampling was done every one hour until 8 hours and followed by sampling at the 24 hours.

2.3.8. SEM-EDX analysis
SEM-EDX analysis was performed on the MIM membranes before and after the 24 hours template removal.

2.3.9. TGA-DTA Analysis
TGA-DTA analysis was carried out on polyeugenol polymers, and polieugenoxy acetic acid, as well as the MIM membrane washed with HCl and ethanol, and unwashed MIM.

3. Results and discussions

3.1. Synthesis of polyeugenol and polyeugenoxy acetic acid
It appears in the FTIR analysis (figure 1) that the loss of eugenol vinyl spectra at wavenumbers of 915cm^{-1}, 997 cm^{-1} and 1648-1638 cm^{-1} indicates that the polymerization has occurred into polyeugenol. The C-O alcohol group (1600 cm^{-1}) transforms into an acid carbonyl group which is indicated by the peak at 1720 cm^{-1}. The alcohol group (O-H) of eugenol and polyeugenol turn to acids at around 3400 cm^{-1}.

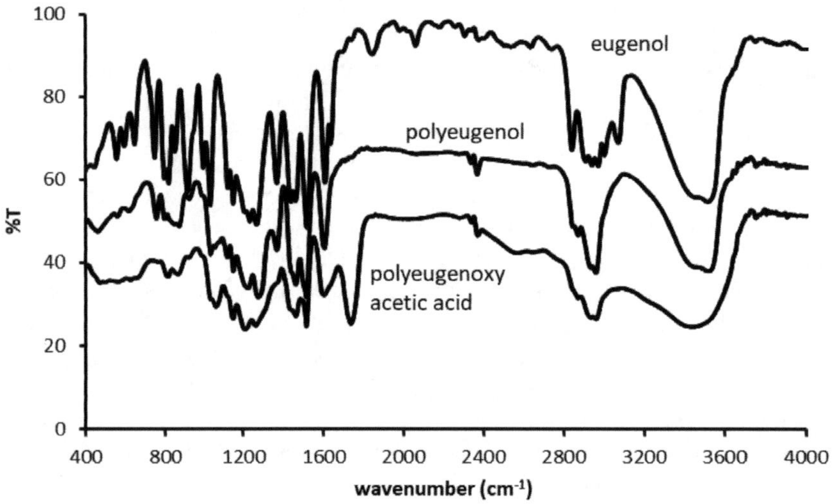

Figure 1. Graph of eugenol IR spectra and their derivatives.

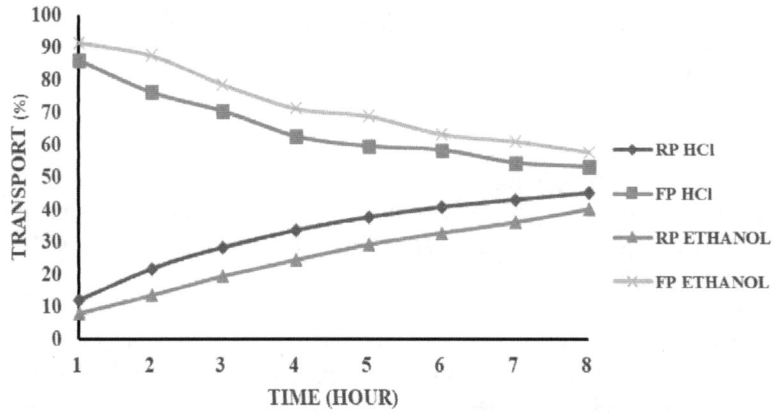

Figure 2. Urea transport in MIM with template release eluent variations.

3.2. Transport of 300 ppm urea with the variation of template release eluent
The performance of the membrane can be seen by transporting 300 ppm urea. It appears in figure 2 that the MIM shows good transport performance, especially at the beginning of transport (up to 8 hours of transport), both in the receiving phase and the feed phase with transport capability almost

half of the initial concentration up to the 8th hour. Thus, it is possible that the presence of the imprinted area on the membrane gives a large percentage of transport value in the urea transport process, and also shows the difference in eluent template release at urea transport performance.

In figure 2, a high yield percentage of transport using a 1 M HCl as eluent was indicated by the surface of the membrane that have a more regular and uniform pore as stated by the results of the characterization with SEM instruments and indicates that the release using HCl can improve the process of releasing the urea template because of its ability to dissolve the urea template is better than ethanol.

3.3. Transport of 25 ppm creatinine with a variation of the template release eluent

Figure 3 shows a high yield of transport percentage using a 1 M HCl eluent. This is indicated by the surface of the membrane having a more regular and uniform pore as stated by the results of the characterization with the SEM instrument and indicates that the release using HCl can improve the template release process better than ethanol.

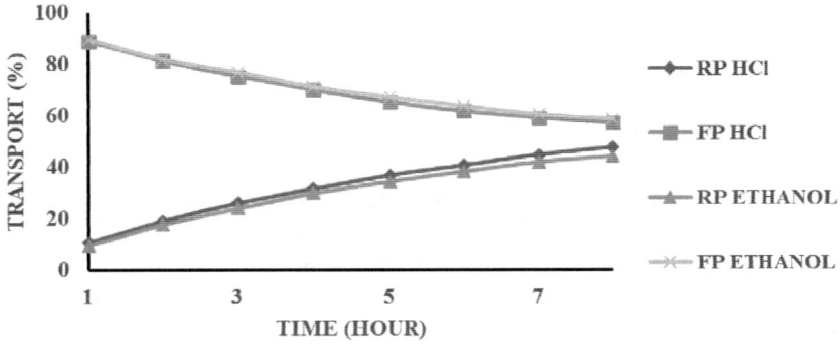

Figure 3. Creatinine transport on MIM with eluent variation release template.

3.4. Transport of 20 ppm vitamin B_{12} with a variety of template release eluents

Figure 4 shows that the membrane with HCl eluent cannot transport Vitamin B_{12} properly (not in large quantities) which can be seen from the graph. The feed phase (FP) has a high percentage of transport indicating that the HCl eluent has the ability to maintain vitamin B_{12} during the transport process and is not able to transport properly due to the template or the imprinted area when compared with ethanol.

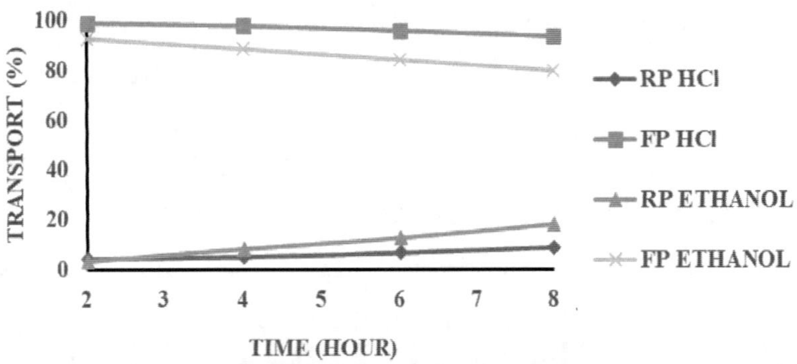

Figure 4. Vitamin B_{12} transport on MIM with eluent variation release template.

IOP Conf. Series: Materials Science and Engineering **959** (2020) 012023 doi:10.1088/1757-899X/959/1/012023

3.5. Urea, creatinine and vitamin B_{12} transport

It appears in figure 5 that the MIM is also capable of transporting creatinine and vitamin B_{12} even though it is not as large as urea, and is very small in transporting vitamin B_{12}. This is caused by creatinine and vitamin B_{12} sizes which are larger than urea. Creatinine has a size of (60 and 113 daltons) [10] while vitamin B_{12} has a size of 8,5 A or 1.35 kDa [11].

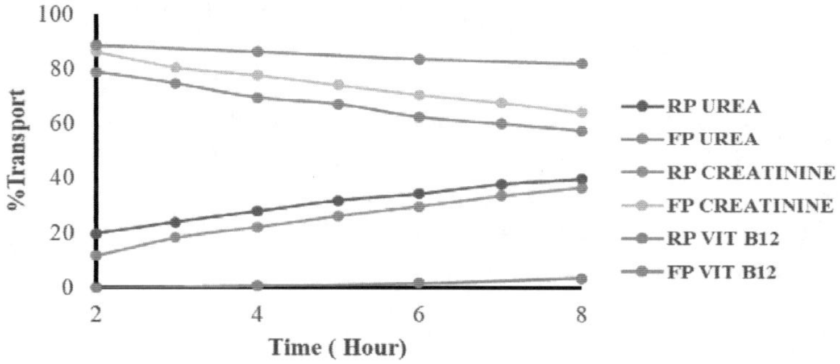

Figure 5. Transport of creatinine, vitamin B_{12} and urea with MIM membrane.

3.6. Membrane characterization using SEM-EDX

Table 1 below shows that washing with HCl for 24 hours can reduce the level of urea which is characterized by the composition of element N after leaching proved by EDX with an element composition of 0% and also shows that the composition of the element N in NIM is 0% because there is no urea template in the NIM. So, in this case, it shows that the urea MIM has been successfully formed and the levels of urea in the feed and receiver phases are not derived from the leaching process but indeed from the urea transport from the feed phase.

Table 1. EDX comparison of MIM and NIM 24-hour washing with HCl.

Element			Weight Concentration		
Number	Symbol	Name	MIM before and after washing HCl		NIM after washing HCl
			Before washing	After washing	After washing
6	C	Carbon	45.07	63.83	65.40
8	O	Oxygen	26.32	36.16	34.59
7	N	Nitrogen	28.57	-	-

While table 2 below showed that washing with ethanol for 24 hours, the composition of element N after leaching on EDX with an elemental composition of 32.62%. So, in this case, the level of urea present in the feed and receiver phases may come from the leaching process.

TABLE 2. EDX comparison of 24-hour washing with ethanol

Element			Weight Concentration	
Number	Symbol	Name	MIM before and after ethanol washing	
			Before washing	After washing
6	C	Carbon	45.07	44.75
8	O	Oxygen	26.32	22.62
7	N	Nitrogen	28.57	32.62

3.7. SEM analysis

Figure 6 shows that the obtained membrane is porous with a size of about 1–5 μm. The size of the pore membrane is suitable for use as hemodialysis membranes [12]. The appearance surface morphology in NIM showed that there are no pores and it showed differences in surface morphology when compared to MIM. Thus, the presence of templates gives a different effect on membrane surface morphology, and will also affect the transport performance.

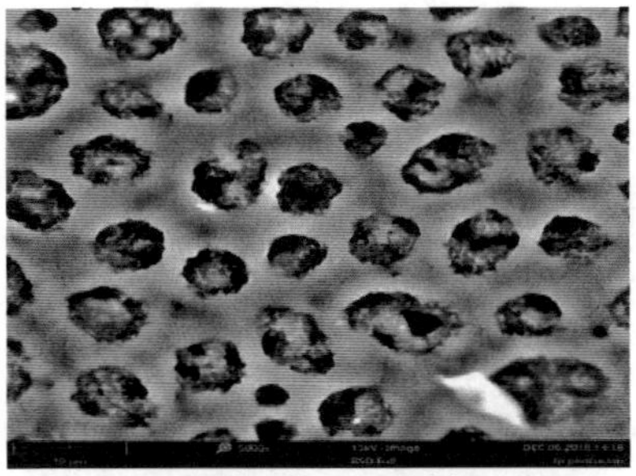
MIM before washing

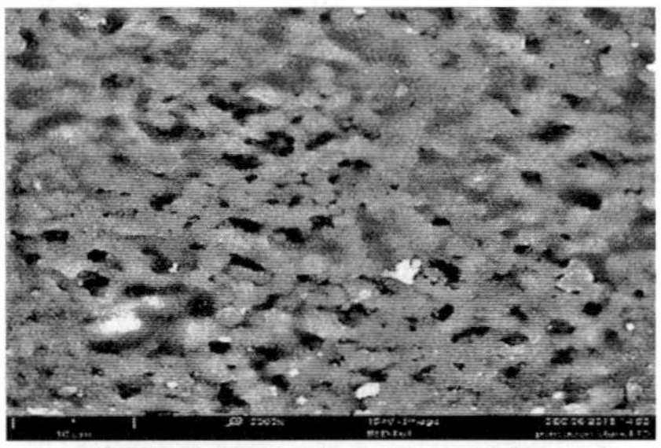
MIM after ethanol washing

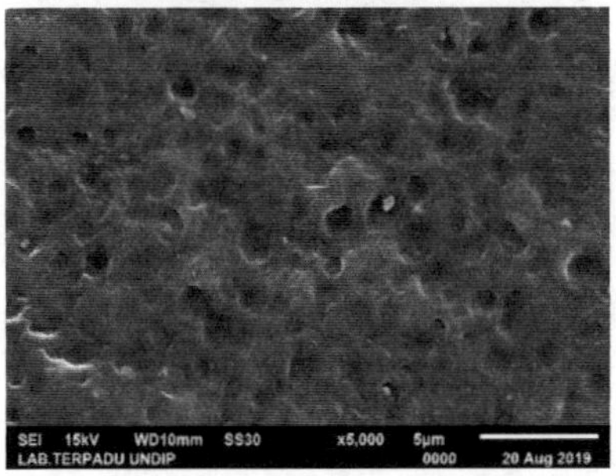

NIM after HCl washing

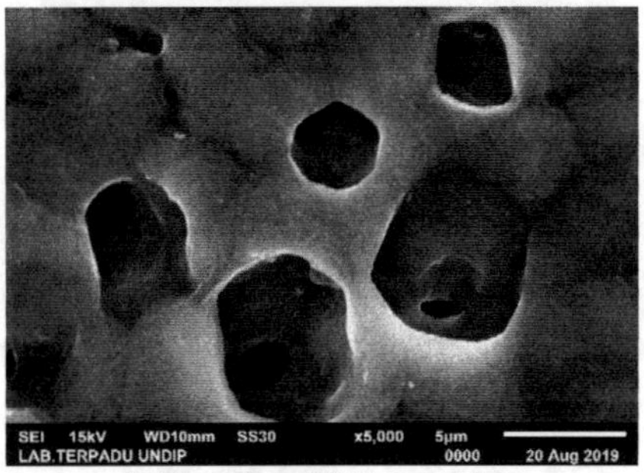

MIM after HCl washing

Figure 6. Comparison of SEM results before and after washing.

3.8. Thermal analysis with TGA-DTA

From the results of the analysis of a weak exothermic p at a temperature range of 50–100°C it is possible to have a dehydration reaction or release of water from water molecules that are absorbed in the material [13–14]. Endothermic peak p no temperature range 303–424°C (figure 7 (A)) and temperature range 321–446°C (figure 7 (B)), the temperature of 220°C and a wide endothermic peak at a temperature of 350–390°C (image C and D in figure 7), and temperatures 275–501°C (figure e) significantly further decomposition with respect to the TGA curve caused by the double bond on the benzene aromatic ring which breaks into aliphatic chains due to high temperatures [15]. A decrease in mass of about 20–60% (image A and B in figure 7), a decrease in mass of about 20–80% (image C and D in figure 7), and a decrease in mass of about 10–80% (figure e) is due to the breakdown/ depolymerization of the benzene ring which is abundant in both of the polymers [13].

The material polyeugenol and polyeugenoxy acetic acid only showed two steps: dehydration and polymer decomposition. This is different from MIM and NIM, because their composition has been cross-linked with PEGDE, PVA, and polyeugenoxy acetic acid polymers, showing that many stages occur at higher temperatures. The increase in chemical stability is due to the crosslinking and also caused by the bonding between molecules which requires more energy for decomposition [16] and the chemical structure that becomes more stable when crosslinked [17-19]. In the TGA-DTA result of polymer polyeugenol and polyeugenoxy acetic acid, it showed that the change of group in the polymers are capable of increasing the thermal resistance of the polymer. Also, the results of MIM and NIM TGA-DTA analysis showed that the presence of template urea in the membrane has no significance influence in the membrane thermal resistance (stability).

IOP Conf. Series: Materials Science and Engineering **959** (2020) 012023 doi:10.1088/1757-899X/959/1/012023

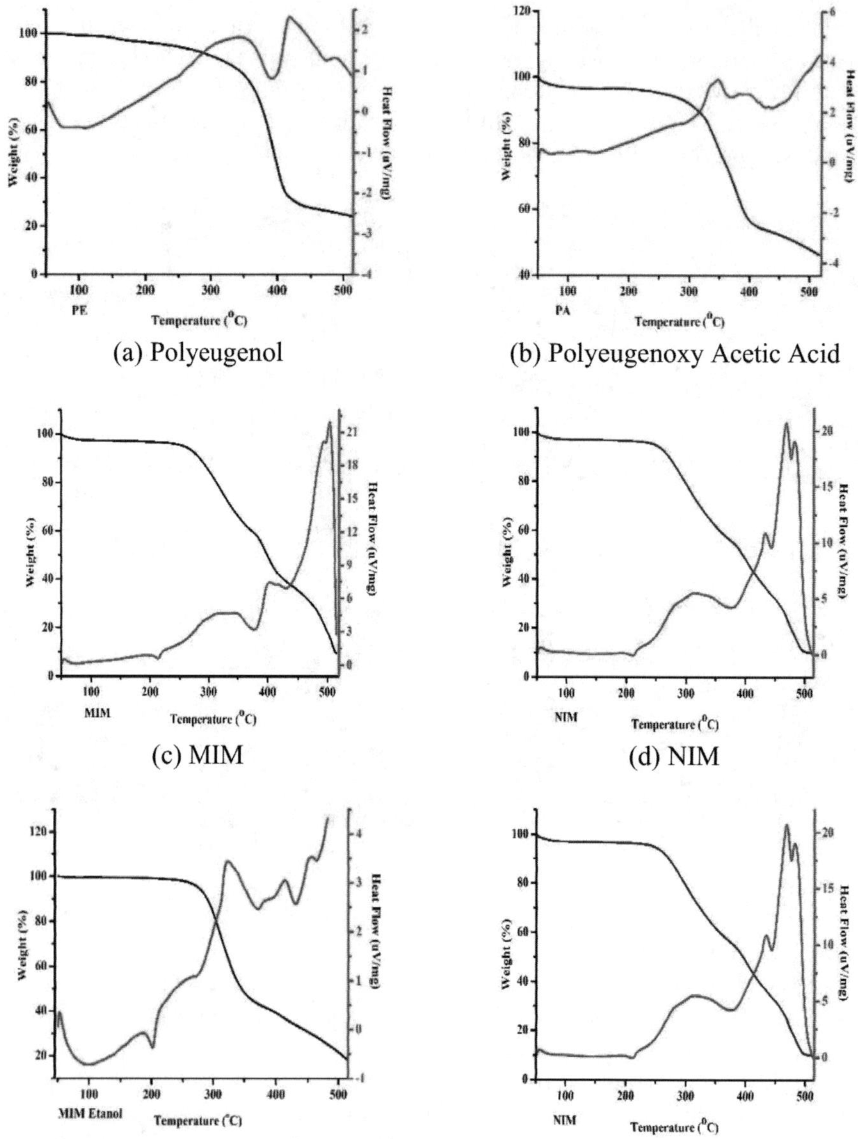

(a) Polyeugenol (b) Polyeugenoxy Acetic Acid

(c) MIM (d) NIM

(e) MIM washed with ethanol (f) MIM washed with HCl

Figure 7. Thermal analysis of polymers and membranes.

4. Conclusions

Membrane synthesis with active compounds derived from polyeugenol in the form of polyeugenoxy acetic acid has been successfully synthesized. Molecularly imprinted urea membrane has been successfully synthesized. The urea imprinted membrane is unable to transport creatinine and vitamin B_{12} properly. The eluent variations of HCl 1 M, Ethanol 10 % for template release produce differences in the results of urea, creatinine, and vitamin B_{12} transport. The HCl 1 M template release effluent can optimize the results of the urea, creatinine and vitamin B_{12} transport percentage. Changing the polymers group from originally polyeugenol into polyeugenoxy acetic acid increases its thermal resistance and there is no significant difference in thermal resistance on the imprinted and the non-imprinted membranes.

IOP Conf. Series: Materials Science and Engineering **959** (2020) 012023 doi:10.1088/1757-899X/959/1/012023

References

[1] Ministry of Health of the Republic of Indonesia 2017, *Pusat Data dan Informasi Kementrian Kesehatan RI* (Jakarta: Ministry of Health of the Republic of Indonesia)

[2] Badan Penelitian dan Pengembangan Kesehatan Kementrian RI 2018 *Riset Kesehatan Dasar (Riskesdas)* (Jakarta)

[3] Kee C M and Idris A 2010 *Sep. Purif. Technol.* **75** 102

[4] Ayuni N P S, Siswanta D and Suratman A 2017 Wahana Matematika dan Sains: Jurnal Matematika, Sains, dan Pembelajarannya **8** 77

[5] Wenten I G, Aryanti P T P, Khoiruddin, Hakim A N and Himma N F 2016 *J. Membr. Sci. Res.* **2** 78

[6] Lusiana R A, Siswanta D, Mudasir and Hayashita T 2013 *Int. J. Chem. Eng. App.* **4** 229

[7] Algieri C, Drioli E, Guzzo L and Donato L 2014 *Sensors* **14** 13863

[8] Djunaidi M C 2016 *Adsorpsi dan Transpor Fe (III) dan Fenol Menggunakan Polieugenol Tercetak Ion dan Molekul* Ph.D. Thesis (Gadjah Mada University)

[9] Lee M H, Tsai T C, Thomas J L and Lin H Y 2008 *Desalination* **234** 126

[10] Hosten A O *BUN and Creatinine* in: Walker H K, Hall W D and Hurst J W 1990 *Clinical Methods: The History, Physical and laboratory Examinations. 3rd edition* (Boston: Butterworths) chapter 193

[11] Talmard C, Guilloreau, Coppel Y, Mazarguil H and Faller P 2007 *ChemBioChem* **8** 163

[12] Djunaidi M C and Wenten I G 2019 *IOP Conf. Ser. Mater. Sci. Eng.* **509** 012069

[13] Djunaidi M C, Khabibi and Ulumuddin I 2016 *IOP Conf. Ser. Mater. Sci. Eng.* **172** 012032

[14] Nayakama F S and Bucks D 1991 *Irrigation Sci.* **12** 187

[15] Kiswandono A A, Siswanta D, Aprilita N H, Santosa S J and Hayashita T 2014 *J. Environ. Friendly Process* **2** 57

[16] Merle G, Hosseiny S S, Wessling M and Nijmeijer K 2012 *J. Membr. Sci.* **409** 191

[17] Djunaidi M C, Jumina, Siswanta D and Ulbricht M 2015 *Asian J. Chem.* **27** 4553

[18] Djunaidi M C, Jumina, Siswanta D and Ulbricht M 2015 *Indones. J. Chem.* **15** 305

[19] Djunaidi M C, Jumina, Siswanta D and Ulbricht M 2015 *Orient. J. Chem.* **32** 77

IOP Conf. Series: Materials Science and Engineering **959** (2020) 012024 doi:10.1088/1757-899X/959/1/012024

Formulation of the Lemongrass (*Cymbopogon citratus*) essential oil-based eco-friendly diffuse solution

T T K Ngan [1,2], T T Hien [2,3,4], P H Danh[3], L T H Nhan[4,5] and L X Tien[4,5,*]

[1] Center of Excellence for Biochemistry and Natural Products, Nguyen Tat Thanh University, Ho Chi Minh City 700000, Vietnam;

[2] NTT Hi-Tech Institute, Nguyen Tat Thanh University, Ho Chi Minh City 700000, Vietnam

[3] Falcuty of Food and Environmental Engineering, Nguyen Tat Thanh University, Ho Chi Minh City, 70000, Viet Nam

[4] Department of Chemical Engineering, Ho Chi Minh City University of Technology (HCMUT), 268 Ly Thuong Kiet Street, District 10, Ho Chi Minh City 700000, Vietnam;

[5] Vietnam National University Ho Chi Minh City, Linh Trung Ward, Thu Duc District, Ho Chi Minh City 700000, Vietnam

*Correspondence: tien.le@hcmut.edu.vn

Abstract. One of the most challenging issues associated with the use of artificial scent is the harmful effects not always visible to the eyes. To meet the growing consumer demand for the natural active ingredients of household products, the team looked for new, rich ingredients that have ample amount of compound that create scents and exhibit high antibacterial activities. The main objective of this study was to formulate a diffused product by replacing the harmful synthetic ingredients with safe natural ingredients. After applying the hydrodistillation method for the extraction process, lemongrass essential oil obtained with optimum efficiency of 0.29%. The process and formula of diffused products are optimized through the process of influencing parameters such as platform oil, emulsifier, solvent, Tween 80 and essential oil. The results indicate that lemongrass essential oil can be used as an eco-friendly alternative, which can add up the aromatic and deodorizing properties to the synthesized fragrance. The research into the application of lemongrass essential oils in many products will provide insight into the potential development of essential oils in the field of home care products.

1. Introduction

Today, essential oils are the subject of many scientific studies and also attract the attention of the cosmetics and pharmaceutical industries due to their potential as aromatic compounds or natural preservatives [1–5]. The wide variety of natural compounds and biological properties that make them attractive to many new and unexplored industries and applications. Regardless of the sensory properties of essential oils, the antibacterial, antifungal and especially aromatic activities of each essential oil are the goals of the study [6–10]. A promising new area of application of essential oils for flavoring and enhancement of preservatives from nature in cosmetics has been investigated [11–15].

Lemongrass has the scientific name of *Cymbopogon citratus* currently has about 55 species. It is one of the plants with a sweet and attractive scent, is applied in much cosmetic technology (e.g. perfume, perfume and bath soap) and food (e.g. tea, baking spices). Lemongrass is a year-round plant with thin, long leaves, is one of the most widely used medicinal plants and is also the most oil-extracted plant in tropical countries like Asia, Africa, and America. The aroma of citronella is due to the composition of the components in the essential oil, they create a gentle fragrance and long-lasting, an important aroma of high-class cosmetics [16], [17]. A number of previously published reports have revealed the presence the compounds identified in citronella are mainly terpenes, alcohols, ketones, aldehydes, and esters. Essential oils contain Citral, Citral, Nerol Geraniol, Citronellal, Terpinolene, Geranyl acetate, Myrcene, and Terpinol Methylheptenone [18–23].

Nowadays, the natural lemongrass essential oil is very expensive. Despite the higher prices, many people still switch to using natural essential oil products instead of synthetic essential oils. However, the determination of the essential oil component serves the production technology of essential oil applications has not been focused yet. There is no method to satisfy the requirements set forth about the efficiency and quality of citronella oil.

Lemongrass essential oil has been studied and applied widely for centuries ago, in many countries around the world [24–28]. Particularly in our country, the research of citronella oil, as well as the ability to apply research results, have been almost neglected. In addition, the safety of cosmetics is always of special interest to researchers and manufacturers, as the influence of microorganisms can lead to product degradation and endanger human health. Diffused products now commonly marketed are mainly blended with aromas to create a fragrance for products that contain products mixed with essential oils to create the fragrance, repel mosquitoes, insects, and deodorants. Besides, the price of these products is quite high, it mainly processed in Vietnam or imported from other countries such as Japan, Indonesia and Malaysia.

To create diffused products in the form of a solution of lemongrass essential oils, several factors such as solute/emulsifier of essential oils, the percentage of essential oils in the product to the highest aroma at a low cost should be taken into consideration. Furthermore, large-scale and low-quality production of essential oil may also do harm to consumer health, as well as the environment by disposing a significant amount of waste, toxic chemicals and non-recyclable containers [29]. For these reasons, research and preparation of environmentally friendly diffusing products made from lemongrass essential oils are essential, taking advantage of available raw materials in Vietnam with competitive prices.

2. Materials and methods

2.1. Raw materials and chemicals

Lemongrass materials were collected and selected in Tan Phu Dong district, Tien Giang province, Vietnam in the spring of 2019. The leaves are cut 20 cm away from the root, pick up and remove withered leaves, stored in a cool, convenient place during production for a few days. The extraction of essential oils is carried out by the hydro distillation system with an industrial scale of inputting about 200–750 kg. 710 kg of lemongrass leaves have been steam distilled with temperature of 100°C and steam flow rate of 110–130 L per hour for 3 hours to obtain essential oils. The addition of sodium sulfate has been done to remove exceed water left in the essential oil, then the oil stored in a closed vial.

Chemicals used: Dipropyllen glycol and paraffin are purchased at Nguyen Ba Trading Production Co., Ltd, Tan Binh District, Ho Chi Minh City. Tween 80 and ethanol were obtained from Xilong Group (China). Coconut oil was from Luong Quoi Coconut Co., Ltd and Borges Extra Virgin Olive Oil from Spain.

2.2. Product blend process

Lemongrass essential oil is first dissolved with a vegetable oil, antioxidants and emulsifiers are stirred to form a homogeneous mixed solution in phase 1. Then water, ethanol and isopropanol are separately stirred in phase 2 forming a homogeneous mixture. After a period of 30 minutes, the phase 1 mixture is slowly added to phase 2, continue stirring for 2–3 hours. After adding colors and different flavors, we will get a diffusion solution.

Table 1. Table of formulation of lemongrass essential oil.

Ingredients	%
Tween 80	2–12
Lemongrass essential oil	2–6
Coconut oil	0–4
Ethanol	0–20
BHT	0.1–0.5
Water	just enough

2.3. Evaluation methods

2.3.1. Method of assessing the appearance of products
Homogeneity and clarity: the product is contained in a transparent jar and visually observes against a white background under light to assess its homogeneity and clarity.
Color: visual observation, color identification and uniformity for preliminary evaluation. If the accurate determination is required, a colorimetric method may be carried out using a colorimeter.
Odor: describes the perceived odor.
pH: measured by using a Consort pH meter (Consort C3010, Belgium)

2.3.2. Methods of evaluating product durability
Products are stored in airtight bottles and stored under different conditions: room temperature, acceleration and thermal shock. The diffusion ability of essential oil was evaluated.

2.3.3. Methods of assessing the volatility of products
Place 2 ml of the sample solution on a petri dish, observe the evaporation time. Weigh the samples before and after leaving it at room temperature for 1 h, 2 h and 3 h. Record the clear aroma diffusion content at each time interval.

3. Results and discussion

3.1. The effect of platform oil contents on diffusion solutions
The influence of the oil on the background of the diffusion solution product is shown in figures 1 and 2. The results showed that the physical appearance of the product when using a coconut oil give the best result in terms of clarity, uniformity, fragrance, pH of the solution. For olive and paraffin oils, the solution is cloudy and small particles of oil appear on the surface. Similarly, when dipropyllen glycol is used as a platform oil, the solution shows a clear separation and the oil layer will float up. Figure 1 shows the effect of the oil on the diffusion of the mosquito repellent solution. The results showed that the use of coconut oil as a platform oil resulted in a much higher degree of diffusion than when using paraffin oil, in which, with a survey time of 3 hours, the obtained diffusion level was the highest (1.045). Figure 2 shows the effect of coconut oil content on diffused products. The results showed that with the content of coconut oil increased from 0–1%, the level of product diffusion increased from 0.871 to 0.925. However, when the content of coconut oil increased to 1.5 and 2%, the level of product diffusion decreased from 0.853 to 0.783. This could be explained by the fact that coconut oil is enriched in triglycerides; thus its addition of at high concentration would increase the solution viscosity, limiting the diffusity of the product [30]. Similar to the conductive oil survey experiment, when the time increases from 1–3 hours, the diffusion of the product increases. In particular, the level of product diffusion at 3 hour intervals soared compared to 1 h and 2 h. Therefore, using a sample of coconut oil with a content of 1% for the next survey.

Table 2. The influence of platform oil on diffusion solutions.

Platform oil	Coconut oil	Olive oil	Paraffin	Dipropyllen glycol
Physical appearance	Slightly turbid	Turbid	Turbid	Turbid
Uniformity	Uniformity	Small oil particles float on the surface of the solution	Oil floating on the surface of the solution	Oil floating on the surface of the solution
Fragrance	Pleasant	Pleasant	Pleasant	Pleasant
pH	6–7	6–7	6–7	6–7

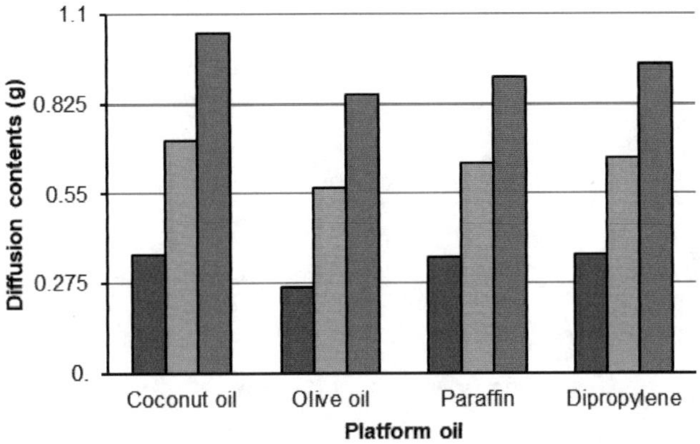

Figure 1. The effect of platform oil on the background of the diffusion solution.

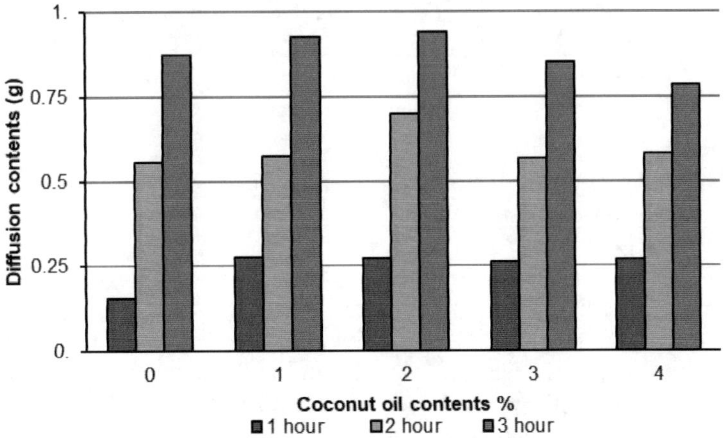

Figure 2. The effect of coconut oil contents on the background of the diffusion solution.

3.2. The effect of Lemongrass essential oil contents on diffusion solutions

The physical appearance of the product using different concentrations of essential oils is presented in table 2. The analysis showed that the content of lemongrass oil is proportional to the turbidity of the product, besides the high content of lemongrass also affects the smell of the product, causing a strong odor when use. The influence of the oil content on the evaporation of the solution is shown in figure 3. Lemongrass essential oil has many attractive aromatic compounds (myrcene and limonene) and some compounds capable of deodorizing and strong antibacterial activity (citral and geraniol) should be used as the main ingredient in mosquito repellent solution. The volatility of the solution increases with the concentration of essential oil from 2 to 4% after 1 hour and 2 hours and 3 hours, when the concentration of 5 and 6% increases, the sample changes to a slightly cloudy state, slightly lower than the 4% sample. Therefore, the content of lemongrass essential oil using 4% is suitable for diffusion products.

Table 3. The influence of lemongrass essential oil contents on diffusion solutions.

Lemongrass essential oil contents	1%	1.5%	2%	2.5%	3%
Physical appearance	Slightly turbid	Slightly turbid	Slightly turbid	Turbid	Turbid
Uniformity	Uniformity	Uniformity	Uniformity	Uniformity	Uniformity
Fragrance	Very light	Light	Pleasant	Pungent	Very pungent
pH	6–7	6–7	6–7	6–7	6–7

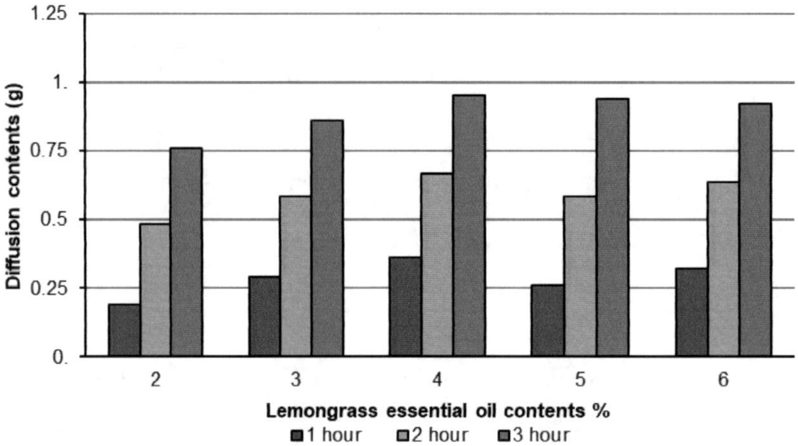

Figure 3. The effect of lemongrass essential oil contents on the background of the diffusion solution.

3.3. The effect of emulsifier on diffusion solutions

The physical appearance of the product when using different types of emulsifiers is shown in table 4. The results showed that, when using emulsifiers such as tween 20, tween 80, the solution was more opaque when using the emulsifier type PEG-40. Regarding the degree of homogeneity, observing the separation of the essential oil in the product base when using emulsion is Tween 20 and when using the other two emulsifiers, the phenomenon does not appear. The effect of emulsifier on the evaporation of the product is shown in figure 4. The results show that the emulsifiers Tween 20 (1.45) and Tween 80 (1.43) have the best evaporation than PEG-40 (1.16). However, in terms of appearance, Tween 20 appears the separation of essential oils. Therefore, Tween 80 is selected as the main emulsifier in the solution for the next survey.

The influence of the content of Tween 80 on the background of the sample solution is shown in figure 5. The results showed that Tween 80 surfactant accounted for 4% of the obtained solution with high stability and low viscosity. The reason is that the amount of surfactant just enough to break the emulsion up to micrometer size [31], while 10% and 12% of the large emulsifier makes the spray of odorant liquid difficult to use. Conversely, when using 2, 4, and 6% low emulsion content leads to unstable system, there is phenomenon of delamination, so use 8% content to investigate the next factor.

Table 4. The influence of emulsifier on diffusion solutions.

Emulsifier	Tween 20	Tween 80	PEG-40
Physical appearance	Turbid	Turbid	Slightly pure
Uniformity	Oil floating on the surface of the solution	Uniformity	Uniformity
Fragrance	Pleasant	Pleasant	Pleasant
pH	6–7	6–7	6–7

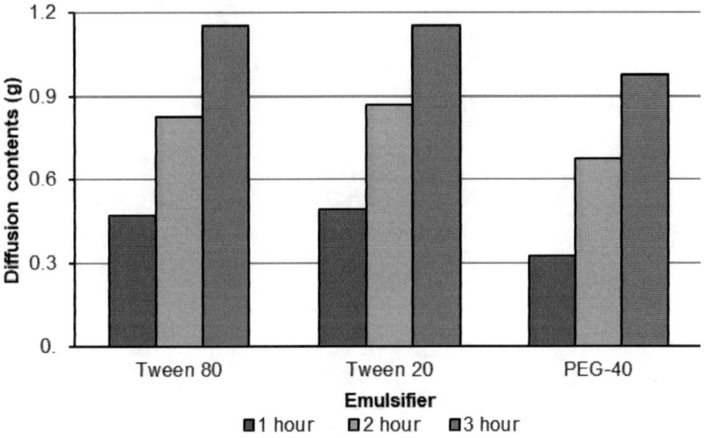

Figure 4. The effect of emulsifier contents on the background of the diffusion solution.

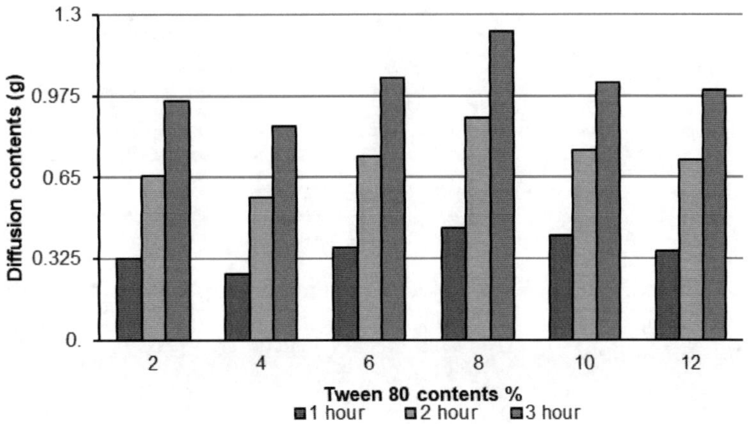

Figure 5. The effect of Tween 80 contents on the background of the diffusion solution.

IOP Conf. Series: Materials Science and Engineering **959** (2020) 012024 doi:10.1088/1757-899X/959/1/012024

3.4. The effect of solvent on diffusion solutions

The physical appearance of the product using different types of solvents is presented in table 4. The results showed that with ethanol the diffusion solution obtained was more transparent when using isopropanol. The diffuse solution used ethanol has a pleasant aroma, while with the isopropanol, the diffuse solution has a strong odor caused by the solvent vapor. The evaporation of diffusion solutions using different types of solvents is shown in figure 6. The results showed that the evaporation of the diffusion solution using ethanol is much higher than that of isopropanol. After a period of 2 hours and 3 hours, the evaporation of ethanol is 0.768–1.053 compared with isopropanol 0.666–0.913 respectively. At the same time, with the specific characteristics of isopropanol when used as a solvent in the solution, it will have a slightly pungent odor, reducing the aroma of essential oils. Therefore, ethanol is used as a solvent for diffusion products.

The evaporation rate of the solution depends mainly on the amount of solvent used, the difference between the concentrations is shown in Figure 7. According to the observation results, the percentage 5% and 10% volatility after 3 hours 1.253-1.179, respectively. Additional amount of solvent content to 15%, 20% of the evaporation gradually increases with the solvent content of 1.289–1.315 after 3 hours. Although, the ethanol content of 15% and 20% has a higher diffusion than the remaining sample, but the solution appears a separation state between the two phases of oil and water, the oil layer separates and floats on surface of solution. Therefore, use 5% ethanol for the subsequent survey of factors.

Table 5. The influence of solvent on diffusion solutions.

Solvent	Ethanol	Isopropanol
Physical appearance	Slightly turbid	Turbid
Uniformity	Uniformity	Uniformity
Fragrance	Pleasant	Very pungent (smell of solvent)
pH	6–7	6–7

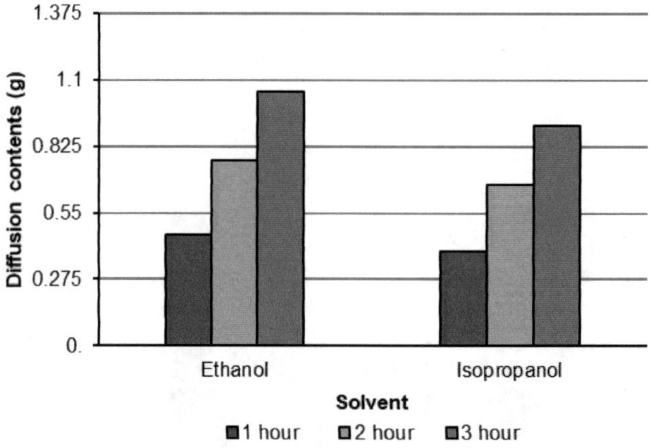

Figure 6. The effect of solvent contents on the background of the diffusion solution.

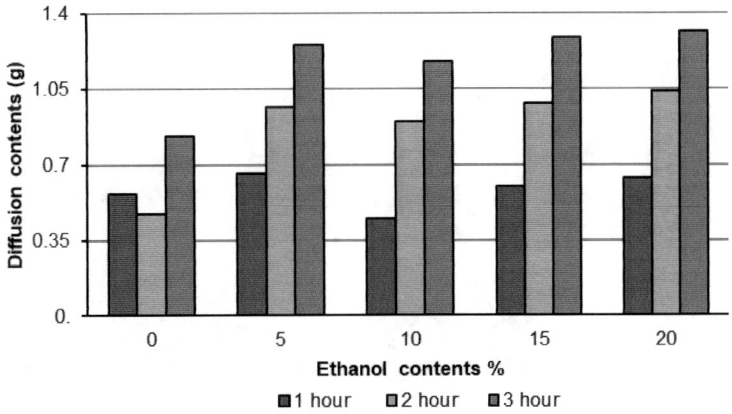

Figure 7. The effect of ethanol contents on the background of the diffusion solution.

3.5. The effect of antioxidants on diffusion solutions

The appearance of the product when using different antioxidants is shown in table 5. The results showed that, with BHA preservative the solution obtained was clear but separated. In contrast, with the anti-oxidant BHT, the solution is cloudy without separation. The influence of the antioxidants used in the sample to the diffusion is shown in figure 8. In terms of evaporation after 3 hours, the BHA 0.995 sample and the BHT 1.247 sample, proving that the BHT model evaporates a lot better. Although the properties of BHT are similar to BHA's, they are more heat-resistant. However, BHT has a lower antioxidant effect than BHA because the space structure of BHT is more bulky than BHA (because in the molecule of BHT, there are 2 groups of tert - butyl around the group - OH). Therefore, use BHT as an antioxidant in solution samples.

The influence of antioxidant content on the background of the product of diffusion solution is shown in figure 9. The volatility between different concentrations does not have much difference, the sample reaches the highest value at the content of 0.1% (1.145). For antioxidant content at 0.2, 0.3, 0.4, 0.5%, the diffusion ability of lemongrass essential oil is also relatively reduced. However, when high levels of antioxidants are used, the sample of the solution becomes more opaque than the rest of the samples, due to its unique properties. Therefore, the use of an antioxidant BHT with a concentration of 0.1% is suitable for diffusion products.

Figure 10 shows the effect of storage conditions (room temperature, acceleration, thermal shock) on the amplification of the solution. The solution sample, when stored under three conditions does not have much difference in the diffusion ability of the sample. On the other hand, samples stored under accelerated conditions and heat shocks turn into a more opaque state than samples under normal conditions. However, the component that accounts for the highest content of lemongrass essential oil is Citral (accounting for more than 80% of the content of lemongrass oil), which is very unstable, easily oxidized and denatured by external conditions such as light, heat which lead to changes in the quality of the diffusion solution [32,33]. This is the disadvantage that we do not expect of lemongrass essential oil because when the Citral component is denatured it will create another structure that leads to changes in the odor or loss of the characteristic smell of citronella oil in the sample.

Table 6. The influence of antioxidants on diffusion solutions.

Antioxidants	BHA	BHT
Physical appearance	Turbid	Slightly turbid
Uniformity	Oil floating on the surface of the solution	Uniformity
Fragrance	Pleasant	Pleasant
pH	6–7	6–7

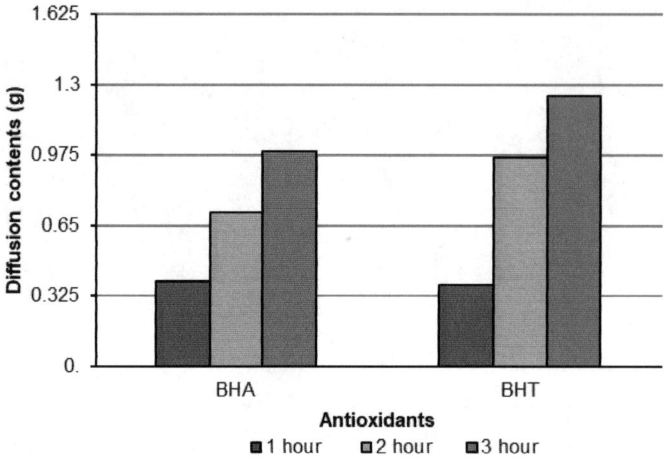

Figure 8. The effect of antioxidants contents on the background of the diffusion solution.

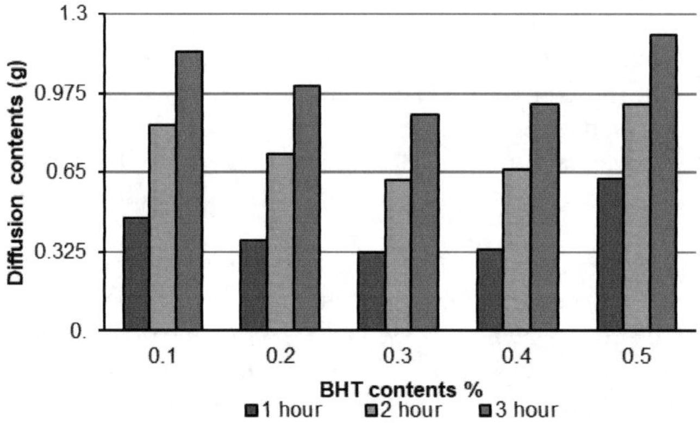

Figure 9. The effect of BHT contents on the background of the diffusion solution.

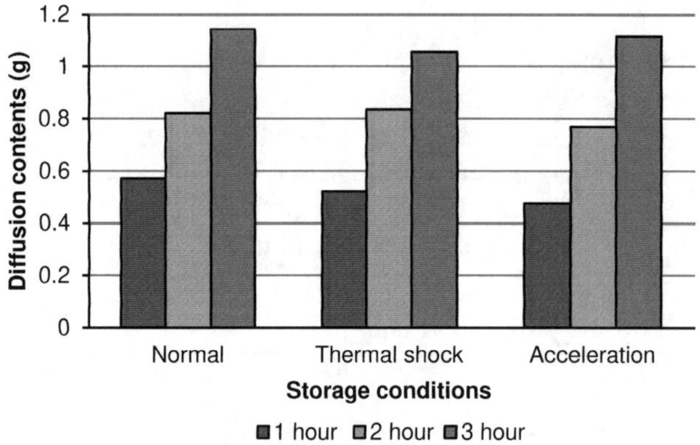

Figure 10. The influence of storage conditions on the background of the diffusion solution.

4. Conclusion

Research has proven that the replacement of citronella essential oil to diffused liquid formulations is an important factor that contributes to the quality of the product, significantly reducing the irritation of scents and improve the safety of products used. The ingredients in the formula are investigated at different levels and select the appropriate content to add to the product. The result is a diffuse product with main ingredients: Coconut oil (1%), Lemongrass essential oil (4%), Ethanol (5%), Tween 80 (8%) as emulsifier and BHT antioxidant (0.1%). The resulted lemongrass essential oil requires room temperature and minimal temperature change to maintain its diffusion quality.

Acknowledgements
This study was supported by Tien Giang Department of Science and Technology, Tien Giang province, Vietnam (the grant number is 103/HĐ-QPTKH&CN).

References

[1] Kim Ngan T T, Hien T T, Le X T, Anh T T, Quan P M, Cang M H, Le Ngoc T T, Danh V T, Yen Trung L N and Toan T Q 2019 *Asian J. Chem.* **31** 2855
[2] Dao T P, Tran T H, Quyen Ngo T C, Kieu Linh H T, Yen Trung L N, Danh V T, Le Ngoc T T, Yen Pham N D, Quan P M and Toan T Q 2019 *Asian J. Chem.* **31** 2827
[3] Ngan T T K, Huong N C, Le X T, Long P Q, Toan T Q, Hoang Vo D M, Danh V T, Yen Trung L N and Trieu T A 2019 P *Asian J. Chem.* **31** 2759
[4] Kim Ngan T T *et al.* 2019 *Asian J. Chem.* **32** 36
[5] Nhan N P T, Hien T T, Nhan L T H, Anh P N Q, Huy L T, Nguyen T C T, Nguyen D T and Bach L G 2018 *SSP* **279** 235
[6] Tran T, Nguyen H, Nguyen D, Nguyen T, Tan H, Nhan L, Nguyen D, Tran L, Do S and Nguyen T 2018 *Processes* **6** 206
[7] Mai H C, Le T T T, Diep T T, Le T H N, Nguyen D T and Bach L G 2018 *Asian J. Chem.* **30** 293
[8] Mai H C, Le T T T, Diep T T, Le T H N, Nguyen D T and Bach L G 2018 *Asian J. Chem.* **30** 293
[9] Dao T P, Nguyen D C, Nguyen D T, Tran T H, Nhan Nguyen P T, Hong Le N T, Le X T, Nguyen D H, N. Vo D V and Bach L G 2019 *Asian J. Chem.* **31** 977
[10] Hien T T, Nhan N P T, Trinh N D, Ho V T T and Bach L G 2018 *SSP* **279** 217
[11] Dreger M and Wielgus K 2013 *Herba Pol.* **59** 142
[12] Hutadilok-Towatana N, Chaiyamutti P, Panthong K, Mahabusarakam W and Rukachaisirikul V 2006 *Pharm. Biol* **44** 221
[13] Menut C, Bessiére J M, Samaté D, Djibo A K, Buchbauer G and Schopper B 2000 *J. Essent. Oil Res.* **12** 207
[14] Mirghani M E S, Liyana Y and Parveen J 2012 *Int. Food Res. J.* **19** 569
[15] Bassolé I H N, Lamien-Meda A, Bayala B, Obame L C, Ilboudo A J, Franz C, Novak J, Nebié R C and Dicko M H 2011 *Phytomedicine* **18** 1070
[16] Okoh O O, Sadimenko A P and Afolayan A. J 2011 *African J. Biotechnol.* **10** 4207
[17] Tran T K N, Tran T H, Le T H N, Xuan T L, Phan N Q A, Cam T I and Long G B 2020 *IOP Conf. Ser.: Mater. Sci. Eng.* **736** 022066
[18] Alam P N, Husin H, Asnawi T M, and Adisalamun 2018 *IOP Conf. Ser.: Mater. Sci. Eng.* **345** 012022
[19] Schaneberg B T and Khan I A 2002 *J. Agric. Food Chem.* **50** 1345
[20] Pino J A and Rosado A 2000 *J. Essent. Oil Res.* **12** 301
[21] Nakahara K, Alzoreky N S, Yoshihashi T, Nguyen H T T and Trakoontivakorn G 2013 *JARQ* **37** 249
[22] Mohamed Hanaa A R, Sallam Y I, El-Leithy A S and Aly S E 2012 *Ann. Agric. Sci.* **57** 113
[23] Desai M A and Parikh J 2015 *ACS Sustainable Chem. Eng.* **3** 421
[24] Brophy J J, Davies N W, Southwell I A, Stiff I A and Williams L R 1989 *J. Agric. Food Chem.* **37** 1330
[25] Ranitha M 2012 *Faculty of Chemical & Natural Resource Engineering* (Universiti Malaysia Pahang)

[26] Abegaz B, Yohannes P G and Dieter R K 1983 *J. Nat. Prod.* **46** 424
[27] Ganjewala D 2009 *Int. J. Essent. Oil Ther.* **3** 56
[28] Lorenzetti B B, Souza G E P, Sarti S J, Santos Filho D and Ferreira S H 1991 *J. Ethnopharmacol.* **34** 43
[29] Antunes M D C and Cavaco A M 2010 *Flavour Fragr. J.* **25** 351
[30] Bezard J, Bugaut M and Clement G 1971 *J Am Oil Chem Soc* **48** 134
[31] Kralova I and Sjöblom J 2009 *J of Disper Sci Technol* **30** 1363
[32] Nguyen H, Campi E M, Roy Jackson W and Patti A F 2009 *Food Chem.* **112** 388
[33] Djordjevic D, Cercaci L, Alamed J, McClements D J and Decker E A 2007 *J. Agric. Food Chem.* **55** 3585

IOP Conf. Series: Materials Science and Engineering **959** (2020) 012025 doi:10.1088/1757-899X/959/1/012025

Application of coconut oil from Ben Tre Province (Vietnam) as the main detergent for body wash products

T T K Ngan[1,2], T T Hien[1,2], N T C Quyen[1,2], P N Q Anh[3,4], L T H Nhan[3,4], M H Cang[5], D D Nhat[6], N D Phuc[6] and L G Bach[1,2,*]

[1] Center of Excellence for Biochemistry and Natural Products, Nguyen Tat Thanh University, Ho Chi Minh City 700000, Vietnam;
[2] NTT Hi-Tech Institute, Nguyen Tat Thanh University, Ho Chi Minh City 700000, Vietnam
[3] Department of Chemical Engineering, Ho Chi Minh City University of Technology (HCMUT), Ho Chi Minh City 700000, Vietnam
[4] Vietnam National University Ho Chi Minh City, Linh Trung Ward, Thu Duc District, Ho Chi Minh City 700000, Vietnam
[5] Department of Chemical Engineering & Processing, Nong Lam University, Ho Chi Minh City, Vietnam
[6] Falcuty of Food and Environmental Engineering, Nguyen Tat Thanh University, Ho Chi Minh City, 70000, Viet Nam

* E-mail: blgiang@ntt.edu.vn

Abstract. Coconut oil is considered the active ingredient as a moisturizer used by many people in tropical regions. At the same time, coconut oil can improve the symptoms of skin disorders by moisturizing and soothing the skin. A body washes formula that contains coconut oil based on the saponification process is formulated through an appropriate content survey. The combination of auxiliary substances such as detergent, foaming agent, moisturizing agent, preservative was also conducted to evaluate different concentrations. Products are assessed and visually observed for appearance, viscosity, pH value, stability, and durability when stored. The body wash formula was found to be white and pearl luster was visually observed. Products using 20% crude soap content, 5% detergent (SLES), 3% glycerin, etc. have all been tested to show good physico-chemical properties in terms of pH value and viscosity, as well as no irritation, upon skin contact. Research shows that coconut oil body wash formulas with good physico-chemical properties can be developed into new body wash cosmetic products.

1. Introduction

Daily exposure to air pollution and sun rays is known as main cause of women's skin damage. That is why a series of skincare products are regularly being launched, which have been meeting the above demand. On the other hand, there are quite a lot of bath products with different features, and most of them contain primary surfactants such as sodium laureth ether sulfate (SLES), sodium lauryl sulfate, disodium lauryl sulfosuccinate and sodium cocoylisothionate. They all have strong cleansing capabilities that can dry out the skin. The formula of body wash products with main ingredients from nature is currently favorable [1]–[3].

Coconut oil is a type of fat extracted from copra and is widely used in the field of food and pharmaceuticals worldwide [4]–[6]. Because of its high saturated fat content, it can be less oxidized and stored up to two years. Coconut oil is currently used in many fields, especially in cosmetics, which is very popular and attracts consumers' attention. Coconut oil contains high levels of saturated fatty acids such as palmitic acid (7–10%), myristic acid (16–21%), lauric acid (45–56%), caprylic acid, capric acid, stearic acid and oleic acid. In terms of nutritional value, coconut oil has been proven to be very good for the cardiovascular and digestion system [7].

Varmaet *et al.* have demonstrated the highest anti-inflammatory activity of coconut oil by inhibiting inflammatory markers and protecting the skin by strengthening the skin barrier function [8]. Burnett *et al.* mentioned in the research that in cosmetic products, coconut oil is used as a foaming agent, stabilizer, and detergent with concentrations up to 50%. There are no indications that these ingredients are sensitizing, primary stimulants, or toxic photosynthetic compounds after human testing. No skin irritation or any sensitization caused by coconut oil [9].

Besides, coconut oil antioxidant activity has been interested in scientists in the world in recent years [10]–[13]. Nevin *et al.* studied the effects of coconut oil on antioxidant enzymes' activity and fat peroxidation levels in male mice Sprague Daw Dawley [14]. Besides, the coconut oil contains polyphenols, which are antioxidants and prevent lipid peroxidation under both in vitro and in vivo conditions, which leads to improved health benefits related to antioxidants; and research results are published in Food Chemistry [14]. From the research results of scientists published in prestigious scientific journals, coconut oil can be used as an ingredient that can create fragrances, conditioners for hair, or moisturizers for skin and was reported in cosmetics in concentrations ranging from 0.0001% to 70%. Coconut oil is as safe as other cosmetic ingredients in use, and the concentrations are described in safety assessments [9], [13], [15], [16].

Therefore, the investment in research and development of coconut products, taking advantage of available raw materials to research high-quality body wash products, is the increasing demand of the market. However, the product lines from coconut applied in body wash products on the Vietnamese market are currently not many and not diverse. Thus, with the potential of coconut oil available in Ben Tre, the locality can fully take the initiative in the source of raw materials to produce natural cosmetic products from coconut oil. In this study, the leading detergent in the body wash was provided by the saponification process of coconut oil with the right content. The product is blended with several other active ingredients such as foaming, thickening, coloring, and aromatic essential oils to enhance the shower gel's aesthetics Therefore, the research objective is to produce bath products using coconut oil as raw material. This product can enhance the use value and income for local people in Bentre Province. Natural body wash products is safe and highly biodegradable materials will contribute to environmental protection and sustainable economic development.

2. Material and methods

2.1. Raw materials and chemicals

Coconut oil used in this study is virgin coconut oil purchased from VietCoCo products of Luong Quoi Coconut Processing Co., Ltd, Chau Thanh District, Ben Tre Province (Vietnam). Coconut oil meets Vietnamese standards (TCVN 7597: 2013) through the process of evaluating physical and chemical properties.

The chemicals used: Sodium Hydroxide (NaOH), sodium lauryl ether sulfate (SLES) as detergents, Glycerin as humectants, Cocamidopropyl bentaine (CAPB) as foaming agent and Repoly 100 as thickeners were purchased at Nguyen Ba Trading Production Co., Ltd, Tan Binh District, Ho Chi Minh City.

2.2. Saponification process
The process of saponification follows previous studies by the authors [6]. Conducting the saponification process is based on a combination of an alkaline solution (NaOH) with a coconut oil ratio of 1: 5 (g / g), of which the alkali content is 11%. The mixture was stirred well and kept reaction time for 3 hours, at a reaction temperature of 85°C. The crude soap base is checked with basic parameters such as pH value, viscosity, foaming, foam strength, and emulsifying ability and cleaning ability.

2.3. The process of mixing of body wash
First, the saponification process is carried out: the reaction creates soap with alkali to make crude soap. Then weigh the exact amount of SLES and H_2O stir until entirely completely together. Continue adding the obtained natural soap to the stirring mixture until dissolved. The formula's ingredients will be examined at different levels and selected appropriate content to add to the formula. Add additives, including auxiliary cleaners, moisturizers, softeners, preservatives, etc. and part of the water, ~~is~~ are dissolved together and stirred gently, and proceed heat (if necessary). Colors and scent were added to enhance the sensory value and diversify products. Also, add electrolyte and dissolved pH adjustment solution to the solution.

2.4. Methods of determining the physical properties of products
- Viscosity: The viscosity of liquid products can be checked by NDJ-9S viscometer
- pH value: Use the Consort pH meter (C3010)
- Foamability: This study uses the shaking test to measure foaming. The liquid is diluted 100 time, and then 5 ml of solution is placed in a stoppered tube, shaking with a moderate force until the maximum amount of foam generated (foam volume does not change).
- Foamability was calculated by the formula:

$$\varepsilon_f = \frac{V_{foam} - V_{liquid}}{V_{foam}}$$

  with ε_f: foamability; V_{foam}: foam volume after shaking; and V_{liquid}: initial liquid volume.
- Emulsion durability: The cleaning effect is shown by the time of emulsification, with selected paraffin oil (simulated for dirt).

2.5. Sensory evaluation method of users
Surveying any 20 people (10 men + 10 women), the evaluation factors are expressed on a scale according to the sensory perception method. In particular, the largest scale is 5 points for the most popular product and the descending scale from the favorite, and lowest in the level of dislike for the zero scales. Factors are assessed as product background, foam level, feeling of use, and product preference level.

3. Result and discussion

3.1. Effect of crude soap
The content of crude soap was surveyed from 10 to 50%. The survey results are shown in figure 1. The foam volume increases with increasing the soap content from 10 to 20% and reaches the highest value at 20% (foamability 0.612, foam durability 0.569). Then decrease gradually when increasing the range to 50%. Therefore, simple soap will be selected with ~~a~~ 20% content to investigate the next factor. In terms of the time of emulsification, samples containing 20% crude soap content had the most extended delamination period under the influence of paraffin oil and tended to decrease when increasing the range from 20%–50%.

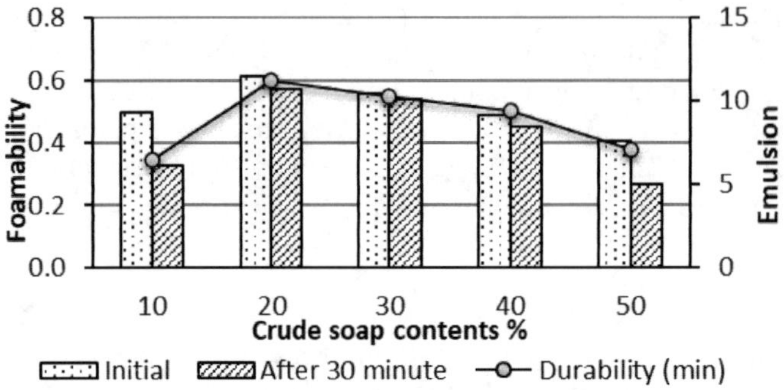

Figure 1. The effect of crude soap on foamability and emulsion durability.

3.2. Effect of Sodium Lauryl Ether Sulfate (SLES) detergent

The surfactant used is Sodium lauryl ether sulfate (SLES). The content of SLES is investigated from 0 to 10%. From the results of figure 2, foam volume did not change when increasing SLES content from 0 to 5%. But when the content increased to 5%, the foamability increased significantly to reach the highest value of 0.612 and also the content that the sample had the most durable time of 12.05 minutes. After that, the foamability decreases with increasing content from 7.5–10% (foamability from 0.593 to 0.573). At the same time, adding EDTA (0.1%) to isolate heavy metal ions, creating stability, also helps prevent the product from being affected by chemical reactions between metals and other active substances. Therefore, the selected SLES content is 5% for the follow-up survey.

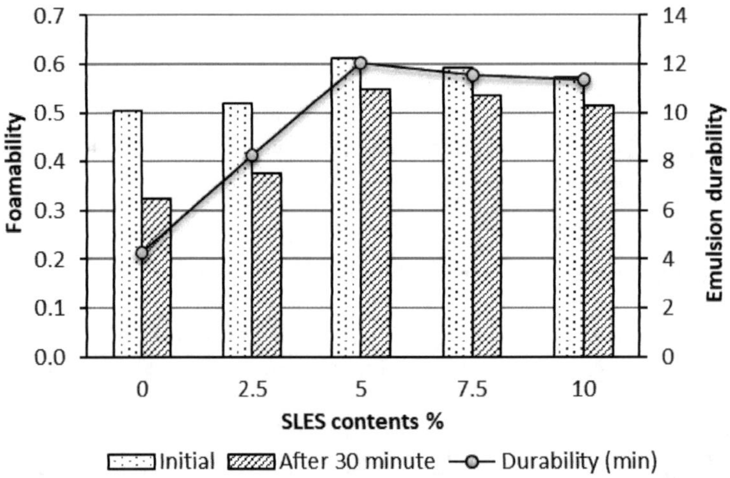

Figure 2. The effect of SLES on foamability and emulsion durability.

3.3. Effect of glycerin humectants

The study results on glycerin content using from 1 to 5%, were obtained, as shown in figure 3. However, there was an apparent change in the range of 3% (formability 0.580). Simultaneously, when evaluating the time of emulsion durability of the sample at 3%, it is the most durable compared to the other content. In comparison, larger values tend to reduce the foam and their stability. On the other hand, glycerin is a benign active substance, suitable for all skin types. It has ~~a~~ high strength besides good foaming ability, so the use of this agent in the product is appropriate for the work's original purpose. Therefore, to ensure the ability to cleanse, and moisturize the skin after use, glycerin is selected to be 3% content to investigate the next factor.

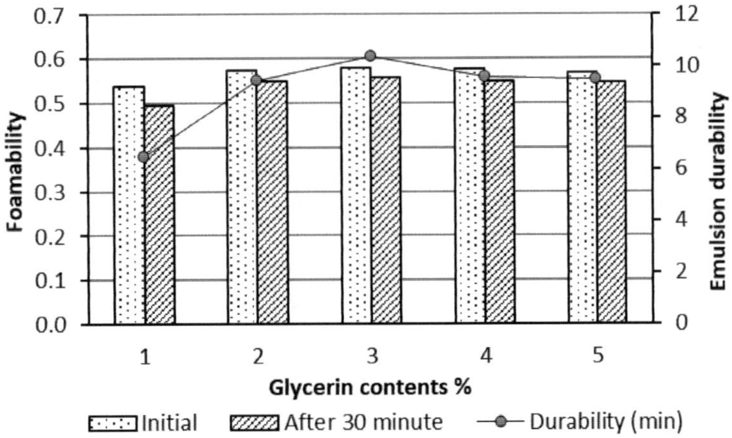

Figure 3. The effect of Glycerin on foamability and emulsion durability.

3.4. Effect of Cocamidopropyl Betaine (CAPB) foaming agent

Cocamidopropyl betaine (CAPB) works to help foam and stabilize the foam, while thickening the product and reducing skin irritation. CAPB content was ranged from 0 to 10%. The results are shown in figure 4. From figure 4, when increasing the CAPB content from 0 to 5%, the foam volume increases and tends to decrease when increasing the range from 5 to 10%. The highest foamability at CAPB 5% content with 0.593 foamability and 0.573 foam durability. On the other hand, the samples increased gradually from 0% to 7.5% in terms of emulsion durability. They tended to decrease at the content of 10%, reaching the highest value at the content of 7.5% with 11.43 minutes. However, at 5% with a stable time of 11.32 minutes, there was not much difference, combined with the ability to create foam, the CAPB content was selected as 5% to investigate the next factors.

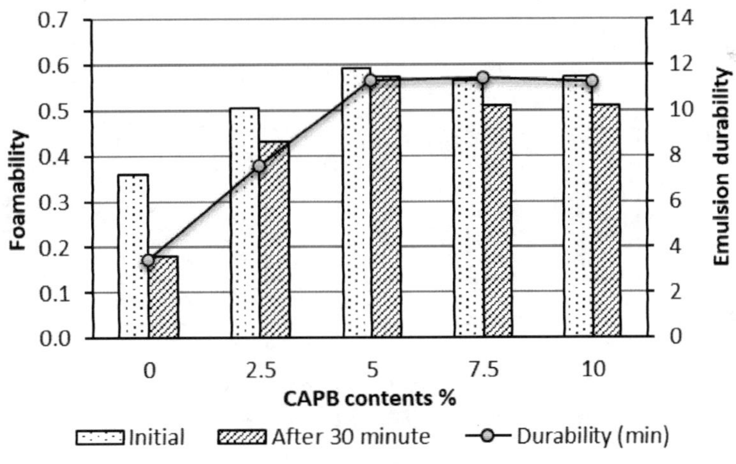

Figure 4. The effect of CAPB on foamability and emulsion durability.

3.5. Effect of P2000 UP and PEG-7 auxiliary cleaners

The two auxiliary cleaners selected for the survey were plantacare 2000 UP and PEG-7. The total amount of chosen additional detergentis 5%. The survey results are shown in figure 5.

According to the results of figure 5, in general, when increasing plantacare 2000 UP content and decreasing PEG-7 content, the foam volume increased. Therefore, use the plantacare 2000 UP auxiliary cleanser without the need for combination with PEG-7. Thus, the next section will examine plantacare 2000 UP content to add to the research product.

Plantacare 2000 UP content was surveyed from 0 to 4%. The influence of P2000 on the foaming and the time of emulsification are shown in figure 6. According to the results of Figure 6, the foam volume changes irregularly when the plantacare 2000 UP content is increased. Specifically, when adding 2% of the substance, the product shows the best cleaning ability (foamability 0.6, foam durability 0.573). The time to maintain the durability of emulsion peaks 11.16 minutes and starts to decrease with increasing content. Therefore, plantacare 2000 UP content in the formula was selected with a of 2%.

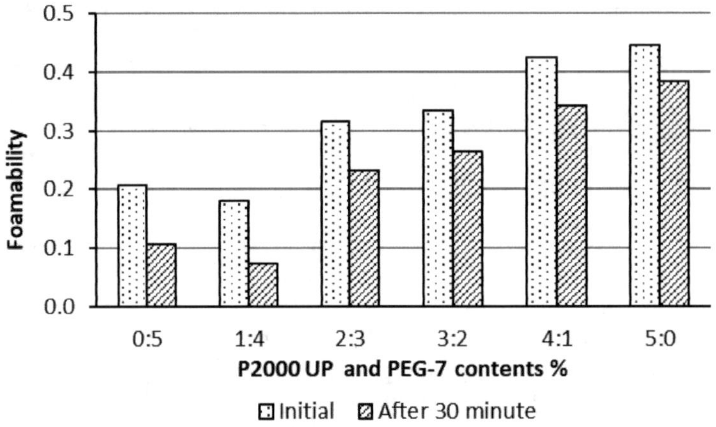

Figure 5. The effect of P2000 UP and PEG-7 on the foamability and emulsion durability.

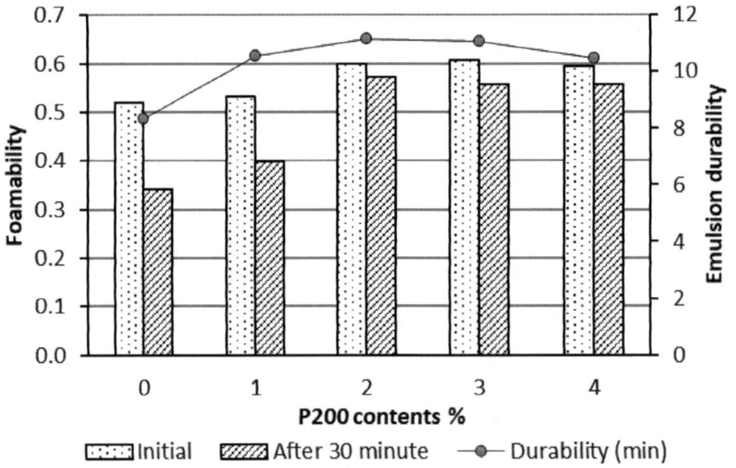

Figure 6. The effect of P2000 on the foamability and emulsion durability.

3.6. Effect of Repoly 100 thickener

About bonded evaluate, repoly 100 is satisfactory, but it does not reach the desired viscosity value. Therefore, to improve viscosity and ensure product appearance, the next survey will combine repoly 100 and NaCl.

After selecting repoly 100 as a thickener, conducting a survey on the content of repoly 100 shows that 4% has foamability 0.583 and foam durability 0.558.

Therefore, we chose repoly 100 and NaCl to conduct a detailed survey on the content used. The result of foaming of repoly 100 at 4% has the highest foamability 0.583 and foam durability 0.558, along with the highest emulsion time at 10.25 minutes at 4% compared to other patterns are shown in figure 7. After optimizing the repoly content of 100, continuing to investigate the content of NaCl agent, the result is the best foaming level of 0.583, and the best emulsifying time reaches 10.25 minutes at the of 5%.

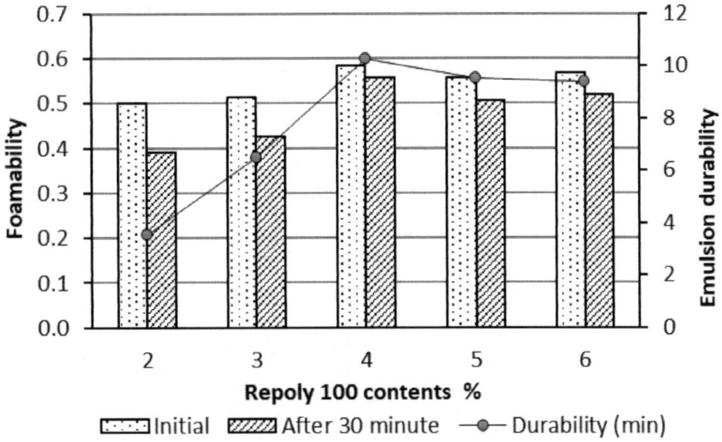

Figure 7. The effect of Repoly 100 on the foamability and emulsion durability.

3.7. Evaluate product properties
Through the assessment results, it can be seen that coconut oil body wash products do not have much difference compared to lifebuoy body wash samples on the market. The Body wash is a kind of wash-away product, it does not stay on the skin for too long, so the pH value ranges quite wide from 5 to 10. For the research, body washes (pH 8.26) because the soap is the leading detergent, so the pH value will range from 8 to 10 to ensure the best soap dish to wash away. Therefore, it can be predicted that the Lifebuoy body washes with lauric acid content need to adjust to high pH value, which will show good cleaning ability. The foam durability of the product above does not change significantly compared to the foaming level, after 30 minutes, the foam volume decreases significantly.

Table 1. Basic specifications of body wash products.

Parameter	Coconut oil body wash	Market sample
Appearance	White, pearl luster product background	White, pearl luster product background
pH	8.26	9.32
Viscosity (P)	>11	>11
Foamability	0.537	0.444
Foam durability *(after 30 min)*	0.519	0.392
Sensory	Small foam, smooth	Small foam, smooth

3.8. Assess the possibility of market penetration
According to the results of figure 8, Hazeline and St. Ives are appreciated, and the level of popularity is outstanding, and the foamability of St. Ives-is the most popular. Although not as superior as the two products above. The of the topic also has a positive effect from the user. The structure and feel on the

two products' skin are well evaluated, but limited to the foaming. Figure 8 shows the average result of Hazeline and St. body wash. Ives is a high-end product, so it scored a high rating around 4/5. Lifebuoy and Safeguard body wash belong to the mid-range product, the price is not high, so it has a lower score. Athough it is a cheap product for research body wash products, the effectiveness is quite positive, according to the user. Therefore, research body wash products are competitive and penetrate the market for cleaning products.

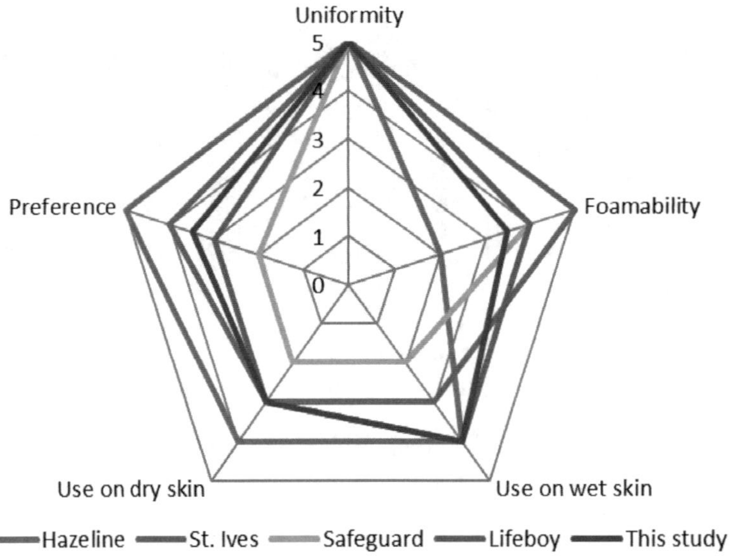

Figure 8. Evaluate senses of body wash users.

4. Conclusion

The crude soap mixture is created by the saponification reaction with the nature of liquid soap, has good detergency, and is suitable to add to the product formula as the detergent to replace a detergent commercial washing, reducing the amount of synthetic detergent. The ingredients in the formula's ingredients are investigated at different levels and select the appropriate content to add to the product. Foamability and foam durability are set targets to examine the influence of ingredients in the formula. The formula of 20% crude soap and 5% SLES, combined with auxiliary detergents, moisturizers, thickeners was suitable to appy for body wash formulation.

Acknowledgements
This study was supported by Ben Tre Department of Science and Technology, Ben Tre Province, Vietnam under the grant number 1136/HĐ-SKHCN.

References

[1] Tran T K N, Tran T H, Le T H N, Xuan T L, Phan N Q A, Cam T I and Long G B 2020 *IOP Conf. Ser. Mater. Sci. Eng.* **736** 022066
[2] Mai H C, Le T T T, Diep T T, Le T H N, Nguyen D T and Bach L G 2018 *Asian J. Chem.* **30** 293
[3] Mai H, Nguyen T, Le T, Nguyen D and Bach L 2019 *Processes* **7** 90
[4] Aburjai T and Natsheh F M 2003 Plants used in cosmetics *Phytother. Res.* **17** 987
[5] Agero A L C and Verallo-Rowell V M 2004 *Dermatitis (Am. J. Contact Dermat.)* **15** 109
[6] Nhan N P T, Hien T T, Nhan L T H, Anh P N Q, Huy L T, Nguyen T C T, Nguyen D T and Bach L G 2018 *SSP* **279** 235
[7] Fife B 2013 *Agro. Food Ind. Hi. Tech.* **24** 7
[8] Varma S R, Sivaprakasam T O, Arumugam I, Dilip N, Raghuraman M, Pavan K B, Rafiq M and

Paramesh R 2019 *J. Tradit. Complement. Med.* **9** 5

[9] Burnett C L, Bergfeld W F, Belsito D V, Klaassen C D, Marks J G, Shank R C, Slaga T J, Snyder P W and Andersen F A 2011 *Int. J. Toxicol.* **30** 5S

[10] Ahmad Z., Sarmidi M.R., and Hasham R. 2017 *Chem. Eng. Trans.* **56** 1657

[11] Shilling M, Matt L, Rubin E, Visitacion M P, Haller N A, Grey S F and Woolverton C J 2013 *J. Med. Food* **16** 1079

[12] Intahphuak S, Khonsung P and Panthong A 2010 *Pharm Biol* **48** 151

[13] Oyi A R, Onaolapo J A and Obi R C 2010 *Res. J. Appl. Sci. Eng. Technol.* **2** 133

[14] Nevin K G and Rajamohan T 2006 *Food Chem.* **99** 260

[15] Mukti R A and Misbahuddin A R 2020 *Proc. 2nd Int. Conf. Soc. Appl. Sci. Technol. Home Econ. (ICONHOMECS 2019)* (Bareng: Atlantis Press)

[16] Satheesh N 2015 *Ann. Food Sci. Technol.* **16** 115

IOP Conf. Series: Materials Science and Engineering **959** (2020) 012026 doi:10.1088/1757-899X/959/1/012026

Development and evaluation of shampoo products based on coconut oil source from Ben Tre Province (Vietnam)

T T K Ngan[1,2], T T Hien[1,2], L T H Nhan[3,4], M H Cang[5], P H Danh[6], N D Phuc[6] and L G Bach[1,2,*]

[1] Center of Excellence for Biochemistry and Natural Products, Nguyen Tat Thanh University, Ho Chi Minh City, Vietnam;
[2] NTT Hi-Tech Institute, Nguyen Tat Thanh University, Ho Chi Minh City, Vietnam
[3] Department of Chemical Engineering, Ho Chi Minh City University of Technology (HCMUT), Ho Chi Minh City 700000, Vietnam
[4] Vietnam National University Ho Chi Minh City, Linh Trung Ward, Thu Duc District, Ho Chi Minh City 700000, Vietnam
[5] Department of Chemical Engineering & Processing, Nong Lam University, Ho Chi Minh City, Vietnam
[6] Falcuty of Food and Environmental Engineering, Nguyen Tat Thanh University, Ho Chi Minh City, 70000, Viet Nam

* E-mail: blgiang@ntt.edu.vn

Abstract In order to enhance the use of coconut oil through its application in personal care products, shampoos formulated with simple soap act as the leading detergent with a 20% content and ingredients other commercial components such as Sodium Lauryl Ether Sulfate, Plantacare 2000, Glycerin, Cocamidopropyl Betaine, etc. A combination of 20% crude soap and a synthetic active substance synthesizing sodium lauryl ether sulfate (SLES) 5% make up the excellent shampoo recipe with cleansing feature. Ingredients of complementary detergents, moisturizers, and smoothing softeners are supplemented with appropriate levels to improve raw soap's disadvantages. Foamability and foam durability are selected targets to examine the influence of ingredients in the formula. The sensory elements of the product reach the set goals with an average evaluation score of about 4/5. Thereby, shampoo products penetrate the cosmetic market and attract many consumers in the future.

1. Introduction

Today, hair care products are expected to provide additional benefits such, as hair beauty or damage care. In addition to basic hair care formulas, there are select products to treat hair problems like dandruff. A large part of the hair care market is focused on products which include hair styling, coloring, and products that change the natural properties of hair fibers. Hair is an essential factor in the outside of the individual. Hair is easily damaged by weather, chemicals, heat, and abrasion; Therefore, suitable products should be used to achieve the best results. Therefore, hair care products are also essential some hair care products such as hair dyes, hair curlers, sprays, shampoos, and conditioners.

On the market today there, are many different types of shampoos, but the composition is mainly synthetic surfactants, so products of natural origin are the tastes of consumers [1]–[5]. Products with synthetic surfactants often have strong cleansing properties, not suitable for people with sensitive skin. In the present time the creation of shampoos product with ingredients of natural origin is a trend in research on personal care products.

Coconuts are grown a lot in Vietnam, especially in Ben Tre, and coconut oil is the most valuable component of coconut. With various features and good effects on the human body, especially skin and hair, coconut oil is a lot of manufacturers' interest and research and development [6]–[8]. Taking advantage of cheap coconut oil in Ben Tre to create cleaning products with good cleaning power, safe and gentle, taking advantage of the inherent benefits of this material will be studied. The application of coconut oil into a variety of products will give more insight into the development potential of coconut oil in the field of cosmetics, thereby widely developing in the cosmetic market. The effect of coconut oil have been studied before, but the scope of application is quite limited, and the product is not diverse. Currently, coconut oil is researched and widely used in many fields such as food, cosmetics, pharmaceuticals and other industries depending on the composition and properties of coconut oil. In cosmetics, coconut oil is used in skin and hair care, showing the ability to treat dandruff very well and help hair soft, firm. Besides, coconut oil can be used instead of moisturizers, preventing dry skin, effective acne and skin infection, excellent anti-aging, helps healthy, smooth skin. Lip balm and sunscreen products from coconut oil have been developed widely to improve the effectiveness of coconut oil and diversify the range of personal care products [9]–[11].

Coconut oil contains approximately 92.1% saturated fatty acids, 6.2% monounsaturated fatty acids, 1.6% polyunsaturated fatty acids [12]. Coconut oil can to kill bacteria thanks to the presence of monolaurin. Coconut oil has been studied and used as a lotion, safe, and effective in moisturizing the skin thanks to the effect of monolaurin [13]. Z. Ahmad and colleagues have shown coconut oil caninduce fibroblast growth and proliferation to enhance wound healing. The optimal concentration of coconut oil obtained in the study was 1.0 mg / mL for the best wound healing of about 56.89 ± 2.35% after 24 hours. Research results show that coconut oil several active ingredients for wound healing [14]. Several studies have shown that coconut oil is widely used for healthy hair growth. Traditionally, coconut oil is used to treat dandruff and lice, while it helps to soften the skin and reduce flaking and dryness. Coconut oil is also used as a protective agent against the damaging effects of UV radiation from the sun [15]. Coconut oil can penetrate the hair cuticle as a hair protective agent in grooming untreated or damaged hair [16]. With the good effects of coconut oil, this is a potential source of raw materials and has been researched and developed in many different applications.

Therefore, based on the inherited results of the research team, from the source of coconut oil material will be made saponification and leading the main detergent in the formula. Besides, focus on completing the procedure of shampoo products, researching on adding active additives to the product. Conduct analysis, evaluate the activity of the products (bonded properties; physical properties; sensory properties when used), and assess the durability of the products (tracking the product parameters when Stored in normal conditions and condition of accelerated aging) Therefore, the implementation of research topics related to coconut oil will expand the understanding and help to exploit better the uses and medicinal properties of local coconut oil.

2. Material and methods

2.1. Raw materials and chemicals

Virgin Coconut Oil is produced from an all-natural raw material and an international-standard production process, by VietCoCo products of Luong Quoi Coconut Processing Co., Ltd, Chau Thanh District, Ben Tre Province (Vietnam). For the primary raw material, coconut oil is tested, identified, and evaluated the oil-specific parameters such as the saponification index, iodine index, peroxide index, and so on.

The chemicals used: Detergents, foaming agents, humectants, thickeners and Sodium Hydroxide (NaOH), etc. are purchased at Nguyen Ba Trading Production Co., Ltd, Tan Binh District, Ho Chi Minh City.

2.2. Saponification process

The saponification reaction is the reaction between oil (triglycerides) and alkali (KOH, NaOH) to form glycerol and salts of fatty acids (soaps). The saponification reaction plays an essential role in the cleansing effect of the product because the properties of the raw soap are affected by the reaction process. Based on the previous research of the authors, the optimal conditions for conducting the saponification reaction to create liquid soap are by examining parameters such as temperature 80ºC, the reaction time of 3 hours, oil ratio coconut/alkali and proper alkali concentration of 11%.

2.3. The process of mixing of face wash

Mixing stage: Select the appropriate soap content and add additives to the system to complete the product's features. Additives may need to be added: auxiliary cleaners (Sodium Lauryl Ether Sulfate), emollients (Glycerin), foam enhancers (Cocamidopropyl Betaine), thickeners (Repoly 100).

Finishing the product: Sensory additives such as color, aroma is added to increase the aesthetics of the product, can use natural flavor and color to enhance the properties of the product. The finished product will be checked and evaluated the appearance and sensory parameters when used.

2.4. The method to assess the bonded properties of products

The state of the product: the product is contained in a transparent jar and visually observed on a white background under light D65 created by VERTILUX 200 projection chamber to assess the uniformity and transparency.

Color: visual observation, color identification, and uniformity for preliminary evaluation. If accurate determination is required, a colorimetric method may be carried out using a colorimeter.

Odor: describes the perceived odor.

2.5. Methods of determining the physical properties of products

- Viscosity: The viscosity of liquid products can be checked by NDJ-9S viscometer
- pH value: Use the Consort pH meter (C3010)
- Foamability: This study uses the shaking test to measure foaming. The liquid is diluted 100 times and then 5 ml of solution is placed in a stoppered tube, shaking with a moderate force until the maximum amount of foam generated (foam volume does not change).

  Foamability was calculated by the formula: $\varepsilon_f = \frac{V_{foam} - V_{liquid}}{V_{foam}}$, with: ε_f: foamability; V_{foam}: foam volume after shaking; and V_{liquid}: initial liquid volume.

3. Result and discussion

The formulation of shampoo products will be based on the shower gel surveyed by the authors, with the effect of simple soap, SLES, glycerin, auxiliary cleaners, foaming agents, and thickeners. However, the surface of skin and hair is not the same, so the amount of moisturizing and softening hair will affect the sensory use. Therefore, next only the sensory survey is used on the hair and is assessed on a scale of 1 to 5.

3.1. Effect of crude soap

Simple soap is the leading detergent in the formula. However, it needs to be combined with a part of commercial detergent to increase the cleaning effect and the sensory use. The results in figure 1 show the apparent change in foam when changing the crude soap content. Based on foaming and foam resistance, samples containing 20% dry soap content reached the highest values with 0.612 and 0.569. In terms of emulsifying time, there is the longest separation time under the influence of paraffin oil (11.18 minutes) and tends to decrease when increasing the content from 20%–50%.

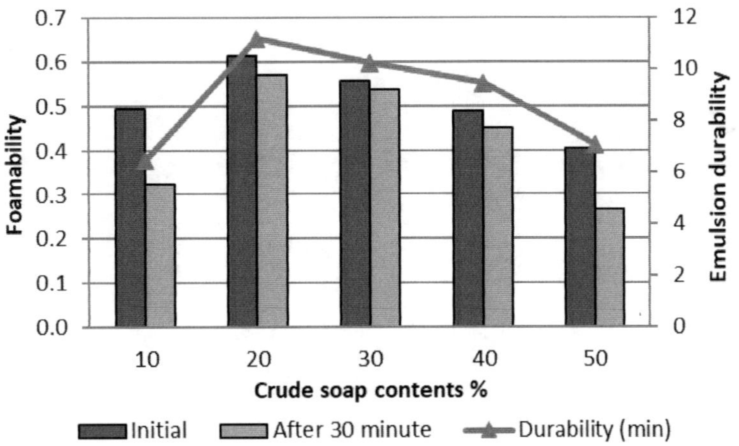

Figure 1. The effect of crude soap on foamability and emulsion durability.

3.2. Effect of Sodium Lauryl Ether Sulfate (SLES) detergent

Crude soap for large foam systems reduces sensory and make a dry skin sensation. Therefore, the need to add synthetic surfactants to create a better sense of use. The surfactant used is SLES; the effect of the agent and detergent content on the foamability and emulsion of the durability of the shampoo is shown in figure 2. The best concentration at 5% with the longest time of 12.05 minutes. Respectively, the foaming and foam resistance is 0.612, and the foam strength is 0.55.

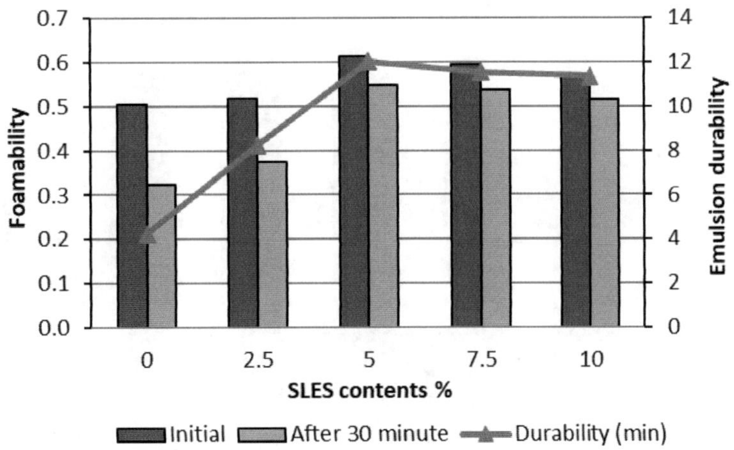

Figure 2. The effect of SLES on foamability and emulsion durability.

3.3. Effect of glycerin humectants

Because the raw soap initially contains glycerin and is not separated from the mixture. It can be predicted that the change in glycerin content affects the foam properties of the product. Therefore, glycerin content was investigated from 1 to 5%. Therefore, to ensure the ability to cleanse, and moisturize the skin after use, glycerin is selected for the 3% content to be the highest with 0.580 foamings and 0.558 foam resistance.

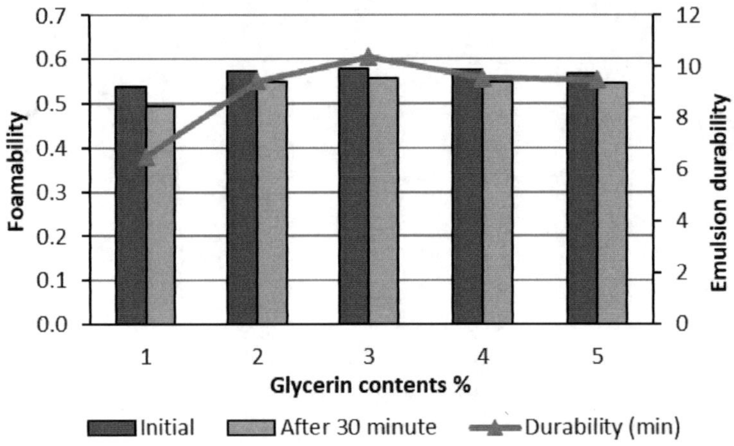

Figure 3. The effect of Glycerin on foamability and emulsion durability.

3.4. Effect of Cocamidopropyl Betaine (CAPB) foaming agent

CAPB helps to foam and stabilize the foam while thickening the product and reducing skin irritation. CAPB content was surveyed from 0 to 10%. The survey results are shown in figure 4. The highest foamability at 5% CAPB content with 0.593 foamings and 0.573 foam strength. However, at 5% content with a stable time of 11.32 minutes, there was not much difference compared to other content, so combined with the ability to foamability, the CAPB content was selected as 5%.

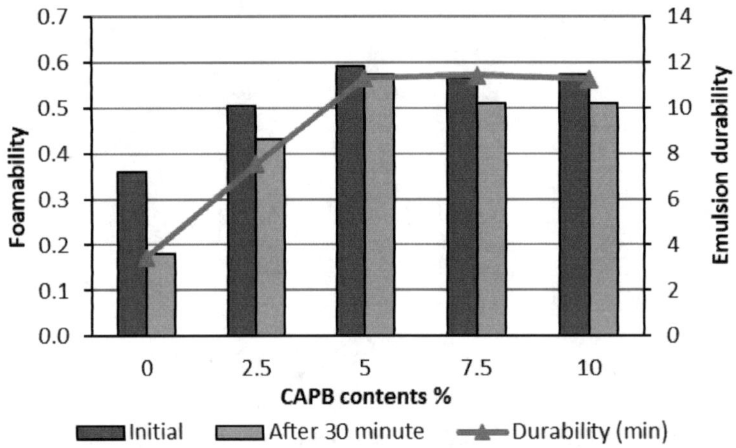

Figure 4. The effect of CAPB on foamability and emulsion durability.

3.5. Effect of P2000 UP and PEG-7 auxiliary cleaners

Adjuvant cleaners added to plantacare 2000 UP and PEG-7 were selected for the survey. The result is shown in figure 5, with a total of detergents being 5% (without the combination of PEG-7).

Specifically, when adding 2% of the plantacare 2000 substance, the product shows the best cleaning ability (foaming capacity 0.6, foam resistance 0.573). The time to maintain the durability of emulsion peaks 11.16 minutes and starts to decrease with increasing content. Therefore, plantacare 2000 UP content in the formula was selected with a content of 2%.

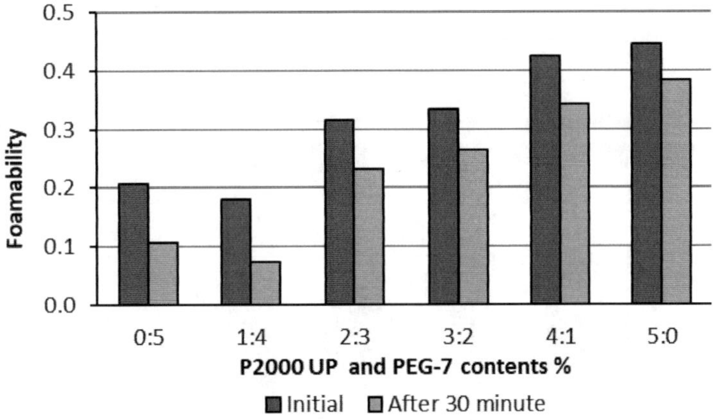

Figure 5. The effect of P2000 UP and PEG-7 on foamability and emulsion durability.

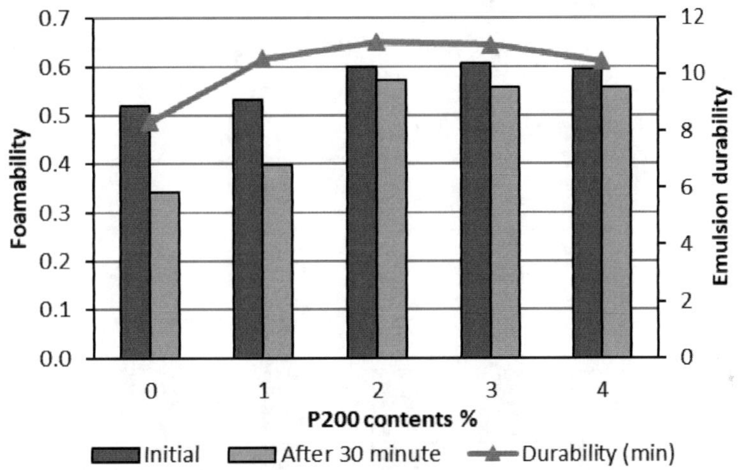

Figure 6. The effect of P2000 on foamability and emulsion durability.

3.6. *Effect of Repoly 100 thickener*

Visually, only Repoly 100 is satisfactory, but it does not reach the desired viscosity value. Therefore, to improve viscosity and ensure product appearance, the next survey will incorporate Repoly 100 and NaCl's addition.

The result of foaming of repoly 100 at 4% has the highest foaming (0.583) and foam strength (0.558), along with the durability of the emulsion at 10.25 minutes at 4% content compared to other Other patterns are shown in figure 7.

3.7. *Effect of hair smooth*

In addition to moisturizing the skin, polyquaternium-7 also has a hair-softening effect and is useful when used in combination with other emollients. Therefore, polyquaternium-7 was investigated at concentrations from 0 to 5%. The evaluation results are shown in figure 8.

With substance content from 0 to 5%, the change of evaluation point is shown in figure 8. Products, when not using polyquaternium-7, feels dry and fibrous hair. But when increasing the content to 1%, the efficiency of use changed significantly and reached the maximum rating. The similar results were shown from 2 to 4%. At 5% content, when used, it creates a greasy, unclean feeling. Therefore, selecting polyquaternium-7 1% content added to the formula to investigate the next factor.

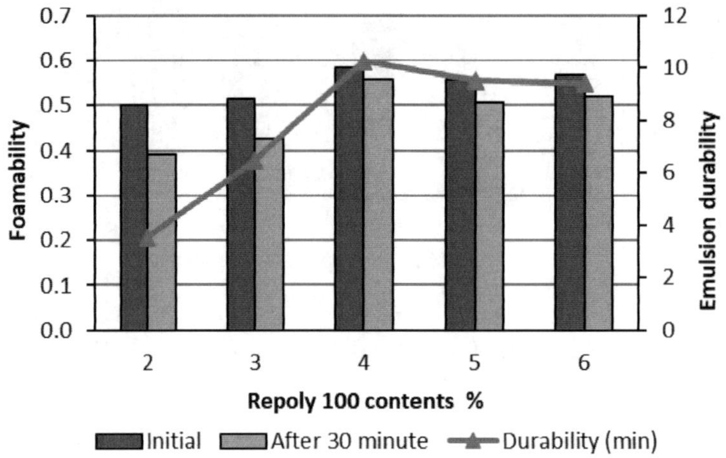

Figure 7. The effect of Repoly 100 on foamability and emulsion durability.

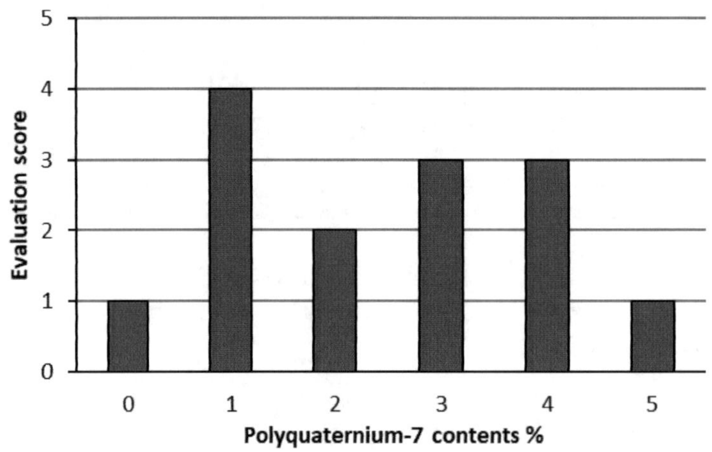

Figure 8. The effect of polyquaternium-7 content on sensory use of shampoo.

3.8. Effect of hair softener

The ingredients of serashine 301-A are the active ingredients that soften hair, and the small amount of this ingredient in the product formula also significantly changes the effectiveness of use. Therefore, the content of serashine 301-A added to the formula does not exceed 3%. The evaluation results are shown in figure 9.

From the results of figure 9, when adding silicone with the content improved from 1 to 3%, the evaluation score increased from 1 to 3. When adding 0.5% serashine 301-A, the highest evaluation score creates a sense of using the best. Therefore, it can be seen with a small amount of silicone, forming a surface layer that covers and softens hair. Therefore, serashine 301-A content is selected at 0.5% to add to the product.

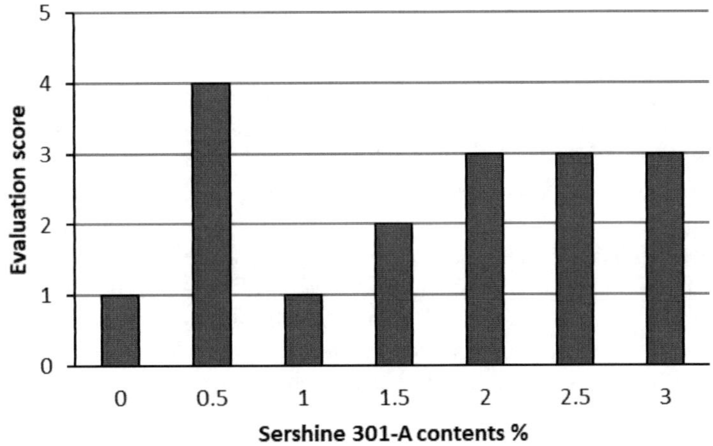

Figure 9. The effect of 301-A content on sensory use of shampoo.

Table 1. Basic specifications of shampoo products.

Parameter	Coconut oil shampoo	Market sample
Appearance	White, pearl luster product background	Light blue, opaque product background
pH	8.24	6.68
Viscosity (P)	>11	>11
Foamability	0.485	0.412
Foam durability *(after 30 min)*	0.462	0.355
Sensory	Small foam, smooth	Big foam, lots, dry hair

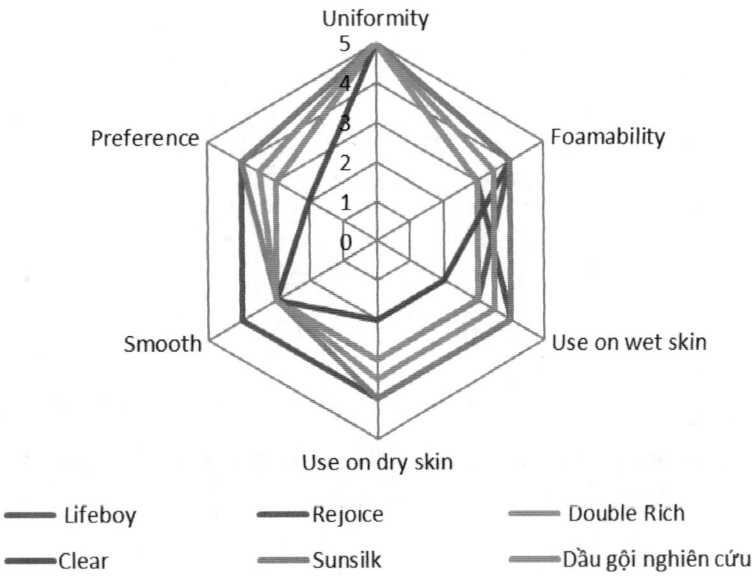

Figure 10. Evaluate senses of body wash users.

IOP Conf. Series: Materials Science and Engineering **959** (2020) 012026 doi:10.1088/1757-899X/959/1/012026

The evaluation results show that coconut oil shampoo products are not much different from the shampoo samples on the market. In terms of pH value, the main product is raw soap, so it needs to adjust the pH in the range of 8 to 10. Specifically, study shampoos have pH 8.24, and market shampoos have pH 6.68, both products meet the pH value requirements. Similarly, with a product viscosity > 11 P, the foaming properties and foam strength meet the original standards. In terms of foamability and foam durability in both products, there is no significant difference.

The result in figure 10 shows the higher the product rating, the further away it is from the center. Therefore, shampoo products in clear and opaque formulations have shown quite satisfactory results.

The rating in figure 10 has a clear difference between the surveyed products, Sunsilk and Rejoice shampoos have a relatively high rating of about 4/5. Lifebuoy, Double Rich and Clear shampoos scored lower, especially Clear, which made the hair feel dry and fibrous, so it had the lowest rating (3 points). With an average score higher than some market products after performing user reviews, the average score of 3.7 researched shampoo products is competitive and penetrates the cleaning products market.

4. Conclusion

From the survey of market products and the introduction of target parameters, the research product aims to have a dry matter content of about 20–35%, a pH value of 8–10, meeting the criteria for ability detergency as well as durability and have an average sensory evaluation score above 3.67. Products after the survey used 20% simple soap combined with other detergents such as Sodium Lauryl Ether Sulfate, Plantacare 2000, besides Cocamidopropyl Betaine, moisturizers, soft skin, and some other substances. To form the complete recipe. After that, the product is checked for appearance and basic parameters products meet the requirements. Although parameters of dry matter content, pH, foam volume, etc. have changed but not significantly, they still met the initial target requirements.

Acknowledgements

This study was supported by Ben Tre Department of Science and Technology, Ben Tre Province, Vietnam under the grant number 1136/HĐ-SKHCN.

References

[1] Punyoyai C, Sirilun S, Chantawannakul P and Chaiyana W 2018 *Cosmetics* **5** 43
[2] Halith S M, Abirami A, Jayaprakash S, Karthikeyini C, Pillai K K and Firthouse P U M 2009 *Der Pharmacia Lettre* **1** 68
[3] Sharma R M, Shah K and Patel J 2011 *Int. J. Pharm. Pharm. Sci.* **4** 402
[4] Singh S, Phurailatpam A and Senjam P 2014 *Afr. J. Trad. Compl. Alt. Med.* **11** 135
[5] Vijayalakshmi A, Sangeetha S and Ranjith N 2018 *Asian J Pharm Clin Res* **11** 121
[6] Agero A L C and Verallo-Rowell V M 2004 *Dermatitis* **15** 109
[7] Aburjai T and Natsheh F M 2003 *Phytother. Res.* **17** 987
[8] Nhan N P T, Hien T T, Nhan L T H, Anh P N Q, Huy L T, Nguyen T C T, Nguyen D T and Bach L G 2018 *Solid state Phenom.* **279** 235
[9] DebMandal M and Mandal S 2011 *Asian Pac. J. Trop. Med.* **4** 241
[10] Dayrit F M 2015 *J. Am. Oil Chem. Soc.* **92** 1
[11] Intahphuak S, Khonsung P and Panthong A 2010 *Pharm. Biol.* **48** 151
[12] Life B 2013 *Agro Food Ind. Hi. Tech.* **24** 7
[13] Shilling M, Matt L, Rubin E, Visitacion M P, Haller N A, Grey S F and Woolverton C J 2013 *J. Med. Food* **16** 1079
[14] Ahmad Z., Sarmidi M.R., and Hasham R. 2017 *Chem. Eng. Trans.* **56** 1657
[15] Aggarwal B, Lamba H S, Sharma P, and Ajeet 2017 *Am. J. Pharmacol. Sci.* **5** 25
[16] Rele A S and Mohile R B 2003 *J. Cosmet. Sci.* **54** 175

IOP Conf. Series: Materials Science and Engineering **959** (2020) 012027 doi:10.1088/1757-899X/959/1/012027

Study on the formula to produce shampoo derived from Lemongrass (*Cymbopogon citratus* (D.C.) Stapf.) essential oil

T T Hien[1,2,3], P T Nhut[3,4], L G Bach[3,4], N P T Nhan[3,4] and L T H Nhan[1,2,*]

[1] Department of Chemical Engineering, Ho Chi Minh City University of Technology (HCMUT), 268 Ly Thuong Kiet Street, District 10, Ho Chi Minh City 700000, Vietnam
[2] Vietnam National University Ho Chi Minh City, Linh Trung Ward, Thu Duc District, Ho Chi Minh City 700000, Vietnam
[3] NTT Hi-Tech Institute, Nguyen Tat Thanh University, Ho Chi Minh City 700000, Vietnam
[4] Center of Excellence for Biochemistry and Natural Products, Nguyen Tat Thanh University, Ho Chi Minh City 700000, Vietnam

* E-mail: lthnhan@hcmut.edu.vn

Abstract. Facing the proliferation of unnatural fragrance personal care products by their convenience and affordability. However, facing the immediate benefits is the insecurity of prolonged exposure to odors, the vast majority of which may exceed the allowable or unknown uses. Therefore, we research the production process of shampoos derived from natural essential oils. In this study, we use lemongrass oil originating from Tan Phu Dong commune, Tien Giang province to conduct steam distillation with optimum efficiency of 0.29%. The general formula for making a shampoo product is based on basic factors such as detergent, humectant, foaming agent, etc. The product evaluation is based on the results of the assessment of foaming ability, degree foam stability, and product stability under different storage conditions. Results showed that shampoo products derived from lemongrass oil gave good evaluation results, scalable, and developed to conduct commercialization.

1. Introduction

Currently, the demand for personal care in men and women is increasing. At the same time, people are also increasingly understanding the close connection between health and beauty. Therefore, the gentle and friendly ingredients is becoming the keyword that is more concerned than ever. Skin-care products are from natural sources, so they get more and more attention from the beauty community, especially valuable and rare ingredients [1]. In the past, beauty technologies were favored with cosmetics due to the rapid effects [2]. Recently, the trend of beauty with natural extracts is the most preferred. Chemical compounds have been gradually replaced by pure and natural materials such as Essential oils of herbs, fruits, pearls, seaweeds, honey, and so on. Natural extracts without chemical treatment do not contain artificial ingredients, colorings, alcohol should ensure safety for users [3].

Natural oils are a complex mixture of many precious and unique compounds [4]. Essential oils are extracted from several plant organs including flowers, leaves, branches, seeds, fruit, roots, wood, or bark are valuable natural products [5]. Besides, concerns about the production of natural, high-value, chemical-free products and residues of solvents or additives, for these reasons, are used as essential oils. Ingredients for many products including food flavors, additives, and flavorings used in the tobacco industry and a mixture of cosmetics and perfumes [6]. They are also used in air cooling and deodorization and in all medical disciplines such as pharmaceuticals, balneology, massage, and homeopathy. In addition to being safe, some naturally derived scents also have beneficial biological effects, such as essential oils extracted from natural herbs. The most popular are essential oils extracted from lemongrass, lemon, tea tree, wind Melaleuca, eucalyptus. Citral has the antibacterial and deodorizing effect [7–14].

Lemongrass essential oil is an essential oil extracted from lemongrass components such as lemongrass leaves, lemongrass root. The scientific name is Cymbopogon citratus. Currently, there are about 55 species [5]. The main lemongrass product is the essential oil accumulated in the leaves, with the oil content varying from 0.4 to 2.0% depending on the variety, ecological conditions and care [14–16]. Lemongrass essential oil has many attractive aromatic compounds that can relieve pain due to minor pain and fight cancer (myrcene and limonene) and some compounds have strong deodorizing, antibacterial properties (citral and geraniol). Lemongrass has been applied in many types of food, medicine, and industry [17–20]. Lemongrass essential oil is a pale yellow, strong smell of lemon odor, usually extracted by distillation directly with an aqueous solvent. The main components of lemongrass essential oils are citral, geraniol, citronellol, nerol, limonene, geranyl, and acetate, linalool, lemongrass have high index [21–25]. We endeavored to formulate a herbal shampoo from Lemongrass essential oil. The prepared formulas were evaluated for foam volume and stability in comparison to a commercial brand.

2. Materials and methods

2.1. Raw materials and mhemicals

Lemongrass essential oil used in this study was extracted by steam distillation. The material used to remove the essential oil is lemongrass leaves, which originated from Tan Phu Dong district, Tien Giang province, Vietnam. Lemongrass oil is removed with a capacity of 300 kg to 750 kg of raw material per batch for 3 hours, with a distillation efficiency of 0.29%. Next, the mixture of essential oils and water is filtered to obliterate water. Finally, lemongrass oil is dried on anhydrous sodium sulfate and stored in a sealed glass container. Chemicals used: Sodium Lauryl Ether Sulphate (SLES), EDTA, PEG-7 Glycerin Cocoate (PEG-7), Coco Amido Propyl Betaine (CAPB), Coco Diethanolamide (CDE), Carbopol thickener, Glycerin, Polyquaternium-10, D-panthenol are purchased at Nguyen Ba Trading Production Co., Ltd, Tan Binh District, Ho Chi Minh city.

2.2. The process of mixing shampoo from lemongrass oil

The process is as follows: Main detergent, copper detergent, hair conditioner, preservative, and a part of water are dissolved together, stirring and heating. At the same time, the viscosity enhancer is dispersed in water, lightly heated to swell, and mixed with the mixture. Continue to increase foaming, opaque substance, lemongrass oil, emulsifier, antioxidants emulsified with water, and then slowly turn into the mix. Add electrolyte and dissolved pH adjustment solution to the system. Finally, conduct cooling to obtain shampoo products from lemongrass oil.

2.3 The evaluation and test methods

Foamability: The sense of use is expressed by the ability to foam the product. This study uses the shaking test to measure foaming. The liquid is diluted 100 times, and then 2 ml of solution is put into a stoppered tube, shaking with a moderate force until the amount of foam generated is maximum (foam volume does not change).

The formula calculates foamability: $\varepsilon_f = \frac{V_{foam}-V_{liquid}}{V_{foam}}$, where ε_f: foaming level; V_{foam}: foam volume after shaking; and V_{liquid}: original volume of liquid.

The durability of the emulsion: The cleaning effect is expressed through the time of emulsification, that is, the emulsifying ability of the product with selected paraffin oil (simulated for dirt). A volume of 2 ml of diluent is added to 2 g of paraffin oil, then shake to produce an emulsion. The use of a stopwatch is used to determine the lifetime of the system when a 1 ml volume of oil is separated.

A number of several influencing factors during storage and transportation of products are also simulated for evaluation. During the storage process, it is inevitable that the denaturation of the product is under the influence of environmental agents, so it is necessary to find the best solution to preserve the product. Products are packed in sealed bottles and stored in different conditions: room temperature, temperature 45°C, and conditions of thermal shock. Carry out observation of the following characteristics: state, color, smell.

3. Result and discussion

3.1. The effect of detergent content on the foaming ability and foaming strength

Based on the reference formula and the goal of creating products with appropriate foam strength and durability, we surveyed selected main active ingredients: SLES 70 and EDTA replaced with SLS and lauric acid. Based on figure 1a, we found that when SLES was the main cleaning agent, the foam was higher (0.4505) than SLS (0.3590) and lauric acid (0.3421). The difference in foam strength is also evident with the most effective agent, SLES (25.3 minutes). The results were similar when the EDTA agent used. Besides, the comparison of experimental samples with market products (TT sample) also shows the foaming ability and foam durability more effectively of the model. It is understood that, for the same type of surfactant, the lower the critical micelle concentration (CMC), the higher the foaming capacity, the longer the foam lasts. Bisedes, compared to the effectiveness of using SLES, it makes the product feel smoother and less irritating to the skin and eyes, creating a soothing and parallel feeling, that is also the cheaper active ingredient compared to SLS lauric acid. The combination of SLES with EDTA and the increase in cleaning ability, also helps stabilize emulsions and isolate metal ions so that they do not affect other compounds, thereby contributing to increase durability for the product. We selected the main detergents in the product, SLES, and EDTA, for subsequent experiments from the above results. The experiment further assessed the foaming and foam strength at different concentrations of active substances SLES and EDTA. Continue based on the results in figure 1; in general, we see that the foaming rate and foam resistance time are different when changing the content using SLES and EDTA. For SLES, foaming reached the highest value at 18% (0.4565) compared to the lowest of 12% (0.3590), and durability reached 24.95 minutes. Similarly, 0.15% of EDTA content had the best foam level of 0.4624 (24.85 minutes), while at 0.05% it was only 0.1935 (17.36 minutes). From the above results, SLES, EDTA with 18% and 0.15% contents, respectively, were used for the next experiments.

3.2. Evaluate the influence and content of copper detergent on foaming ability and foam strength

Evaluation of the foaming ability and foam strength of surfactants is carried out through a comparison test of the active ingredients Minocol and PEG-7. In terms of foaming, PEG-7 showed significantly higher results (0.4363) than Minocol (0.1667) (figure 2(a)). Besides, the foam durability time of PEG-7 is also superior to (24.93 minutes) minocol (15.13 minutes) and products on the market (TT sample) (22.05 minutes). In terms of properties, PEG-7 is neutral, non-volatile, and well compatible with cationic surfactants (SLES), thereby increasing and stabilizing the foaming level of the model. PEG-7 is an easy-to-use, hypoallergenic, smooth-feeling but non-greasy feeling, suitable for general personal care products. From the above survey, select PEG-7 to conduct an appropriate content evaluation to use the base formula. The survey results showed that the content of 1.2%, the foaming reached the best value (0.4589) compared to 1.8% (0.3243). Foam retention time peaks at 23.18 minutes and begins to decrease with increasing content. From here, PEG-7 with 1.2% content was selected to conduct the next experiments.

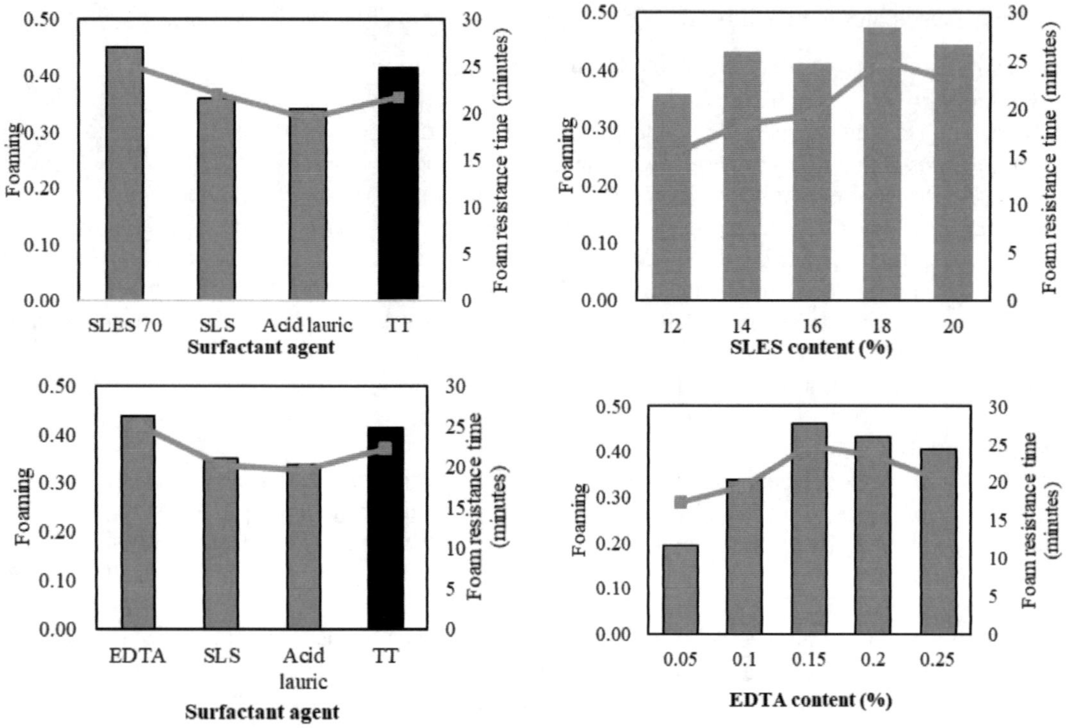

Figure 1. The effect of surface-active agents on the foam generation and foam resistance time.

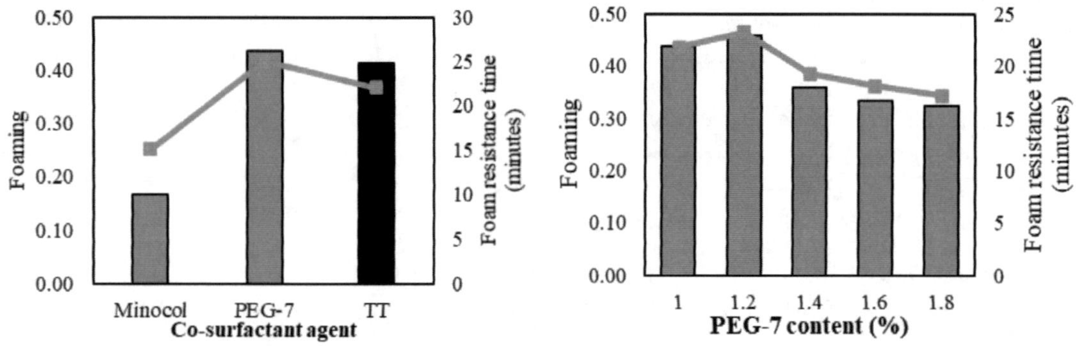

Figure 2. Influence of surface-active co-agents, the content of use on foaming and foam durability time (a) PEG-7 and other surfactants; (b) PEG-7 content.

3.3. The assesment effect of foaming agents concentration on the foaming ability and foam stability

Experimental results are shown in figure 3, showing that using the CAPB and CDE mixture gives the best foaming (0.4559) compared to using only CAPB (0.3289) and CDE (0.3865). The foaming effect of CAPB and CDE blends is also higher than that of market products (TT sample) (0.4069). Considering the foam durability time, the CAPB and CDE mixture proved its ability to maintain foam stability for 24.95 minutes, higher than the product's value on the market. Based on the above results, the experiments to survey the content of CAPB and CDE use were conducted at different values. For CAPB (figure 3(b)), the foaming level peaked when the active ingredient content was 7% (0.4635), in parallel, the CDE agent also showed the best foaming effect at 2%. (0.4540) (figure 3(c)). Corresponding results were recorded at foam resistance time. In fact, CAPB is a commonly used raw material because of its foaming properties and its ability to reduce surface tension, combined with CDE, which helps stabilize foam, increase emulsification and sensitization. Pleasing sensation to skin and hair. Thereby increasing the sensory value as well as the effective use of the product. Thus, the use of CAPB and CDE with 7%, 2% content is suitable for the next survey.

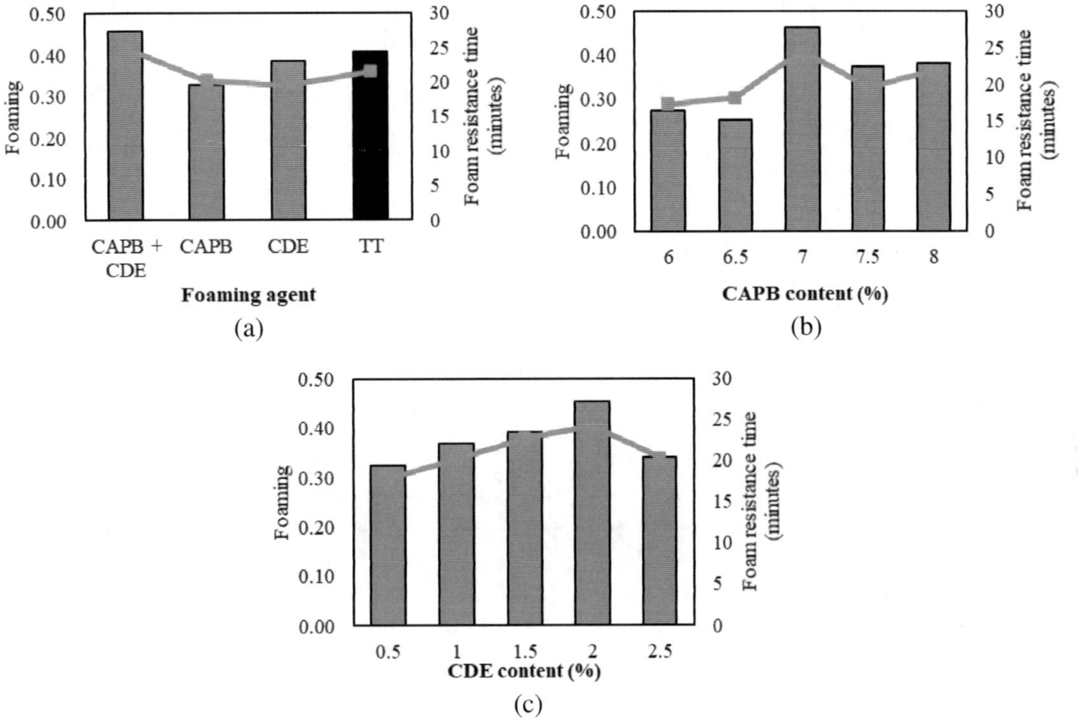

Figure 3. Influence of foaming agents, use content to foaming and foam strength of: (a) Mixtures of CAPB, CDE and other surfactant co-agents; (b) CAPB content; (c) CDE content.

3.4. The effect of foaming agents' concentration on the foaming ability and foam resistance time

The results of the solid agent survey are presented in figure 4(a). Easily found, the foaming and foam durability time when using carbopol is higher (0.4624 / 24.38 minutes) than CMC (0.3243 / 14.23 minutes), HEC (0.4013 / 18.66 minutes) and Xanthangum (0.2857 / 16.36 minutes) respectively. Besides, the foaming effect when using carbopol as the main thickening agent also gives better results than products on the market (TT sample). Considering the actual production, the active ingredients CMC, HEC, and xanthangum tend to form blocks in high humidity. Hence, they are easy to clump when working with water, resulting in a lot of time preparing of raw materials. Therefore, we choose carbopol to conduct a detailed survey of the content used in the background formula. The result shows that the foam content of carbopol is in amount of 22% (0.4553), which is the largest compared to 20% (0.2857) (figure 4(b)). The results of the foam strength evaluation showed similar results. With the ability to thicken the surface-active system, enhance viscosity, increase the components' cohesion, prevent silicon deposition, thereby limiting layer separation and retaining durability for the product, Carbopol was selected as the key solidity agent for the next experiments.

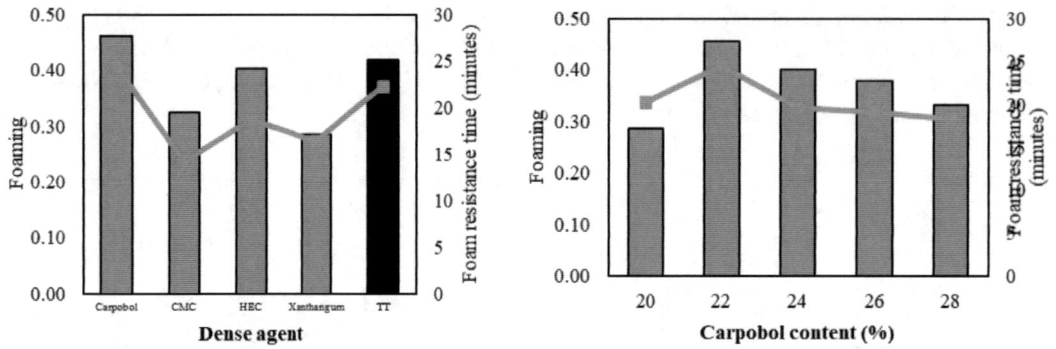

(a) (b)

Figure 4. The effect of the thickening agent concentration on the foaming and foaming strength: (a) Carbopol and other surfactants; (b) Carbopol content.

3.5. The effect of humectants on the foaming ability and foam strength

The ability and durability of foam in different humidifying agents are assessed and shown in figure 5(a). Based on the figure, we see that glycerin has a higher level of foaming (0.4635) than sodium lactate (0.3954), dipropylene glycol (0.3498), and products on the market (TT sample). In parallel, the longest foam resistance when using glycerin (24.21 minutes) and the lowest for dipropylene glycol (16.31 minutes). When evaluating the content of use, 1% glycerin gives the best foaming and foam resistance (0.4898 / 24.49 minutes), while at higher values tends to reduce the foam and their stability. In terms of cost, sodium lactate and dipropylene glycol are less economical than glycerin. On the other hand, glycerin is a benign active substance suitable for all scalp types. It has high stability and good foaming ability, so the use of this agent in the product is appropriate for the original purpose of the research.

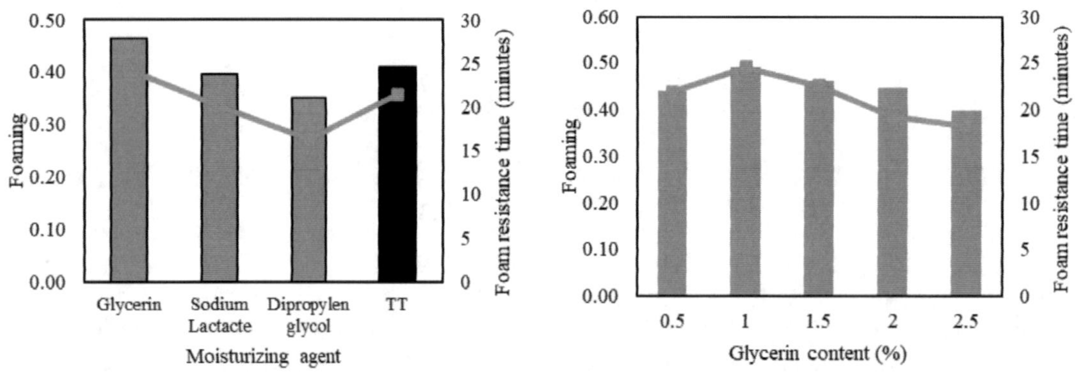

(a) (b)

Figure 5. Effect of moisturizing agents, the content used on the foaming and foam strength of: (a) Glycerin and other surfactants; (b) Glycerin content.

3.6 The effect of smoothing concentration on the foaming ability and foam strength

Based on the survey results presented in figures 6a, it is found that the smoothing agents have a significant effect on foaming and foam strength, most notably the two PEG- 10 (0.4253 / 20.08 minutes) and D-panthenol (0.3976 / 21.50 minutes). However, the combination of all three active ingredients PEG-10, D-panthenol, and 3140E gave better results (0.4505 / 24.32 minutes) compared to a single active ingredient and compared to the product market (TT sample). Conduct experiments to survey the content of active ingredients used each active ingredient's formula, obtaining the highest foaming and foam stability results at 0.35% (figure 6(b)), 0.2%. (figure 6(c)), 2% (figure 6(d)) correspond to PEG-10, D-panthenol and 3140E. Selecting all three active ingredients based on their specific properties,

PEG-10 is compatible with protein-based surfaces (skin, hair, etc.), reducing static electricity, thereby improving the texture of fibers. Hair is damaged by physical or chemical causes. Besides, 3140E and D-panthenol are added to protect, moisturize, and increase smoothness, helping to overcome inflammation and irritation, helping to promote the proliferation of fibroblasts in the class. The epidermis, where damage to the skin and hair forms. Therefore, with the smoothing agent group, we chose the combination of PEG-10, D-panthenol, and 3140E for the next experiment.

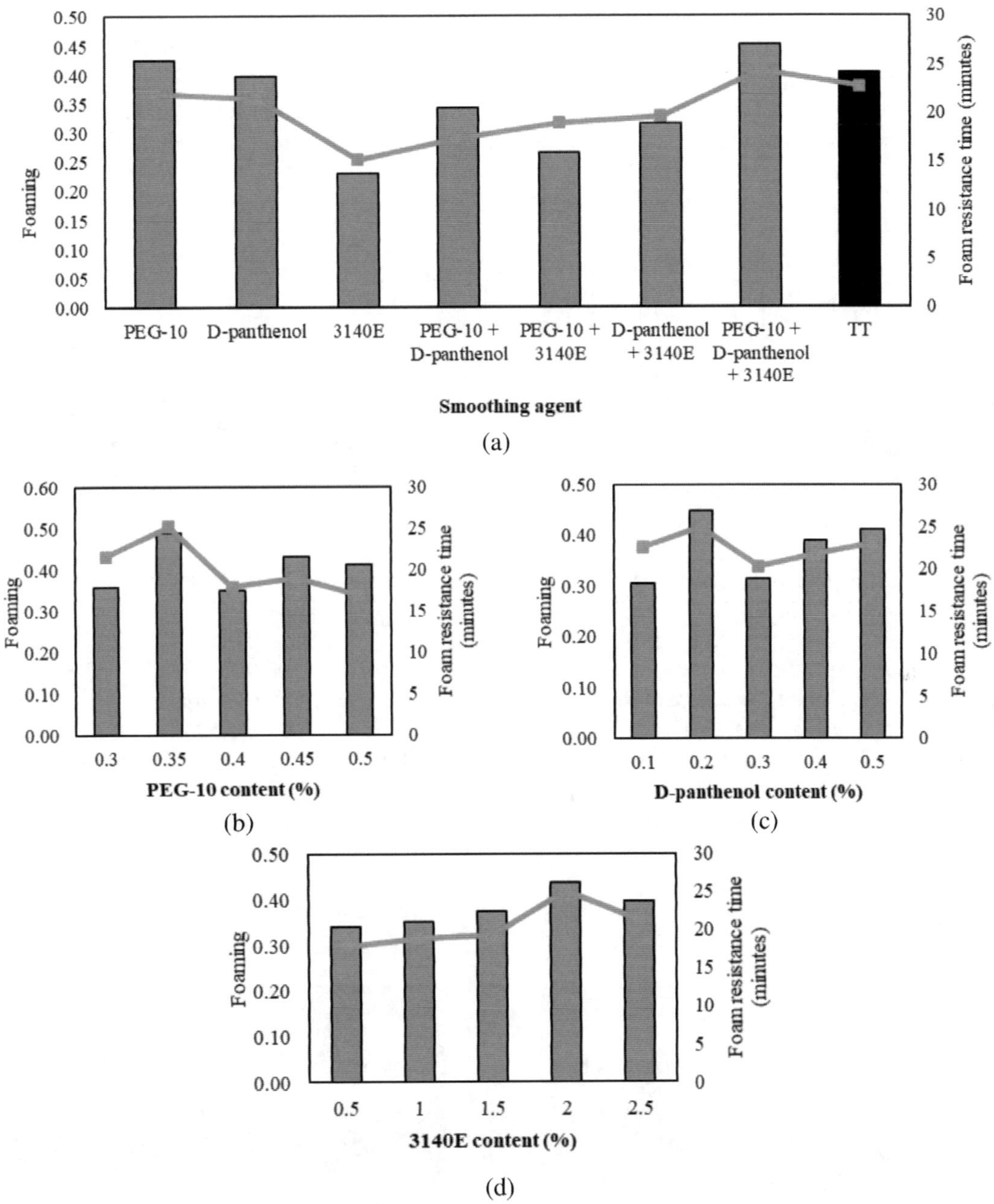

Figure 6. Effects of smoothing agents and levels on foaming and foam strength: (a) Smoothing agents; (b) PEG-10; (c) D-panthenol; (d) 3140E.

IOP Conf. Series: Materials Science and Engineering **959** (2020) 012027 doi:10.1088/1757-899X/959/1/012027

3.7 The effect of antioxidants and Content on the foaming ability and foam strength

The effects of antioxidants are shown in figures 7. Accordingly, we surveyed in two different conditions (nano and non-nanochemical) using two preservatives, BHT, BHA, and a mixture of BHT and BHA. The results showed that all 8 studied samples achieved better foaming and durability than market samples (TT sample). It is easy to see that, for 4 non-nano parrticles, the sample with BHT is the main antioxidant which has little change compared to the remaining samples, the same perception was also seen in 4 nanosized samples. Therefore, we selected samples containing BHT with nanoization to continue evaluating their appropriate usage. Figure 3.7 shows the influence of antioxidants on shampoo products from lemongrass oil. As we can see, the foaming results achieved the best at concentration of 0.2% (0.4623), tended to decrease to close to the market sample gradually, the durability of the sample at the concentration of 0.2% was assessed to meet the requirements. For the purpose of commercializing shampoo products. Therefore, we choose the BHT active ingredient with 0.2% content to be nanosized for further research.

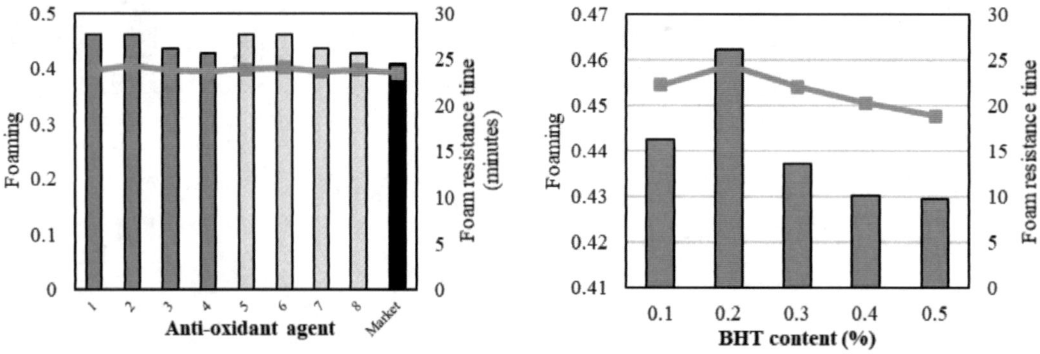

Figure 7. Effect of antioxidants and content on foaming and foam durability (1. Essential oil; 2. BHT; 3. BHT; 4. BHA + BHT; 5. Essential oils + Nanoization; 6 BHA + BHT + Nanoization; 7. BHT + Nanoization; 8. BHA + BHT + Nanoization)

3.8. The effect of Lemongrass Essential oil in the foaming ability and fpam durability

The influence of the oil content is shown in figure 8. Thereby, we surveyed 5 levels of content ranging from 0.1% to 0.5%. The results showed that all samples had better foaming and durability than market samples (TT sample). Therefore, it shows that increasing the content of lemongrass oil in the range of 0.1 -0.5 (%) does not affect the product's properties. However, when deploying a larger scale, it is necessary to assess in detail the economics of use. In terms of appearance, there is no big difference in the milestones using essential oils. Besides, when assessing the scent, at 0.4% content for a pleasant aroma, no harshness, no discomfort, and good fragrance. In a nutshell, we use lemongrass oil with a 0.4% content for the general formula of the product.

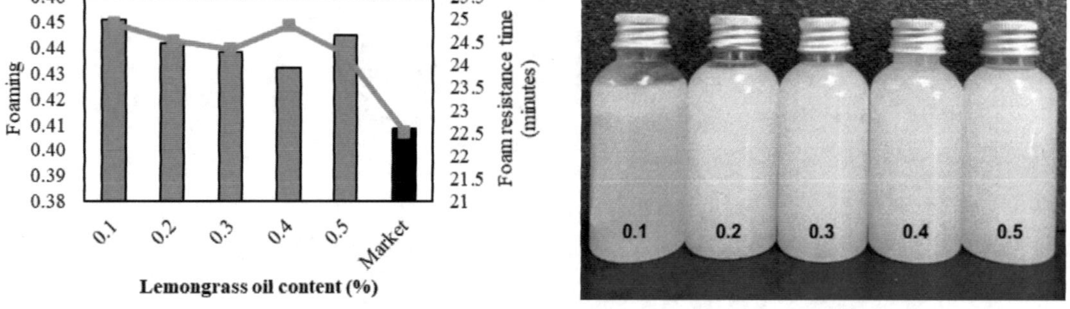

Figure 8. Influence of lemongrass oil content on product appearance.

3.9. The effect of storage conditions on the scent of lemongrass essential oil in products

Figure 9 shows the effect of changing storage conditions on the foaming properties and the foam retention time of shampoo products. The results showed that, when changing the storage conditions did not affect the foaming and foam durability time of the product, the survey samples did not have a big difference compared with the market sample (TT sample). Assessing the appearance, the accelerated sample has a slight yellowing, but does not affect too much the product's appearance. This can be handled thoroughly after the product is colored.

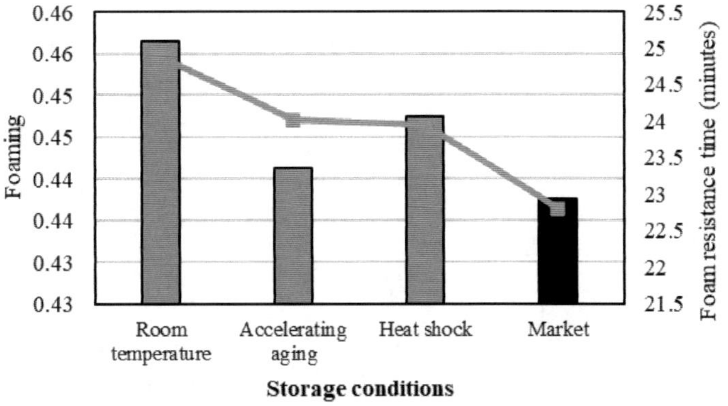

Figure 9. The effect of storage conditions on foam generation and foam strength.

4. Conclusion

Factors affecting the production process are found and optimized for the process including: detergents (EDTA, 70% SLES), co-cleaning agents (PEG-7 glyceryl cocoate), humectants (Glycerin), foaming agents (CAPB, CDE), thickeners (Polygel 1.4%), smoothing agents (Polyquaterium-10, D-panthenol, 3140E), antioxidants (BHT), additives (citric acid 30%) , plantacare 1200, NaOH, NaCl), lemongrass oil. From the results of the experiment, it has been shown that the shampo derived from lemongrass oil has high cleaning efficiency, there is no significant difference compared to existing products on the market. From this, the results can be used to conduct in-depth research and apply products to commercialization in practice.

Acknowledgements

This study was supported by Tien Giang Department of Science and Technology, Tien Giang province, Vietnam, under grant number 103/HĐ-QPTKH&CN.

References

[1] Ahmad M, Khan M A and Zafar M 2008 *Indian J. Tradit. Knowl.* **7** 421
[2] Santucci B, Cristaudo A, Cannistraci C and Picardo M 1987 *Contact Derm.* **16** 93
[3] Bester K 2009 *J. Chromatogr. A* **1216** 470
[4] Adlard E R 2010 *Chroma* **72** 1021
[5] Tajidin, N. E 2012 *Afr. J. Biotechnol.* **11**
[6] Gurib-Fakim A and Demarne F 1995 *J. Essent. Oil Res.* **7** 105
[7] Lee S-H and Kim J-H 2019 *Process Biochem.* **76** 187
[8] Kim T W and Kim J-H 2016 *Korean J. Chem. Eng.* **33** 3175
[9] Milojević S Ž, Stojanović T D, Palić R, Lazić M L and Veljković V B 2008 *Biochem. Eng. J.* **39** 547
[10] Stanojevic L, Stankovic M, Cakic M, Nikolic V, Nikolic L, Ilic D and Radulovic N 2011 *Hem. Ind.* **65** 455

[11] Lota M-L, de Rocca Serra D, Tomi F, Jacquemond C and Casanova J 2002 *J. Agric. Food Chem.* **50** 796
[12] Ganjewala D 2009 *Int. J. Essemt. Oil. Ther.* **3** 56
[13] Pino J A and Rosado A 2000 *J. Essent. Oil Res.* **12** 301
[14] Koul V K, Gandotra B M, Koul S, Ghosh S, Tikoo C L and Gupta A K 2004 *Indian J. Chem. Technol.* **11** 135
[15] Singh A and Kumar A 2017 *Int. J. Innov. Res. Sci. Eng. Technol.* **6** 13989
[16] Shakir I K and Salih S J 2015 *Iraqi J. Chem. Pet. Eng.* **16** 11
[17] Miyazaki Y 1969 *J Agr Res Quart* **4** 37
[18] Nhan N P T, Hien T T, Nhan L T H, Anh P N Q, Huy L T, Nguyen T C T, Nguyen D T and Bach L G 2018 *Solid State Phenom.* **279** 235
[19] Kim Ngan T T, Thu Thuy D T, Tuyen T T, Inh C T, Bich H T, Long P Q, Chien N Q, Kieu Linh H T, Yen Trung L N, Tung N Q, Nguyen D C, Bach L G and Toan T Q 2019 *Asian J. Chem.* **32** 36
[20] Ngan T T K, Huong N C, Le X T, Long P Q, Toan T Q, Hoang Vo D M, Danh V T, Yen Trung L N and Trieu T A 2019 *Asian J. Chem.* **31** 2759
[21] Kim Ngan T T, Hien T T, Le X T, Anh T T, Quan P M, Cang M H, Le Ngoc T T, Danh V T, Yen Trung L N and Toan T Q 2019 *Asian J. Chem.* **31** 2855
[22] Tran T H, Nguyen P T N, Pham T N, Nguyen D C, Dao T P, Nguyen T D, Nguyen D H, Vo D V N, Le X T, Le N T H and Bach L G 2019 *IOP Conf. Ser. Mater. Sci. Eng.* **479** 012002
[23] Tran T H, Nguyen P T N, Pham T N, Nguyen D C, Dao T P, Nguyen T D, Nguyen D H, Vo D V N, Le X T, Le N T H and Bach L G 2019 *Processes* **7** 56
[24] Mai H C, Le T T T, Diep T T, Le T H N, Nguyen D T and Bach L G 2018 *Asian J. Chem.* **30** 293
[25] Mai H, Nguyen T, Le T, Nguyen D and Bach L 2019 *Processes* **7** 90

IOP Conf. Series: Materials Science and Engineering **959** (2020) 012028 doi:10.1088/1757-899X/959/1/012028

The effect of chemical refining process on the physicochemical and fatty acid composition of water melon seed oil (*Citrulus lanatus* L.)

H Soetjipto[1], C P Septiyarini[1] and S Hartini[1]
[1] Program Studi Kimia, Fakultas Sains dan Matematika,
Universitas Kristen Satya Wacana, Salatiga

E-mail: hartati.sucipto@uksw.edu

Abstract. A lot of watermelon seed waste is wasted, generally its only used to make "kuaci", a kind of salted saline a snack made from watermelon seeds. As a member of *Cucurbitaceae* family, watermelon seeds can be used as a source of vegetable oil that predicted contains squalene inside like other *Cucurbitaceae* members. This study aims to determine the effect of refining process on the physicochemical properties and composition watermelon seed oil. Watermelon seed oil which is obtained by using Soxhlet extractor in n-hexane solvent was refined in 3 steps, which was degumming, neutralization, and bleaching, respectively. Degumming process used 3 concentrations of H_3PO_4, which were 0,15%; 0,20%; and 0,25%, respectively; whereas 3 zeolit concentrations of 2%; 4%; and 8%, respectively were used in bleaching steps. Data were analyzed by using 3x3 Factorial design and it was laid out with Randomized Completely Block Design, 3 replications and as the block is time analysis. The optimum yield of the watermelon (*C. lanatus* L.) seed oil is obtained in the amount of 48.47±0.91 % at 0.25% phosphoric acid concentration and 4% Zeolite concentration. The result of physicochemical analysis before and after the refining process have changed. The density, peroxide value and acid value show a significantly decrease, while only the saponification value shows an increase. Fatty acid composition (profile) of WSO before and after refining process does not change, comprises linoleic acid, palmitic acid, stearic acid and squalene (unsaturated hydrocarbon). Linoleic acid is the main compound of WSO (76.69%), followed by palmitic acid (14.60%), stearic acid (6.53%) and squalene (2.18%), respectively.

1. Introduction

Watermelon (*Citrulus lanatus* L.) is classified as a member of *Cucurbitaceae* family. Usually, people are only using the fruit flesh and the skin is preserved as a sweet snack while the seeds are removed. Watermelon seeds are only used to make "kuaci", a kind of dry salty snacks made from watermelon seeds. These seeds contain high protein and unsaturated fatty acids as linoleic and oleic acid [1,2]. Watermelon seeds contains plant oil in amount of 16.79 %, also contains linoleic acid 75.71%, palmitic acid 18.10%, and stearic acid 6.19% [3].

Oil extracted generally contains phospholipid, free fatty acids, pigment and sap which affects the oil quality such as aroma, shelf life, and clarity of the oil [4]. The refining process will improve the shelf life of M.oleifera oil and the chemical refining process can increase the quality of kenaf (*Hibiscus cannabinus*) seed oil [5,6]. Therefore, improving the oil quality has to be done. Refining oil aims is to avoid oil damage which usually occurs due to the hydrolysis and oxidation reactions. The presence of free fatty acid can trigger oil damage, which it affects the stability of vegetable oil [7].

The first step of the refining process is degumming, followed by neutralization, and then bleaching. In the degumming process, the main compound to be separated is the phosphatide. This compound will cause unwanted colors , flavors, and reduce the shelf life of the oil [8]. The sap or mucus, gum which contains phosphatide, protein, carbohydrate, residues, water, and resin will be separated to increase the quality of vegetable oil [9]. The neutralization step using sodium hydroxide (NaOH) is mostly done on an industrial scale because it is more efficient than other neutralizations. This step is also able to remove the color and dirt remains in the oil. And bleaching is the next step which will (want) to remove the unwanted colors remain inside the oil. Bleaching, as the last step, can be done using active carbon, zeolite, or bentonite [10].

Refining process using phosphoric acid in degumming step followed by a neutralization step affect the physicochemical properties of kachnar seed oil (*Bauhinia purpurea* L.) [11]. The refined oil obtained show the more brighter color, cleaner than the crude oil before refining and it also has a change in its fatty acid' composition. In this study has focused on the removal of impurity especially of gummy materials and waxes to improve the physicochemical and composition of Watermelon Seed Oil.

2. Materials and methods

Watermelon seed was obtained from a field in Trengguli village, Demak, Central Java Indonesia. All the solvents and chemicals used were purchased of analytical grade (E-Merck Germany, and SMART lab Indonesia). The instruments used in this work were analytical balance to the accuracy 0.0001 g (Ohaus PA124, USA), soxhlet apparatus, waterbath (Memmert WNB 14), rotary evaporator (Buchi R-114, Swiss), Gas-chromatography- Mass Spectrophotometer (GC-MS) perkin-Elmer.

2.1. Sample preparation method [2]

The seeds are washed and dried in oven at 60° C for 24 h and then it is crushed with a grinder.

2.2. Refining process of Watermelon Seed Oil (WSO)

2.2.1. Degumming process [12] *modified*

Degumming process using three concentrations of phosphoric acid (85% conc.) whichater are 0.15%, 0.20%, and 0.25%, respectively.

2.2.2. Neutralization process [13] *modified*

In this process, a concentration of 9.5% sodium hydroxide was used.

2.2.3. Bleaching process [14] *modified*

The neutralized oil added using three zeolite concentrations which are 2%, 4%, and 8%, respectively. The natural zeolite was used in this experiment, before use zeolite was activated by heating in a furnace.

2.3. Physicochemical characteristics of WSO

All of the items were determined according to the AOCS method. The yield of oil extraction obtained was determined by gravimetric method using 4 digits balance [15], density [16], water content [17], acid value (AV) [18], peroxide value (PV) [19], and saponification value (SV) [20], respectively.

2.4. Statistical analysis

Data were analyzed using a 3x3 Factorial design, and it was laid out with a Randomized Completely Block Design (RCBD), 3 replications and as a group is the time of analysis. The first factor is H_3PO_4 concentration consisted of 3 concentrations which are 0.15%; 0.20%; and 0.25%, respectively. The second factor is the zeolite concentration also consisted of 3 concentrations are: 2 %, 4%, 8%, respectively. The comparison between treatment means was carried out with honestly significant difference test with 5% of level significance [21].

3. Result and discussion

Table 1 summarizes the yield of oil obtained after refining process using phosphoric acid 0.15%, 0.20% and 0.25%. The optimal yield of WSO was obtained with the use of 0.15% as much as 33.43 ± 4.35 % because the use of phosphoric acid 0.15% and 0.25% statistically give the same results, while on the use of 0.20% phosphoric acid give lower yield.

Table 1. Yield of WSO in various concentration of 85% phosphoric acid.

	85% H_3PO_4		
Yield (%)	0.15%	0.20%	0.25%
Average ±SE[a]	33.43± 4.35	26.77±11.06	35.45±9.63
W[a] = 5.54	(b)[b]	(a)[b]	(b)[b]

[a] SE= Standard Error of Estimate; W= Honestly Significant Difference (HSD) at 5% level of significance.
[b] Numbers followed by the same letter at the same row or column indicate that the treatment means are not significantly different, whereas the numbers followed by different letter at the same row or column indicate that the different treatment means are significant different. This note is also used for table 3.

According to the previous studies, during oil refining process showed that the oil requires phosphoric acid in unequal amounts depending on the oil type, and generally its ranged between 0.15–0.25%. The addition of phosphoric acid in degumming step will change unhydratable gum to be hydratable and then remove both of them [22,23]. Gum must be removed in the degumming step, because it will inhibit the oil separation due to the soap formation in the saponification process. The gum is also able to act as an emulsifier that's why it will emulsify the oil causing the yield to be decreasing [9].

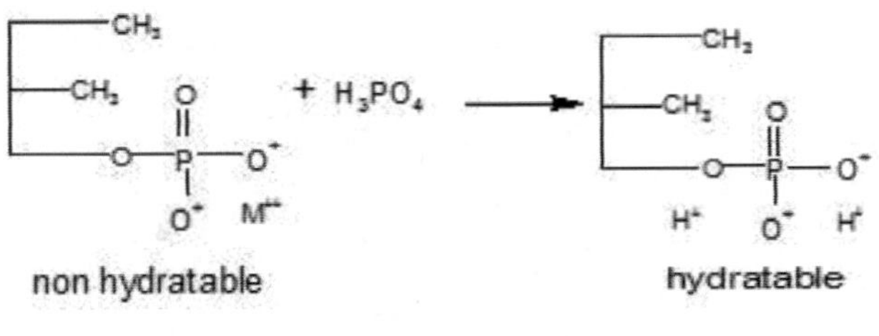

Figure 1. Degumming reaction [24].

After degumming WSO was neutralized by using NaOH until the rinse water neutral, then followed by bleaching step. The yield obtained after bleaching process isshown in table 2.

Table 2. Yield of WSO in various concentration of zeolite.

	Zeolite concentrations		
Yield (%)	2%	4%	8%
Average ±SE	34.75±8.79	31.68±6.0	29.22±10.02
W = 5.54	(a)	(a)	(a)

The optimum yield in the amount of 34.75%±8.79 was obtained in the use of 2% zeolite concentration. The more the use of zeolites up to 8% neither increases nor decreases yield. Zeolite has

3 layers structure that allows triglyceride molecules to be absorbed due to interactions with H^+ ions in the envelope contained in the zeolite structure [25]. Zeolite also has an open pores structure with a large surface area so as to allow the uptake of dye molecules. Zeolite can be used to remove oil colorant because it has a high adsorptive. The addition of 4% zeolites' concentration combined with 0.25% of phosphoric acid (85% conc.) gave the maximum WSO yield in the amount of 48.47±0.91% [26]. The maximum WSO yield obtained is suspected due to the oil is relatively clean that the soap result in the purification process does not absorb many impurities and does not reduce the yield [27].

Table 3. The average yield of WSO resulting from interaction after the refining process.

H_3PO_4 Concentration (%)	Zeolite concentration (%)		
	2	4	8
0.15% W= 9.59	34.79± 4.42 (a) (b)	21.06±11.67 (a) (a)	44.43±0.73 (b) (c)
0.20% W= 9.59	33.51±3.33 (a) (b)	25.50±0.46 (a) (ab)	21.30±8.28 (a) (a)
0.25% W= 9.59	35.96±3.86 (a) (b)	**48.47±0.91 (b) (c)**	21.30±0.99 (a) (a)
	W= 9.59	W= 9.59	W= 9.59

3.1. Physicochemical properties of WSO (C.lanatus) pre- and post- refining.

Physicochemical properties of WSO are shown in table 4. Crude oil of WSO was obtained as much 23.22%, this result is lower than Baboli (50%) [1], but is higher than the Ariani result of 16.79% [3]. This difference is possible due to several reasons, for example: a watermelon varieties, soil conditions, and the environmental conditions in which watermelons grow [28].

Table 4. Physicochemical analysis of WSO.

Parameter	WSO		Reference	
	Before	After	Ariani [3]	Baboli [1]
Yield (%)	23.22	48.47	16.79±0.82	50±0.2
H_2O concentration (%)	9.14	0.49		
Density(g/ml)	0.82	0.80		-
Peroxide value (Mequiv O_2/Kg Oil)	4.50	0.90	73.50±1.59	3.24±0.1
Saponification value (mg KOH/g Oil)	26.92	88.86	77.87±2.26	200±0.1
Acid Value (mg NaOH/g Oil)	4.13	1.55	5.05±0.00	2.4±0.1

After the refining process changes occur in the physicochemical properties of the oil, for example, a decrease in the water content of oil from 9.14% to 0.49%, peroxide value from 4.5 to 0.9. Decrease of water content occur because in the degumming process using phosphoric acid, the water contained in oil will be washed into the acid solution, so that the water content of the oil will drop. The water content in the oil is an important parameter due to the hydrolysis reaction that can reduce the quality of the oil. The decrease of peroxide value cause of the present of sodium hydroxide in neutralization process. This compound will reacts with free fatty acid and peroxide polymers [29]. Oil oxidation is strongly influenced by fatty acid constituent, heating time and storage condition [30]. The higher the peroxide value the worse the oil condition [31].

The density of the oil before and after refining does not differ greatly from 0.82 to 0.80 because the refining process removed the oil impurities so that it affects its density [32]. The saponification value of WSO increased after the refining process. The low saponification value indicate that the oil has large fatty acid molecules and vice versa [33].

The refining process also decreased acid value because when saponification process occured, free fatty acid will react with sodium hydroxide and separated to become soapstock. The higher the acid value of the oil , the worse the quality of the oil because a lot of triglyceride molecule hydrolyzed [34]. The refining process canto increase oil quality [33].

3.2. Fatty acid composition of the WSO

The fatty acid composition of the crude and refined WSO is shown in table 5, while the results of the gas chromatogram are shown in figure 2 and figure 3. Four fatty acids are detected from WSO before refining process, each of the peaks shows time retention is 24.523, 16.069, 15.799, and 13.473 minutes respectively, while after refining process WSO also demonstrated four fatty acid with retention time 24.537, 16.149, 15.914, and 13.523 minutes respectively. Every peak was analyzed using MS to determine the compound then compare with Wiley database figure 2 and figure 3.

Figure 2 is a spectrum of one compound from WSO shows a similar fragmentation with figure 3 spectrum of 9,12-Octadecadienoic acid from Wiley database, so it is believed that spectrum in figure 2 is 9,12-Octadecadienoic acid. By using the similar way all the peaks that appear in WSO spectra have been identified. The result of identification from all the peaks before and after refined are shown in table 5.

After the refining process the composition of fatty acid is similar, a little change in fatty acid composition was observed especially in the concentration of the compounds. The reason for this change is the present of alkali solution in neutralization which causes the saponification reaction.

The linoleic acid content in this study which are 77.80% is higher in comparison with Baboli finding 68.3% [1]; and a little bit higher than linoleic acid of Ariani report is in the amount of 75.71% [3]. Table 5 also shown the presence of a squalene compound in relatively small amount (2.18%) after purification. Squalene is a powerful antioxidant compound so it is widely used in cosmetics and treatment for skin health. This compound is also able to reduce the toxicity of drugs consumed and has antitumor activity [35].

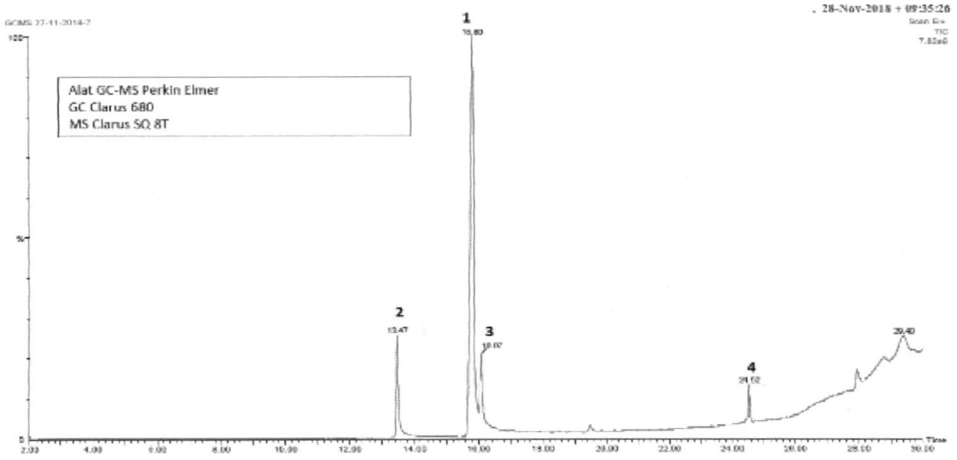

Figure 2. Chromatogram of WSO before refining process,

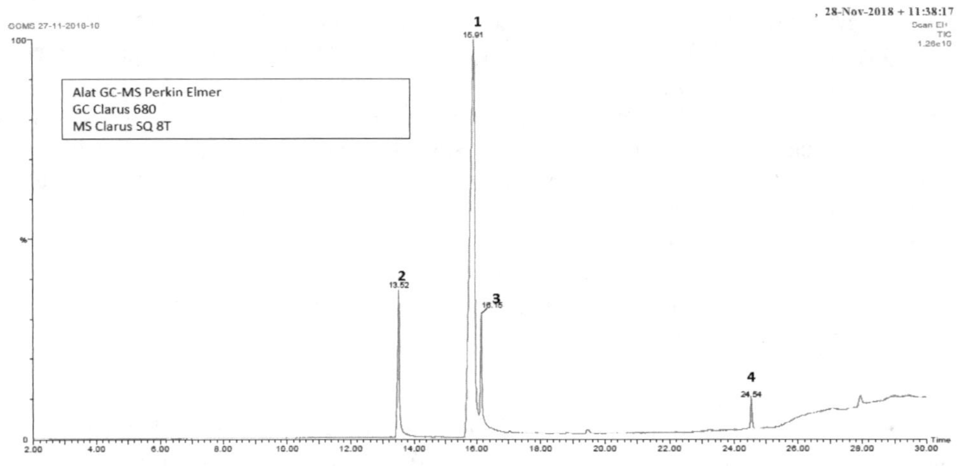

Figure 3. Chromatogram of WSO after refining process.

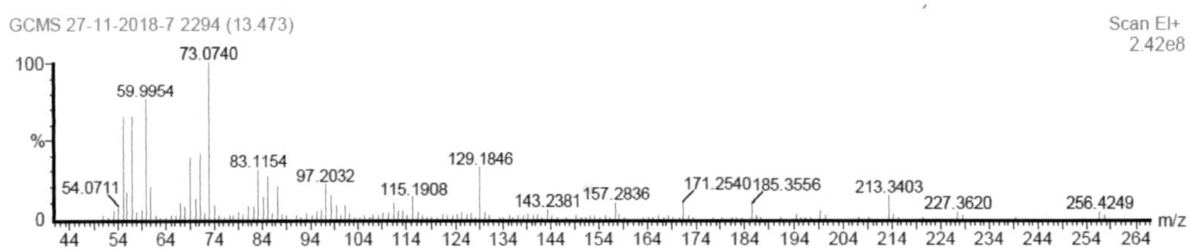

Figure 4. Spectrum of 9,12-octadecadienoic acid from WSO.

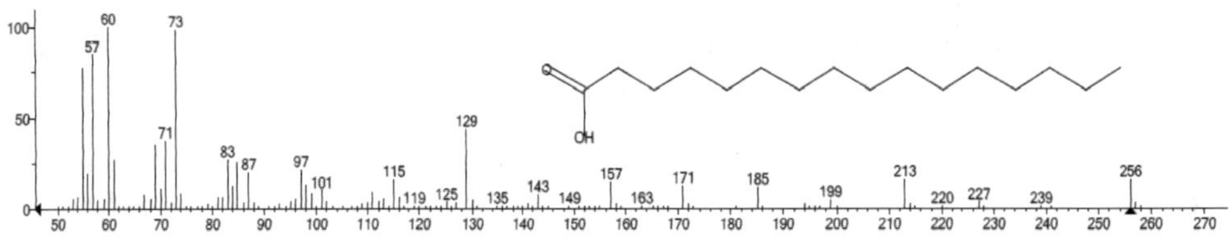

Figure 5. Spectrum of 9,12-octadecadienoic acid from WSO (Wiley database).

Table 5. Fatty acid profiles from all the peaks of WSO before and after refined.

Name of compounds	MW[a]	Molecule structure	Before		After	
			Retent. Time (min)	*Area* (%)	Retent. Time (min)	*Area* (%)
Hexadecanoic acid (Palmitic acid)	256	$C_{16}H_{32}O_2$	13.473		13.04	
9,12-Octadecadienoic acid (Z,Z)- (Linoleic acid)	280	$C_{18}H_{32}O_2$	15.799		77.80	
Octadecanoic acid (Stearic acid)	284	$C_{18}H_{36}O_2$	16.069		5.43	
Squalene	410	$C_{30}H_{50}$	24.523		3.73	

[a] MW = molecule weight.

4. Conclusion

The best results of WSO were obtained with the addition of phosphoric acid 0.25% with 4% zeolite is 48.47±0.91%. Refining process influence physicochemical properties and concentration of fatty acid but it is not influence the profile of WSO. The density, peroxide value and acid value show a decrease, while the saponification value shows an increase. Fatty acid composition (profile) of WSO before and after refining process does not change, comprises linoleic acid, palmitic acid, stearic acid, and squalene (unsaturated hydrocarbon). Linoleic acid is the main compound of WSO (76.69%), follow by palmitic acid (14.60%), stearic acid (6.53%), and squalene (2.18%).

Acknowledgements

The authors acknowledge the financial support by Ministry of Research and Technology for Higher Education [KemenRistekDikti Republik Indonesia], the program is PDUPT 2019–2020.

Conflict of interest

The authors declare they have no conflict of interest.

References

[1] Baboli Z M and Safe Kordi A A 2010 *J. Am. Oil Chem. Soc.* **87** 667
[2] de Conto L C, Gragnani M A L, Maus D, Ambiel H C I, Chiu M C, Grimaldi R and Goncalves L A G 2011 *J. Am. Oil Chem. Soc.* **88** 1709
[3] Ariani P W 2015, *Karakterisasi dan Komposisi Kimia Minyak Biji Semangka* (*Citrulus Lanatus* L) *Varietas Sengkaling* Undergraduate Thesis (Universitas Kristen Satya Wacana)
[4] Verleyen T, Sosinska U, Ioannidou S, Verhe R, Dewettinck K, Huyghebaert A and De Greyr W 2002 *J. Am. Oil Chem. Soc.* **79** 947
[5] Sánchez-Machado D I, López-Cervantes J, Núñez-Gastélum J A, Servín De La Mora-López G, López-Hernández J and Paseiro-Losada P 2015 *Food Chem.* **187** 53
[6] Chew S C, Tan C P, Long K and Nyam K L 2016 *Ind. Crops Prod.* **89** 59
[7] Suryani E, Susanto W H and Wijayanti N 2016 *J. Pangan Agroindustri* **4** 120
[8] Morad N A, Aziz M K A and Mohd Zin R 2006 *Process Design In Degumming and Bleaching of Palm Oil* Project Report (Universiti Teknologi Malaysia)
[9] Prasetyo J and Afriyansah M R 2017 *J. Ilmiah Teknik Kimia* **1**
[10] Kurniati Y and Susanto W H 2015 *J. Pangan Agroindustri* **3** 193
[11] Victoria T 2017 *The Impact of H_3PO_4 Content and NaOH Concentration at Refining of Kachnar (Bauhinia Purpurea L.) Seeds Oil* Undergraduate Thesis (Universitas Kristen Satya Wacana)
[12] American Oil Chemists' Society (AOCS) 1989 *Official Methods And Recommended Practices Of The American Oil Chemists' Society Champaign*
[13] American Oil Chemists' Society (AOCS) 1998 *Official Methods And Recommended Practices Of The American Oil Chemists' Society Champaign* In p. Method Ca 9b-52
[14] American Oil Chemists' Society (AOCS) 1992 *Official Methods And Recommended Practices Of The American Oil Chemists' Society Champaign*. In p. Method Cc 8b-52
[15] American Oil Chemists' Society Champaign 1998 *Official Methods and Recommended Practices of the American Oil Chemists' Society* (p. Aa 4-38)
[16] American Oil Chemists' Society Champaign 2005 *Official Methods and Recommended Practices of the American Oil Chemists' Society* (p. Cc 10A-25)
[17] American Oil Chemists' Society 1989 *Official Methods and Recommended Practices of the American Oil Chemists' Society Champaign* p. Ca 2b-38
[18] American Oil Chemists' Society Champaign 1997 *Official Methods and Recommended Practices of the American Oil Chemists' Society* (p. Cd 8-53)
[19] American Oil Chemists' Society Champaign 1997 *Official Methods and Recommended Practices of the American Oil Chemists' Society* (p. Ca 5a-40)
[20] American Oil Chemists' Society Champaign 2003 *Official Methods and Recommended Practices*

of the American Oil Chemists' Society (p. Cd 3-25)
[21] Steel R G D and J H T 1989 *Analisa Data Statistik Deskrtiptif* (Surabaya: Erlangga)
[22] Hafidi A, Pioch D and Ajana H 2005 *Innov. Food Sci. Emerg. Technol.* **6** 203
[23] Arita S, Anindya A S and Wildayani H 2009 *J. Teknik Kimia* **16** 58
[24] Keraba A A, Soetjipto H and Kristijanto A I 2017 *Pros. Semin. Nas. Kim. Pendidik. Kim.*
[25] Gusti I, Budiadnyani A, Kelautan P and Sorong D P 2017 *Technol. Sci. Eng. J.* **1** 2
[26] Salawudeen T O, Arinkoola A O, Jimoh M O and Akinwande B A 2014 *J. Miner. Mater. Charact. Eng.* **02** 586
[27] Bija S 2017 *Pemurnian dengan Tahapan Degumming dan Netralisasi pada Pembuatan Softgel Minyak Ikan Sardin dan Uji Stabilitasnya* Master Thesis (Institut Pertanian Bogor)
[28] Leilah A A and Al-Khateeb S A 2003 *J. Agr. Sci* **28** 819
[29] Cahyaningtyas R, Soetjipto H and Riyanto C A 2017 *Prosiding SNST ke-8* **1** 1
[30] Lawson H 1995 *Food Oils and Fats: Technology, Utilization and Nutrition* (New York: Springer)
[31] Fauziah A W 2013 *Karakterisasi dan Penentuan Komposisi Asam Lemak dari Pemurnian Limbah Pengalengan Ikan dengan Variasi Waktu Simpan Limbah dan Suhu pada Degumming* Undergraduate Thesis (Universitas Jember)
[32] Apriani R 2008 *Studi Ekstraksi dan Penentuan Sifat Fisiko-Kimia Serta Komposisi Asam Lemak Penyusun Trigliserida dari Minyak Biji Pepaya (Carica papaya)* Undergraduate Thesis (Universitas Indonesia)
[33] Budiman F, Ambari O and Surest A H *J. Teknik Kimia* **18** 37
[34] Wildan A, Inggrid D, Hartati I and Widayat 2013 *Momentum* **9** 1
[35] Insani S A, Suseno S H and Jacoeb A M 2017 *J. Pengolahan Hasil Perikanan Indonesia* **20** 494

IOP Conf. Series: Materials Science and Engineering **959** (2020) 012029 doi:10.1088/1757-899X/959/1/012029

Optimization infusion condition on flavonoid and antioxidant activity of herb beverages using response surface methodology

Y Martono[1,2,*] and J Muninggar[1,2]
[1] Faculty of Science and Mathematics, Universitas Kristen Satya Wacana Indonesia
[2] Fitofarmaka Research Center, Universitas Kristen Satya Wacana Indonesia
Diponegoro street No. 52–60 Salatiga, Central Java, Indonesia 50711

[*] E-mail: yohanes.martono@uksw.edu

Abstract. Herbs have potential to be developed as functional beverages to reduce the risk of Covid-19 infection. One of herb ready to serve form is brewed in a bag. This study optimizes the conditions for herb beverages infusion in a silk bag based on their flavonoid content and antioxidant activity. Optimization is completed using the Response Surface Methodology with 3^3 Central Composite Design models. As the factor X1 is the mass of the herbs in the bag, X2 is the infusion frequency and X3 is the steeping time. Each factor consists of 3 levels. The flavonoid content was determined using the UV-Vis spectrophotometric method. The antioxidant activity was measured based on the DPPH free radical scavenging capacity. The results of this study indicate that the optimal condition is achieved in the herb mass of 1.2 g with an infusion frequency of 18 times and steeping time for 15 minutes. Flavonoids content and antioxidant activity in herb beverages at optimal infusion conditions were 942,353 µg QE / mL (117.79 mg QE / g) and 83.06%, respectively.

1. Introduction

Covid-19 has become a pandemic worldwide according to a statement from WHO [1] In Indonesia, until 6 August 2020, 118,753 people were confirmed positive for Covid-19. The mortality rate for sufferers reached 4.6% of confirmed cases [2]. Every day, confirmed cases continue to increase with a significant increase in number. However, until this writing, there has not been found a drug or vaccine that can cure Covid-19.

The mortality rate for Covid-19 is greatly influenced by comorbidities in patients. In Indonesia, one of the comorbidities that many positive COVID-19 patients suffer from is diabetes mellitus (DM) by 36%. Cases of death in COVID-19 patients who have a history of comorbidities are 16.8% [3]. These facts indicate that DM can increase the risk of COVID-19 infection and death of positive COVID-19 cases. High blood sugar levels in humans can increase the ACE2 receptor as receptor binding domain for COVID-19 virus infection into the cells and organs of the human body. In addition, high blood sugar levels in DM sufferers reduce immunity so that they are easily infected with the COVID-19 virus. Furthermore, the condition of high blood sugar in humans causes the body to produce immune blocking genes such as IL-6 and TNF-α which can trigger cytokine storms, thereby increasing the severity of the condition and the risk of death for COVID-19 positive patients. [4].

Due to no vaccine or specific drug has been found now days that can cure positive cases of COVID-19, so that people are very vulnerable to infection with the COVID-19 virus. One way to reduce the risk of COVID-19 infection is to control blood sugar and cholesterol levels for both people with DM and non-DM [5]. Indonesian people use herb plants to help reduce the risk of impact or prevent Covid-19 infection.

Our previous studies showed that the combination of several herb plants and named Steamori can make the fasting blood sugar levels of streptozotocin (STZ) induced rats from DM to normal in just 28 days of treatment. Furthermore, this herb solution can increase the amount and basal blood insulin sensitivity of DM rats accompanied by low cholesterol levels with better results than the positive control for the drug metformin. These results indicate that the STeaMori herbs combination has great potential to control blood sugar and cholesterol so that it can reduce the risk of COVID-19 infection. Steamori herbs contain flavonoid compounds and antioxidant activity that have health effects on the body [6].

Herb bag is one form of practical packaging for consuming herbal beverages. Bioactivity and health benefit are affected by bioactive compounds contained in herb beverages. The optimal condition of herbal beverages infusion needs to be studied so that herbal beverages contain optimal bioactive flavonoid and antioxidant activity when consumed. The serving conditions that can be optimized include herb mass, infusion frequency and steeping time of herb bag in hot water. The purpose of this study is to optimize infusion condition of herb bag Steamori based on flavonoid and antioxidant activity.

2. Materials and method

2.1. Materials

Steamori herbal samples powder with particle size of 100 mesh were obtained from the herbal industry of PT. Sinar Herba Radix in Tangerang, Indonesia. Moisture content of herbal samples is less than 10%. Quercetin standard with purity > 99% based on HPLC was obtained from Sigma Aldrich (USA). The DPPH, $NaNO_2$ and Al $(NO_3)_3$ reagents were obtained from E-Merck, Germany.

2.2. Method

2.2.1. Optimization of Herb ready to serve infusion conditions

Optimization of infusion conditions for the Steamori herb bag comprise mass of steamori herbs packed in silk bags (0.19 – 2.21 g). Infusion was held by immersed all bag parts into hot water and repeat immersed after previous treatment. Infusion frequencies applied were 2 - 22 times. Steeping treatment was incubating herb bag in hot water used for certain period. Steeping times optimized were 0 – 31.82 minutes. Herb bag is brewed using 150 mL hot water at 75°C. Optimization was carried out based on the flavonoid content and antioxidant activity of the drink.

2.2.2. Determination of flavonoid

A total of 2.0 mL of sample was added with 0.2 mL of $NaNO_2$ 5% then allowed to stand for 6 minutes. The solution was added 0.2 mL of Al $(NO_3)_3$ 10% and let stand for 6 minutes. Subsequently, the solution was added with 2 mL of NaOH 4% and let stand 15 minutes. The absorbance of the solution is measured using a UV-Vis spectrometer (Shimadzu 1240) at a wavelength of 501 nm (A_{501}). Total flavonoid content were measured based on the standard curve between the standard curve of quercetin vs A_{501}. Total flavonoid are expressed as μg quercetin equivalent / mL sample (μg QE / mL sample) [7].

2.2.3. Antioxidant activity

The antioxidant activity was determined based on the DPPH free radical scavenging capacity based on the method [8]. A total of 0.25 mL of sample was added with 1.0 mL of 0.4 mM DPPH solution in ethanol. The solution was homogenized with vortex and incubated for 30 minutes. The solution was then measured for its absorbance at a wavelength of 515 nm (A_{515}). The blank is ethanol solution while the control solution is 0.4 mM DPPH solution. Antioxidant activity is expressed as the percentage of DPPH inhibition according to equation (1):

$$\text{Inhibition DPPH (\%)} = \frac{(A_{515\ \text{control}} - A_{515\ \text{sample}})}{A_{515\ \text{control}}} \times 100\% \qquad (1)$$

2.2.4. Data analysis

Optimization was carried out by using the Response Surface Methodology (RSM) design using the 3^3 Central Composite Design (CCD) model. As factor X1 is the mass of Steamori herbs (0.6; 1.2; 1.8 g), factor X2 is the frequency of infusion (6; 12; 18 times) and factor X3 is the steeping time (5; 15; 25 minutes). Prediction model and lack of fit test used ANOVA with P value = 0.05. The prediction model obtained is used to determine the optimal conditions for infusion condition of herbs bags based on total flavonoid content and antioxidant activity.

3. Result and discussion

Optimization of Steamori herb infusion conditions is carried out to obtain optimal flavonoid content and antioxidant activity. Factors that can influence the flavonoid and antioxidant activity in herb infusion are herb mass/water ratio and steeping time [9]. Bioactive compound contained in herb is flavonoids. In the herb plant matrix system, the bioactive compounds of herb can provide bioactivity both individually and in combination [10]. It is necessary to optimize the condition for herb infusion condition. The results of total flavonoids and antioxidant activity in the optimization design for Steamori herb infusion conditions are shown in table 1.

Table 1. Total flavonoids and antioxidant activity on the optimization of the Steamori herb infusion conditions design

Run	X1, mass (g)	X2, steeping frequent (time)	steeping time (minutes)	Flavonoid (μg QE/mL)	Antioxidant Activity (% inhibition)
1	0.19	12	15	193.33	35.98
2	1.2	12	15	809.20	86.27
3	0.6	6	5	265.88	65.40
4	1.2	12	31.82	410.98	81.74
5	1.2	12	15	865.88	86.27
6	1.2	12	15	863.92	86.66
7	1.2	2	15	644.31	85.10
8	1.8	6	5	185.49	46.69
9	1.2	12	15	763.92	82.76
10	1.2	22	15	789.41	83.66
11	1.8	6	25	483.53	79.43
12	0.6	18	5	465.88	85.97
13	2.21	12	15	624.71	89.15
14	1.2	12	15	797.26	83.98
15	1.8	18	25	489.41	83.62
16	1.8	18	5	420.78	85.25
17	1.2	12	0	422.75	86.85
18	0.6	18	25	483.53	80.25
19	1.2	12	15	822.75	83.88
20	0.6	6	25	444.31	78.86

3.1. Flavonoid

One of the bioactive compounds contained in Steamori herbs are flavonoids. Flavonoids have various bioactivity which are included antioxidant activity. The ability of flavonoid compounds as antioxidants is due to the capacity of flavonoid compounds in donating electrons and hydrogen atoms from their polyhydroxy groups to scavenge free radicals. The capacity of the electron donor and the hydrogen atom of the flavonoid compound is linearly proportional to the concentration of the flavonoid compound [11]. The optimization of Steamori herbs infusion conditions resulted in a quadratic model equation. The ANOVA assay of model prediction and the lack of fit are shown in table 2.

Table 2. ANOVA for response surface quadratic model of flavonoid.

Source	Sum of squares	df	Mean square	F value	p-value Prob > F	
Model	7.517E+005	9	83523.04	58.08	< 0.0001	significant
Lack of fit	2248.03	2	1124.02	0.72	0.5316	not significant

Based on the ANOVA assay, the prediction model shows a significant model. The lack of fit test shows not significant, which means that the predicted and actual data are not significantly different. Factors that had a significant effect of flavonoid content in beverages were infusion frequency (X2) and steeping time (X3). The R^2 of the model achieved was 0.9868 and the adjusted R^2 was 0.9698. The mathematical model for predicting flavonoid content based on infusion conditions is shown in equation (2) as follows:

$$y = -582.47920X1^2 - 0.90256X2^2 - 1.68593X3^2 - 3.28461X1X2 + 1.54257X1X3$$
$$-0.61176X2X3 + 142.36577X1 + 42.44189X2 + 61.42914X3 - 843,60194 \quad (2)$$

Based on the mathematical model, it can be seen that the effect of herb mass, infusion frequency, steeping time and the interaction of infusion frequency and steeping time have an effect on increasing flavonoids. The increase in herb mass can increase the flavonoid source of the sample. The frequency of infusion and steeping time have an effect on the longer contact time of the sample as a source of flavonoids with hot water as a solvent so that the solubility of flavonoids in the solvent increases. Hot water and the processing of herbal plants into powder can cause cells to break easily and metabolite compounds to diffuse into the solvent. The diffusion of the active compound can also increase with the longer contact time between herb cells and hot water. The polarity of a compound greatly affects its solubility in a solvent. The use of hot water with a temperature of 75–80°C is due to increase the solubility of flavonoid compounds. However, the use of temperatures above 50°C needs to be limited in steeping time because longer steeping time of herb in hot water will degrade flavonoid compounds. [12]. The square of steeping time actually has the effect of lowering flavonoid levels. Furthermore, it is necessary to optimize the steeping time which provides higher levels of flavonoids. Surface plot of model prediction is displayed as shown in figure 1.

3.2. Antioxidant activity

The antioxidant activity was determined based on the inhibition of DPPH free radicals by electrons from the active compounds contained in herb beverages. The higher the electron and hydrogen donor capacity by the active compound, the higher the antioxidant activity [11]. One of bioactive compounds that have antioxidant activity is the flavonoid group [10]. Based on the ANOVA assay, the prediction model obtained was significant while the lack of fit test gave not significant results (table 3).

R^2 value of model prediction is 0.9888 and the adjusted R^2 is 0.9719. The ANOVA assay showed that there was a significant interaction between the mass of herbs (X1) and steeping time (X3). Interaction also occurred between the frequencies (X2) and steeping time (X3). The square of each factor also has a significant effect. The mass factor (X1) and infusion frequency (X2) independently had a significant effect, but the steeping time factor (X3) independently had no significant effect on antioxidant activity. The prediction model equation achieved is shown in equation (3).

$$y = 11.62542X1^2 - 0.22668X2^2 - 0.001086X3^2 + 0.15381X1X2 + 0.82771X1X3 \\ -0.20204X2X3 - 49.04170X1 + 10.51477X2 + 1.36732X3 + 32.51162 \quad (3)$$

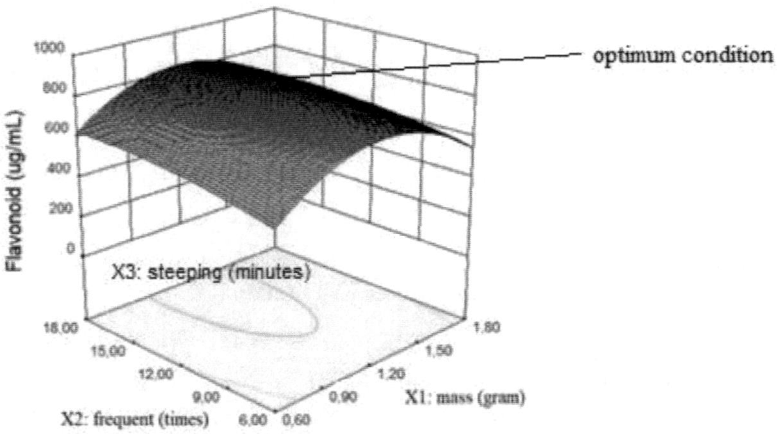

Figure 1. Surface plot of flavonoid prediction quadratic model

Table 3. ANOVA for response surface quadratic model of antioxidant activity.

Source	Sum of squares	df	Mean square	F value	p-value Prob > F	
Model	1625.73	9	180.64	58.68	< 0.0001	significant
Lack of fit	5.18	1	5.18	1.95	0.2214	not significant

The mass herb as independent factor (X1) had the highest effect on lowering antioxidant activity. High mass oh herb will lower the mass to solvent ratio which can reduce the extraction capacity of the solvent to dissolve the active compound. In addition, the higher the mass of herb extracted, the lower the antioxidant activity because it is possible for many non-antioxidant compounds to be dissolved in the beverage matrix. The interaction between mass and infusion frequency (X1.X2) increased the antioxidant activity but the increase did not have a significant effect. In this study, one of the critical conditions is the interaction between mass and steeping time (X1.X3) which has a significant effect on increasing antioxidant activity. The mass ratio to the solvent, temperature and contact time can affect the quantity of dissolved bioactive compounds [13] so that it will affect its antioxidant activity. In this study, it was found that mass factor independently actually reduced antioxidant activity, but squared and its interactions with other factors could increase antioxidant activity. Therefore, it is necessary to optimize the mass of herbs and other factors to optimize the dissolving capacity of compounds that have antioxidant activity. Figure 2 shows the surface plot of the prediction model for antioxidant activity.

A correlation assay was also carried out between the total of flavonoid compounds and antioxidant activity. The results of the correlation test gave a correlation value of 0.76. Correlation indicates that the content of flavonoid compounds with antioxidant activity has a positive correlation. Total flavonoid content is linear proportionally affect antioxidant activity of beverages. However, it needs to be investigated more deeply whether the quality or type of flavonoid compounds or the quantity of flavonoid compounds is more affected to antioxidant activity.

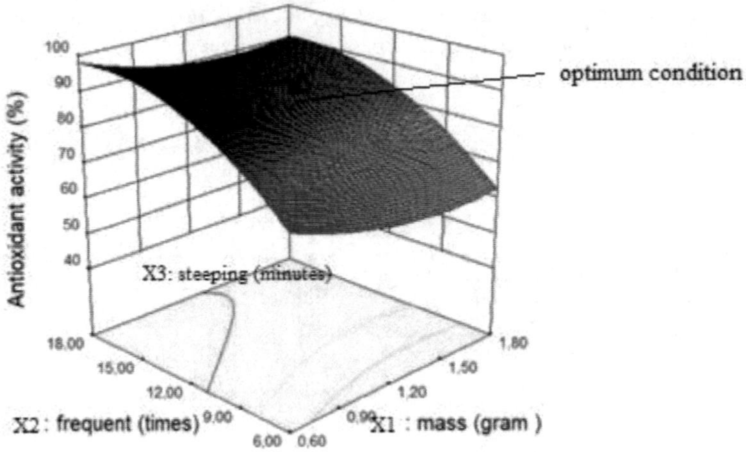

Figure 2. Surface plot of antioxidant activity prediction quadratic model.

3.3. Optimization infusion condition

Optimization of infusion conditions is carried out using targeted optimization. The conditions chosen are those that give desirability> 0.90; flavonoids> 800 µg QE / mL and antioxidant activity> 80%. Based on the optimization, the optimum conditions achieved were 1.2 g of herb mass with a infusion frequency of 18 times and a steeping time of 15 minutes. Total Flavonoid content and antioxidant activity of herb beverages at optimum infusion conditions were 942.353 µg QE / mL (117.79 mg QE / g) and 83.06%, respectively.

4. Conclusion

The optimum infusion condition for Steamori herb beverages based on total flavonoid content and antioxidant activity is a herb mass of 1.2 g with a infusion frequency of 18 times and a steeping time of 15 minutes. The critical factors for optimization were found in the interaction of herb mass and steeping time.

Acknowledgment

We would like to thank to Indonesia Ministry of Research and Technology/National Research and Innovation Agency (Kemenristek/BRIN) who has funded this research through Penelitian Terapan Unggulan Perguruan Tinggi (Applied Research) scheme of 2020 with contract No. 001/LL6/PG/SP2H/PL.I/2020. We also thank to Universitas Kristen Satya Wacana Indonesia which has facilitated of this publication.

References

[1] WHO *Director-General's opening remarks at the media briefing on COVID-19 - 11 March 2020* [Internet]. [cited 2020 Aug 14]. Available from: https://www.who.int/dg/speeches/detail/who-director-general-s-opening-remarks-at-the-media-briefing-on-covid-19---11-march-2020

[2] COVID-19 GTPP. Peta Sebaran | Gugus tugas percepatan penanganan COVID-19 [Internet]. covid19.go.id. [cited 2020 Aug 14]. Available from: https://covid19.go.id/peta-sebaran

[3] Media KC. *Penyakit Bawaan Apa Saja yang Banyak Diderita Pasien Covid-19 di Indonesia?* [Internet]. KOMPAS.com. [cited 2020 Aug 14]. Available from: https://www.kompas.com/tren/read/2020/05/02/161500565/penyakit-bawaan-apa-saja-yang-banyak-diderita-pasien-covid-19-di-indonesia

[4] Pal R and Bhansali 2020 *Diabetes Res Clin Pract* **162** 1

[5] Gupta R, Ghosh A, Singh A K and Misra A 2020 *Diabetes Metab. Syndr.* **14** 211

[6] Martono Y and Muninggar J 2019 *Laporan Penelitian Terapan Kemenristekdikti: Optimasi Proses Produksi Minuman Fungsional Stevia sebagai Obat Herbal Terstandar (OHT) yang Berkhasiat Antidiabetes* Report (Salatiga: Universitas Kristen Satya Wacana)

[7] Lee K W, Kim Y J, Lee H J and Lee C Y 2003 *J. Agric. Food Chem.* **51** 729

[8] Wahyuni O T and Hertiani T 2016 *Maj. Obat Trad.* **21** 86

[9] da Silveira T F F *et al.* 2017 *Plant Foods Hum. Nutr.* **72** 219

[10] Chandrasekara A and Shahidi F 2018 *J. Tradit. Complement. Med.* **8** 451

[11] Vieyra F E M *et al.* 2009 *Cav. Free Radic. Res.* **43** 553

[12] Khoza B S, Dubery I A, Byth-Illing H A, Steenkamp P A, Chimuka L and Madala N E 2016. *Food Anal. Methods* **9** 1480

[13] Pinelo M, Rubilar M, Jerez M, Sineiro J and Núñez M J 2005 *J. Agric. Food Chem.* **53** 2111

IOP Conf. Series: Materials Science and Engineering **959** (2020) 012030 doi:10.1088/1757-899X/959/1/012030

Hydrolysis of sorghum starch to glucose using organic acid catalyst from rosella flower extract (*Hibiscus sabdariffa* L.)

Riniati[1,*], S H Abdulloh[1], R H Fauziyah[1] and Istiqomaturohmah[1]
[1] Chemical Analysis Study Program, Chemical Engineering Department, Politeknik Negeri Bandung, Indonesia

[*] E-mail: riniati@polban.ac.id

Abstract. Highly starch content in sorghum seed has a high potential raw material for the sorghum fluor to be a raw material of liquid sugar known as glucose syrup. The syrup is usually produced through enzymatically or chemically hydrolysis using a strong acid. In this study, sorghum starch is hydrolyzed chemically using a catalyst of organic acids extracted from roselle flower (*Hibiscus sabdariffa* L.). The goal of the study is to produce glucose syrup free from chemical agents so as generally recognized as safe (GRAS). Rosella flowers are known to contain high amounts of organic acids such as citric acid, malic acid, tartaric acid, oxalic acid, and hibiscus acid. Apart from having natural acid content, rosella flowers also contain vitamins (ascorbic acid) and are rich in natural red pigment dyes from the anthocyanins that can be an added value for the glucose syrup products. Organic acid extraction of dried rosella flowers was carried out at a temperature of 100°C for 30 minutes in a water bath and filtered. The resulting acid extract then determined the value of the total titrated acid (TTA) by the acid-base titration method. Sorghum starch is prepared by mixing sorghum flour with distilled water and then cooled at 4°C for 12 hours. Starch is obtained from the dried filtrate to obtain a water content of <14%. The starch hydrolysis process with rosella flower extract was carried out in several variations, namely the hydrolysis time of 1 - 4 hours, the TTA concentration of rosella 4 - 16%. Meanwhile, the temperature and stirring speed were fixed at 96°C and 200 rpm. The resulting glucose levels were analyzed by the spectrophotometric method. From this research, the result shows that the best sorghum starch hydrolysis condition was obtained at the addition of 16% acid extract for 180 minutes which gave 13.83% of glucose yield

1. Introduction

Sorghum (Sorghum bicolor L. Moench) is a cereal crop that has an average starch content of 69.5%. Sorghum starch consists of amylose (20–30%) and amylopectin (70–80%), depending on genetic and environmental factors [1]. Based on these data, sorghum has excellent potential to be processed into other, more useful products based on its nutritional content. The high starch content in sorghum plants makes sorghum widely used as a raw material in various industries such as the beer starch industry, liquid sugar (syrup), jaggery, ethanol, glue, paint, paper, degradable plastics and others[2]. Sorghum seeds can be made into sugar, liquid glucose, or fructose syrup according to the sugar content of the seeds [3].

Several studies of hydrolysis with synthetic organic acids have been carried out. Organic acids, such as citric acid and malic acid, are able to interact with the hydroxyl groups of cassava starch and act in the acid hydrolysis of starch molecules, reducing the starch chain connections [4]. Hydrolysis of durian rind to produce reducing sugar used 0.68 N maleic acid catalyst for 120 minutes at 95 °C to produce reducing sugar levels of 0.119 mg / mL [5]. Meanwhile, in the study of hydrolysis of sweet potato starch with a 5% citric acid catalyst for 1 hour at 45 °C with stirring which still produced a dextrose equivalent value of 2.21%, the DE value also showed an increase along with the increase in acid concentration and hydrolysis time [6].

Modification of starch using citric and stearic acids can reduce the value of swelling power as well as increase the solubility of starch which is thought to be due to the formation of short chains in acid hydrolysis and the effect of increasing the hydration capacity of starch chain hydrolysis[2]. The addition of citric acid to rice flour causes partial hydrolysis, which results in linear and short branched chains of crystalline and amorphous sections [7]. Heating at a higher temperature (> 80 °C) in the presence of citric acid causes hydrolysis of the amylose and amylopectin chain acids, and which affects the gel strength and retrogradation of the corn starch gel [8]. Citric acid is a naturally occurring polycarboxylic acid that can be extracted from various fruits and is generally considered safe [9]. Organic acids have high potential in the production of modified starch for the food and processing industries. Since starch and organic acids are readily available and low-cost materials, modification of starch with organic acids and their derivatives allows much room for further research on the reaction of starch with organic acids [10].

Rosella extract contains a high presentation of organic acids, namely 13–24% hibiscus acid, 12–20% citric acid, 2–9% malic acid, 8% tartaric acid, and 0.02–0.05% ascorbic acid [11]. Fresh roselle petals contain ascorbic acid from 6.7 to 14 mg/100 g while dried petals are 260–280 mg/100g [12]. The presence of organic acids with a high presentation is expected to increase the DE value from hydrolysis. Variables that can affect the hydrolysis process using an acid catalyst include the size of the material to be hydrolyzed, the stirring speed during hydrolysis, the acid concentration, the time and temperature of the hydrolysis [13]. The operating temperature used for the hydrolysis of cellulose can increase glucose levels to the optimum level [14].

2. Material and methods

In general, the research begins with the preparation of sorghum starch and roselle extract, then performs hydrolysis by mixing starch and roselle extract under specific ratios and conditions. Glucose resulting from hydrolysis was analyzed qualitatively using Benedict reagent. To determine glucose levels, quantitatively analyzed using a refractometer and spectrophotometer.

2.1. Making sorghum starch

The sorghum flour is added with distilled water with a ratio of flour: distilled water = 1: 2 (1 kg flour: 2 L distilled water), then the mixture is filtered using a filter cloth. The dregs from the filtering are extracted again with the ratio of dregs: distilled water = 1: 1. Then the starch suspension was stored in the refrigerator at 4 °C for 12 hours. The topwater is removed, and wet starch is obtained. Wet starch is dried to obtain a water content of <14%.

2.2. Rosella extraction

For the manufacture of natural acid catalysts, dried roselle flowers are added with distilled water with a ratio of roselle flowers: aqua dest = 1:10 (100 g of dried roselle flowers: 1 L of distilled water) then the sample is extracted at 100 °C for 30 minutes and filtered until no solids are seen. On the filtrate. The extract obtained was measured the total titrated acid (TTA) by the acid-base titration method using the 848 Titrino Plus auto titrator. In the TTA measurement, a 10 g sample was put into a 100 ml measuring flask and diluted to the mark with distilled water. The 10 mL diluted sample is transferred to a small beaker, and distilled water is added until the electrode is immersed. Titration is carried out using a 0.1 N NaOH solution, which has been standardized until the endpoint of the titration is reached. The titration is done in duplicate. The total titrated acid is calculated using the equation:

$$\%\ \textbf{total titrated acid} = \frac{\textbf{mL NaOHxN NaOHxDilution Factor}}{\textbf{gram of extract}} \times \mathbf{100}.$$

2.3. Hydrolysis of sorghum starch with acid catalyst from roselle extract
One of gram sorghum starch is added with distilled water as much as 50 mL then heated until it dissolves and thickens. The solution was added to the acid extract with a volume variation of 4; 8; and 16 % TTA. Then refluxed at 95°C and stirred 200 rpm for 1, 2, 3 and 4 hours.

2.4. Analysis of glucose result of hydrolysis of starch sorghum

2.4.1. Glucose qualitative test with Benedict's reagent
One without of mL hydrolysis sample was put in a test tube, then added 1 mL of Benedict's reagent and stirred until blended. The test tube is then put into a beaker filled with boiling water, and boil for 5 minutes. After 5 minutes, the test tube and beaker are removed from the hotplate and cooled. The color change of the solution was then observed.

2.4.2. Analysis of glucose levels by Somogyi-Nelson method
One without of mL solution sample was added with 1 mL of Nelson's reagent and heated in a water bath for ± 20 minutes and immediately cooled down by putting it in a beaker filled with cold water so that the tube temperature reached 25°C then added to each tube 1 mL of Arsenomolybdate reagent, shaken until all the Cu_2O precipitates dissolved again. After all Cu_2O precipitates were completely dissolved, 7 mL of distilled water was added and shaken until homogeneous and measured the absorbance (A) of each solution using a Vis spectrophotometer with a wavelength of 540 nm. A standard curve was made showing the relationship between standard glucose concentration and absorbance.

3. Result and Discussion

3.1. The total value of titrated acid (TTA) of roselle extract
To determine The TTA value, a number of roselle extract samples titrated with a standardized NaOH solution. From the calculation results based on the volume of the equivalence point on the titration curve in figure 1, the dry roselle extract with a concentration of 10% (w / v) has a % TTA of 31,46%. The total acid in the sample will correlate with the pH of the sample. The dried roselle extract solution is purple-red and has a pH of 2.25.

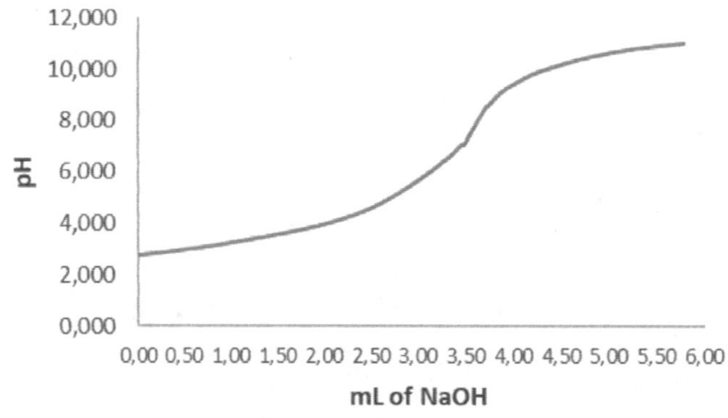

Figure 1. TTA analysis titration curve of rosella extract.

3.2. Glucose qualitative test with benedict's reagent
After hydrolyzing sample for 1 hour, using reagent benedict qualitatively showed positive glucose. The presence of glucose ranging from 0.5% to 1% is indicated by the appearance of yellowish-green colour and cloudy. The other blue colour one of starch sorghum is the solution which no glucose contained as figure 2.

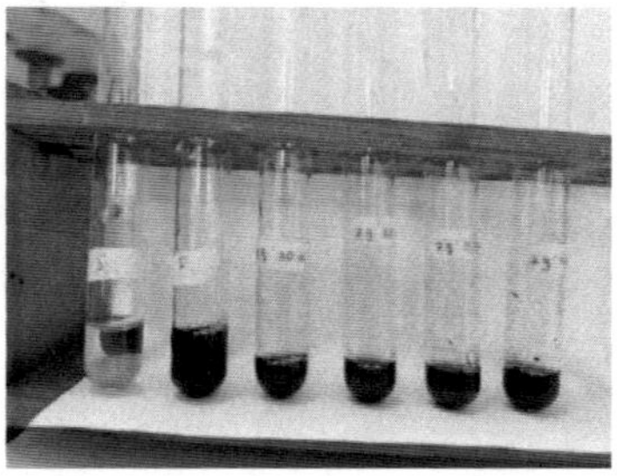

Figure 2. Qualitative test of glucose hydrolysis results

3.3. Effect of TTA concentration and hydrolysis time on glucose levels from hydrolysis

The acid concentration of the roselle extract was varied based on% TTA, by diluting the roselle extract to 4; 8 and 16% were added to the preheated starch (final starch concentration 1 g/100 mL). In the preliminary test to determine glucose levels by measuring% Brix using a refractometer, it appears that the higher the% TTA of organic acids from the roselle extract added, the higher the % Brix, this indicates the higher glucose levels as figure 3. From the time variation of 1–4 hours, it appears that the hydrolysis process decreases after the 3 hours.

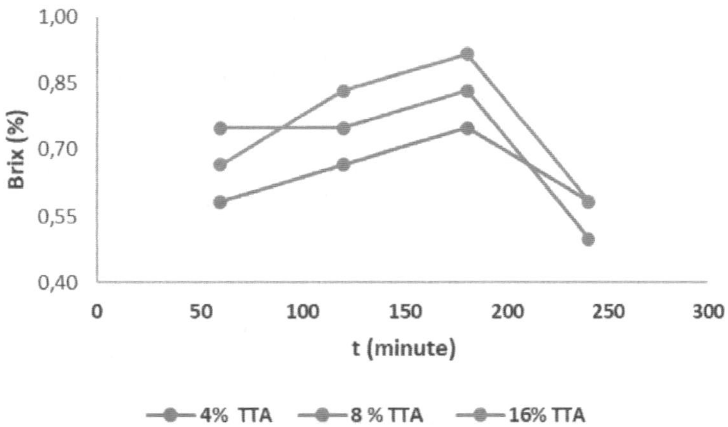

Figure 3. Effect of % TAT and hydrolysis time on the value of % brix.

The results of glucose concentration from absorbance measurements by spectrophotometry which have been converted into % yield of glucose can be seen in figure 4. With an increase in glucose levels in hydrolyzed starch, it can be said that organic acids in roselle extract can affect the starch hydrolysis process. When water-insoluble starch is heated in water, the starch granules expand and burst. At this time, a gelatinization process occurs in which amylose and amylopectin can be removed from the granules, and the starch suspension becomes a thick paste. It is thought that organic acids such as citric acid enter the starch granules, causing hydrolysis of the glucose chains in the starch granules, then causing many of the glucose chains to dissolve into the starch paste.

4. Conclusion

This research can be concluded that the hydrolysis process of sorghum starch using an acid catalyst from rosella flower extract is influenced by the TTA concentration value. With 16% concentration of TTA against 1 gram of starch, it can produce 13,8% glucose at the optimum time for hydrolysis for 3 hours. This study recommends further research in the manufacture of glucose syrup using acids from roselle flower extracts.

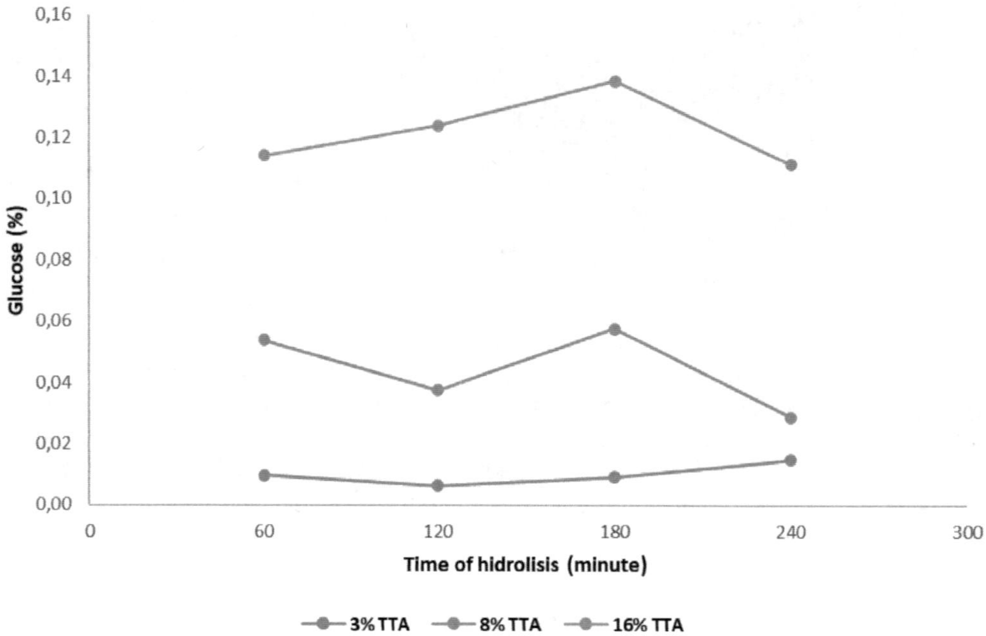

Figure 4. The effect of TTA concentration of rosella flower extract on glucose levels.

Acknowledgment
We gratefully acknowledge support from the POLBAN Research and Community Service Center (PPPM) grant No. B/249.41/PL1.R7/PG.00.03/2020 which has provided funding for this research.

References

[1] Suarni S 2017 *J. Penelit. Pengemb. Pertan.* **35** 99
[2] Singh H, Chang Y H, Lin J H, Singh N and Singh N 2011 *Food Res. Int.* **44** 2949
[3] Irawan B 2011 *Forum Penelitian Agro Ekonomi* **2** 99
[4] Majzoobi M and Beparva P 2014 *Food Chem.* **147** 312
[5] Obed, Alimuddin A H and Harlia 2015 *J. Kim. Khatulistiwa* **4** 67
[6] Shin S I, Lee C J, Kim D I, Lee H A, Cheong J J, Chung K M, Baik M Y, Park C S, Kim C H and Moon T W 2007 *J. Cereal Sci.* **45** 24
[7] Hirashima M, Takashi R and Nishimari K 2012 *Food Hydrocoll.* **27** 390
[8] Olsson E, Menzel C, Johansson C, Andersson R, Koch K and Järnström L 2013 *Carbohydr. Polym.* **98** 1505
[9] Da-Costa-Rocha I, Bonnlaender B, Sievers H, Pischel I and Heinrich M 2014 *Food Chem.* **165** 424
[10] Ačkar D, Babić J, Jozinović A, Miličević B, Jokić S, Miličević R, Rajič M and Šubarić D 2015 *Molecules* **20** 19554
[11] Ismail A, Hainida E, Ikram K, Saadiah H and Nazri M 2008 *Food* **2** 1
[12] Artati E K, Irvina F and Fatimah 2012 *Ekuilibrium* **11** 73
[13] Wahyudi J, Wibowo W A, Rais Y A and Kusumawardani A 2011 *Pros. Semin. Nas. Tek. Kim. Kejuangan* **B09** 1
[14] Subagio H and Aqil M 2013 *Pros. Semin. Nas. Inov. Teknol. Pertan.* 199

IOP Conf. Series: Materials Science and Engineering **959** (2020) 012031 doi:10.1088/1757-899X/959/1/012031

Interaction of mutant PBP2a and bioactive compounds from Streptomyces with anti-MRSA activities

A Laksono, A Asnani* and P Iswanto
Chemistry Department, Faculty of Mathematics and Sciences,
Universitas Jenderal Soedirman, Purwokerto 53123

* E-mail: ari.asnani@unsoed.ac.id

Abstract. Methicillin-Resistant *Staphylococcus aureus* (MRSA) is the leading cause of nosocomial infections in hospitals. Treatment of MRSA infection using ceftaroline has been reported to be resistant due to mutations in the Penicillin Binding Protein (PBP)2a. In silico's approach through virtual screening can analyze the bioactive compounds that can bind effectively to mutant PBP2a. The potential source of bioactive compounds with anti-MRSA activities is Streptomyces, which is the main antibiotic-producing bacteria. Thus, the study aimed to analyze the interactions of PBP2a/mutant PBP2a against ceftaroline and the interactions between mutant PBP2a against bioactive compounds from Streptomyces. The protein receptors were PBP2a (PDB 3ZG0) and mutant PBP2a (PDB 4CPK). The ligands used were ceftaroline (CID 9852981) as control and nine bioactive compounds produced by Streptomyces. Protein preparation and visualization used Discovery Studio, ligand preparation used Marvin, and molecular docking used Autodock4. The alignment results showed that mutant PBP2a has a more extended amino acid sequence (643 amino acids) than PBP2a (641 amino acids). The mutations that occurred in mutant PBP2a caused conformation changes in the active site of mutant PBP2a so that the interaction between ceftaroline and mutant PBP2a decreased. The virtual screening results indicated that 1-acetyl-β-carboline was the most potent compound as anti-MRSA with the lowest binding energy (-7.12 Kcal /mol) compared to ceftaroline (-6.32 Kcal/mol). The amino acids involved in the binding of 1-acetyl-β-carboline with PBP2a mutant were Ser403, Ser461, Asn464, Thr600; Ser462, Tyr446, and Ala642. This result suggests that 1-acetyl-β-carboline has better interaction with mutant PBP2a, hence might serve as a potential anti-MRSA compound.

1. Introduction

Methicillin-Resistant Staphylococcus aureus (MRSA) is *Staphylococcus aureus* bacteria resistant to β-lactam antibiotics such as penicillin and its derivatives, namely methicillin, oxacillin, dicloxacillin, nafcillin, and cephalosporin. Resistance to methicillin is caused by the MecA gene that produces Penicillin Binding Protein (PBP) 2a, which is insensitive to β-lactam antibiotics [1]. PBP has a significant role in the peptidoglycan biosynthetic reaction in the formation of bacterial cell walls [2]. There are four types of PBP, namely PBP1, PBP2, PBP3, and PBP4, which are complex proteins with glycosylase and transpeptidase activities. In MRSA, there is a fifth type of PBP, namely PBP2a, coded by the MecA gene [3]. This PBP2a causes MRSA to be insensitive to β-lactam class antibiotics and causes multiresistant strains.

One of the options for treating MRSA infection is the use of ceftaroline antibiotic, which serves as transpeptidase inhibitor in PBP2a [4]. However, Fishovitz et al. [5] have reported a mutation in PBP2a so that MRSA becomes resistant to ceftaroline. Resistance occurs due to the mutation of two amino acids on the allosteric site of PBP2a, which causes changes in the conformation of the active site. These changes cause a decrease in the effectiveness of ceftaroline antibiotic binding and result in the ineffectiveness of ceftaroline for MRSA therapy [6]. The incidence of MRSA resistance to ceftaroline encourages the development of new bioactive compounds as anti-MRSA. WHO has even classified MRSA in the High-Risk category on the Priority Pathogen List for research and development of new antibiotics [7].

Exploring bioactive compounds that are antagonistic can be done through virtual screening of potentially bioactive compounds [8]. Molecular docking can be performed between the mutant PBP2a receptor and bioactive compounds as ligands to obtain bioactive compounds capable of interacting with mutant PBP2a. Molecular docking is an in-silico approach to pairing a protein molecule on the active side with other molecules with a specific purpose, for example, looking for potential new drugs [9]. It is known that ceftaroline antibiotic binds PBP2a on the active side at the Serin 403 position [10]. Therefore, the active site of PBP2a can be used as a reference for molecular docking to determine the interaction of PBP2a with other bioactive compounds to test the compound's potential as anti-MRSA. Lavanya et al. [11] conducted molecular docking to screen potential compounds from the ZINC database capable of interacting with MRSA that is resistant to ceftaroline. Mohamed et al. [12] reported that indigenous Sudanese natural ingredients, namely isoflavonoids-17, chalcon-21, diphyllin, and coumarin-29, were able to bind effectively with PBP2a in Ser404 using the molecular docking method.

One of the potential sources of bioactive compounds is actinomycetes, especially the genus Streptomyces which is the main antibiotic-producing bacteria. Streptomyces are known to produce antibiotic, anticancer, anti-inflammatory, anti-fungal, protein inhibitor, and other active compounds [13]. Indeed, the genus Streptomyces can produce potential compounds with anti-MRSA activity. Kemung et al. [14] have reviewed bioactive compounds with anti-MRSA activity produced by Streptomyces. Asnani et al. [15] have also reported that Streptomyces from mangrove sediment at Segara Anakan has potential as anti-MRSA. Hence, this study aimed to compare PBP2a protein interaction with mutant PBP2a against ceftaroline and screen bioactive compounds from Streptomyces as anti-MRSA candidates. The bioactive compounds with the most potential to bind to mutant PBP2a are expected to be the most potential anti-MRSA candidates.

2. Materials and methods

The PBP2a receptor protein was PDB 3ZGO, a PBP2a crystal structure that binds to ceftaroline [10]. The mutant PBP2a was PDB 4CPK, which is a crystal structure of PBP2a double clinical mutant N146K-E150K from MRSA [5]. The crystal structures of the two proteins were downloaded from the Protein Data Bank. The alignment of mutant PBP2a and PBP2a used Clustal Omega (1.2.4) multiple sequence alignment to analyze the differences caused by mutations. Protein preparation used Autodock4 [9] version 1.5.6.

The ligands were the antibiotic ceftaroline as control and nine bioactive compounds produced by Streptomyces, namely 1-acetyl-β-carboline, marinopyrrole A, cyslabdan, 4'deacetyl griseucin A, griseucin A, polyketomycin, heliquinomycin, and chaxamycin D. All ligand structures were downloaded from PubChem and optimized with MarvinSketch 20.3.

Interaction of mutant PBP2a and bioactive compounds from Streptomyces was carried out based on the active side of mutant PBP2a, which is Serin 403 coordinates [10] using Autodock4 [9] version 1.5.6. Mutant PBP2a protein was not given a degree of freedom. In contrast, ligands were given a degree of freedom to predict the best position on the active site of the protein. The bond coordinates were x: -27.959; y: -19.845; and z: 6.711. The box's size to predict binding position was 50 on Autodock and docking parameters used LGA (Lamarckian Genetic Algorithm). Visualization of the amino acid bonds involved in the binding position and the interactions between the receptor protein and the ligand used the BIOVIA Discovery Studio Visualizer.

3. Results and discussion

3.1. Alignment of PBP2a and mutant PBP2a

Protein receptors alignment were performed by comparing the crystal structure of PBP2a (PDB 3ZG0), which binds to ceftaroline antibiotics with the crystal structure of mutant PBP2a (PDB 4CPK). Figure 1 shows the differences in the structures of PBP2a and mutant PBP2a. The mutations that occur in mutant PBP2a are on the allosteric site, which causes changes in the protein conformation so that ceftaroline cannot effectively bind to the active site [5]. This mutation causes MRSA to become resistant to ceftaroline.

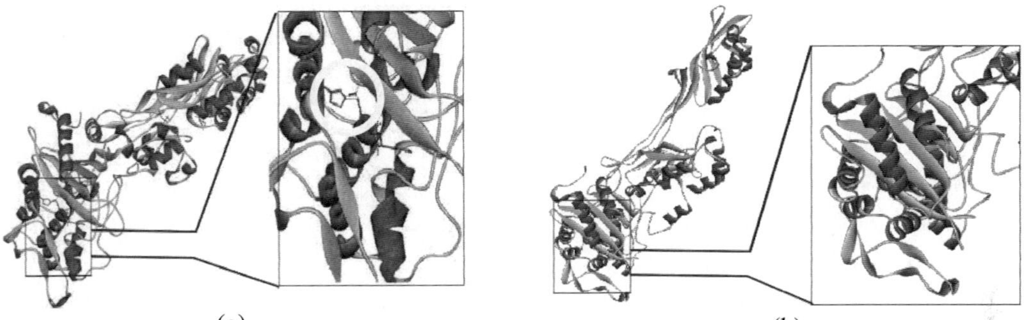

(a) (b)

Figure 1. The crystal structure (a) PBP2a binds to the ceftaroline antibiotic on the active site of the A chain indicates by the yellow circle, (b) mutant PBP2a resistant to antibiotics does not bind to ligands on the active site of the A chain.

In addition to differences in the molecular structure of mutant PBP2a and PBP2a, there are also differences in the amino acid sequences of the two proteins. The results of the alignment analysis of the amino acid sequences of mutant PBP2a and PBP2a showed differences in the number of amino acids. The mutant PBP2a has a more extended amino acid sequence than PBP2a. The longer mutant PBP2a protein presumably is caused by a mutation on the allosteric site that changes the protein conformation. Table 1 shows the differences in amino acids of PBP2a (PDB 3ZG0) and mutant PBP2a (PDB 4CPK).

Table 1. The results of the alignment analysis of 3ZG0 and 4CPK.

PBP2a	Amino acids		Position	
	PDB 3ZG0	PDB 4CPK	PDB 3ZG0	PDB 4CPK
A chain	-	Lysine	-	1
	Asparagine	Lysine	120	121
	Glutamic acid	Lysine	124	125
	-	Glutamic acid	-	643
B chain	-	Lysine	-	1
	Asparagine	Lysine	120	121
	Glutamic acid	Lysine	124	125

3.2. Ligand optimization

The ligands used in this study were ceftaroline antibiotics as control and nine bioactive compounds with potency as anti-MRSA from Streptomyces. Structure optimization of ligands used MarvinSketch. Structure optimization aims to ensure the correct structure of the ligands. It minimizes the energy to stabilize the structures before the ligands are used in molecular docking. The ligand optimization generated structure with energies ranges from 46.47 to 227.48 Kcal/mol (table 2). The compounds that have been minimized were then used as ligands in molecular docking.

Table 2. The Energy minimization from ligands optimization

No	Compounds	CID	Energy (Kcal/mol)
1	Ceftaroline	138756682	127.30
2	1-Acetyl-β-carboline	638667	46.47
3	Frigocyclinone	11476774	109.56
4	Marinopyrrol A	24797083	85.86
5	Cyslabdan	24778014	113.11
6	4'deacetyl griseusin A	56834145	90.09
7	Grieusin A	16102131	89.14
8	Polyketomycin	10395823	227.48
9	Heliquinomycin	101714016	127.30
10	Chaxamycin D	53344649	164.45

3.3. Interaction of mutant PBP2a and ligands

Molecular docking was performed 20 times from the LGA calculation to produce the best position with the lowest binding energy. The smaller binding energy with the receptor protein indicates good results in which the binding is more stable than that with the higher binding energy. The results of the molecular docking of nine bioactive compounds from Streptomyces based on binding energy are presented in table 3. The 1-acetyl-β-carboline compound has the lowest binding energy (–7.12 Kcal/mol) compared to other bioactive compounds, including the control ceftaroline (–6.32 Kcal/mol). This result suggests that the interaction between 1-acetyl-β-carboline and mutant PBP2a is the most effective so that 1-acetyl-β-carboline has the potential as anti-MRSA. The best interactions between ligands and receptor proteins were further visualized using Discovery Studio to analyze the amino acid bonds.

Table 3. The binding energy from the interaction of protein and ligand.

No	Compounds	Binding Energy (Kcal/mol)
1	Ceftaroline	–6.32
2	1-acetyl-β-carboline	–7.12
3	Frigocyclinone	–6.66
4	Marinopyrrol A	–6.27
5	Cyslabdan	–5.67
6	4'deacetyl-griseusin A	–5.10
7	Grieusin A	–4.55
8	Polyketomycin	–4.49
9	Heliquinomycin	–3.55
10	Chaxamycin D	–1.41

The docking result of ceftaroline on the active site of the mutant PBP2a has a binding energy of –6.32 Kcal/mol. The amino acid involved in the binding of ceftaroline with mutant PBP2a are lysine at a position of 430 (Lys430) with hydrogen bonding (figure 2). Amino acids involved in the binding interactions of ceftaroline and mutant PBP2a through hydrogen bonds are Lys430, Tyr446, Asn464, Gln 521, Thr600, Ala601, and Gln613 in the A chain.

Ceftaroline is known to have a high affinity for PBP2a. It binds to the active site of the PBP2a on the amino acid Ser403 [10]. The results of visualization of ceftaroline binding with mutant PBP2a showed different binding to PBP2a. This difference in binding is probably the cause of the different responses of MRSA containing mutant PBP2a to ceftaroline therapy. Allosteric site changes that occur in mutant PBP2a has changed the conformation on the active site of mutant PBP2a [5]. Apparently, the binding affinity of the receptor to ceftaroline decreases due to conformation changes of the active site of the protein, thus resulting in resistance to ceftaroline.

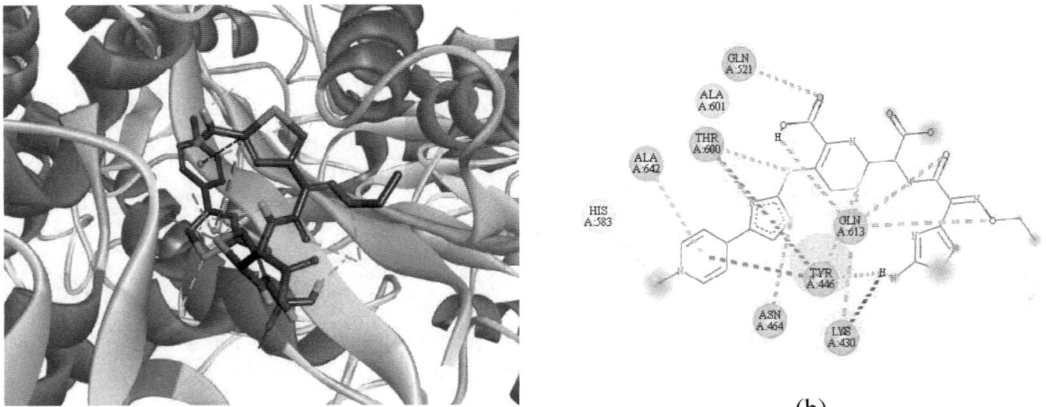

(a)　(b)

Figure 2. (a) 3D visualization of ceftaroline-protein receptor binding interactions; (b) 2D visualization of the amino acids involved.

Nine bioactive compounds from Streptomyces were successfully docking on the active site of mutant PBP2a with the same active site as PBP2a. Based on the docking results, 1-acetyl-β-carboline and frigocyclinone compounds are potential anti-MRSA compounds with bond energies of –7.12 and 6.62 Kcal/mol. The affinity binding of the 1-acetyl-β-carboline and frigocyclinone to mutant PBP2a are stronger than that of ceftaroline.

1-Acetyl-β-carboline is a compound of the alkaloid indole group isolated from *Streptomyces sp.* 04DH52 [16]. This molecule could enter and bind to the active site of mutant PBP2a with the least bond energy compared to other ligands. The small binding energy shows the bond's strength in which the smaller the value, the stronger the bonds formed. Docking of the 1-acetyl-β-carboline with the mutant PBP2a receptor protein gave binding energy of –7.12 Kcal/mol. Amino acids involved in the binding of 1-acetyl-β-carboline with receptor proteins through hydrogen bonding are Ser403, Ser461, Asn464, Thr600; and through the pi-sigma bonds are Ser462, Tyr446, and Ala642 in the A chain (figure 3A).
Frigocyclinone is a bioactive compound isolated from *Streptomyces griseus* strain NTK 97 [17]. Frigocyclinone has good inhibitory activity against Gram-positive bacteria *Bacillus subtilis* DSM 10 and *S. aureus* DSM 20231. Docking of the frigocyclinone molecule with the mutant PBP2a receptor protein gave binding energy of –6.66 Kcal/mol. Amino acids involved in the binding of frigocyclinone with receptor through hydrogen bonds are Lys430, Gln521, Gln613, Glu602; and through phi-sigma bonds is Met641 on the A chain (figure 3B).

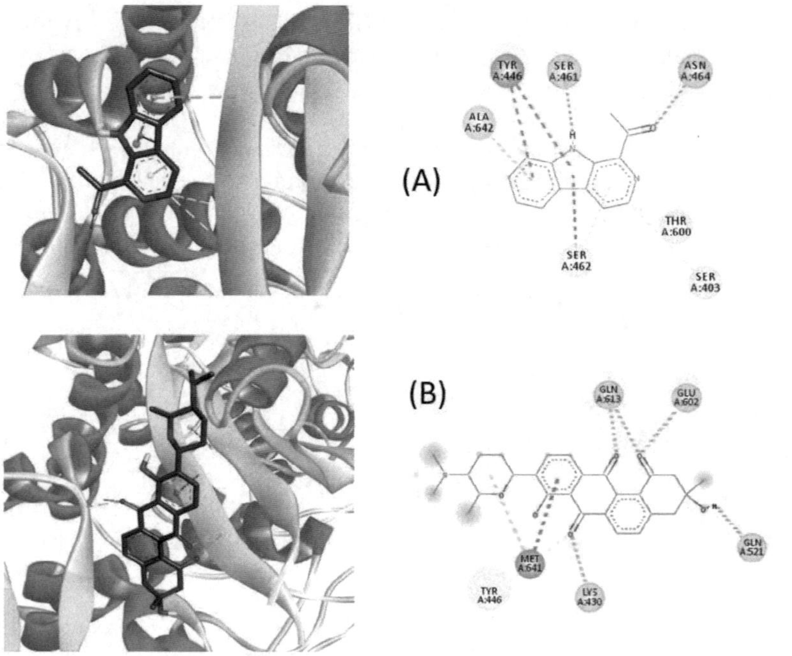

Figure 3. 3D visualization and 2D visualization of the amino acids involved in the interaction of (a) 1-acetyl-β-carboline and mutant PBP2a protein receptor, and (b) frigocyclinone and mutant PBP2a.

4. Conclusion

A virtual screening approach can be used to analyze the potential of bioactive compounds as anti-MRSA. Virtual screening of nine bioactive compounds from Streptomyces using molecular docking showed that 1-acetyl-β-carboline and frigocyclinone are potential compounds as anti-MRSA compared to ceftaroline control based on binding energy. This approach is expected to save time, effort, and cost before drug design and development is carried out for production.

Acknowledgment

This research was supported by the Directorate of Research and Community Service, Ministry of Research and Technology/National Agency for Research and Innovation of the Republic of Indonesia through the Fundamental research grant Contract No. T/1641/UN23.18/PT.01.03/2020.

References

[1] Lim D and Strynadka N C J 2002 *Nat. Struct. Biol.* **9** 870
[2] Sauvage E and Terrak M 2016 *Antibiotics* **5** 12
[3] Fishovitz J, Hermoso J A, Chang M and Mobashery S 2014 *Life* **66** 572
[4] Mendoza N, Hernandez, Tyring and Lema S 2012 *Infect. Drug Resist.* **5** 23
[5] Fishovitz J, Rojas-altuve A, Otero L H, Dawley M, Carrasco-lo C, Chang M, Hermoso J A and Mobashery S 2014 *J. Am. Chem. Soc.* **136** 9814
[6] Long S W, Olsen R J, Mehta S C, Palzkill T, Cernoch P L, Perez K K, Musick W L, Rosato A E and Musser J M 2014 *Antimicrob. Agents Chemother.* **58** 6668
[7] WHO 2017 World Health Organization releases global priority list of antibiotic-resistant bacteria to guide research, discovery, and development of new antibiotics
[8] Malathi K and Ramaiah S 2018 *Biotechnol. Genet. Eng. Rev.* **34** 243
[9] Morris G M, Huey R, Lindstrom W, Sanner M F, Belew R K, Goodsell D S and Olson A J 2009 *J. Comput. Chem.* **30** 2785
[10] Otero L H *et al.* 2013 *Proc. Natl. Acad. Sci. U. S. A.* **110** 16808

IOP Conf. Series: Materials Science and Engineering **959** (2020) 012031 doi:10.1088/1757-899X/959/1/012031

[11] Lavanya P, Ramaiah S and Anbarasu A 2016 *J. Cell. Biochem.* **117** 542
[12] Mohamed S B, Adlan T A, Khalafalla N A, Abdalla N I, Ali Z S A, Munir KA A, Hassan M M and Elnour M A B 2019 *Evol. Bioinforma.* **15** 1
[13] Harir M, Bendif H, Bellahcene M, Fortas Z and Pogni R 2018 Streptomyces secondary metabolites *Basic Biology and Applications of Actinobacteria* ed S Enany (IntechOpen) chapter 6
[14] Kemung H M, Tan L T-H, Khan T M, Chan K-G, Pusparajah P, Goh B-H and Lee L-H 2018 *Front. Microbiol.* **9** 2221
[15] Asnani A, Luviriani E and Oedjijono O 2020 *J. Kim. Sains Apl.* **23** 1
[16] Shin H J, Lee H-S and Lee D-S 2010 *J. Microbiol. Biotechnol.* **20** 501
[17] Bruntner C *et al.* 2005 *J. Antibiot.* **58** 346

The agronomic performance and artemisinin content of colchicine-induced polyploid genotypes *Artemisia cina*

M M Herawati[1,*], E Pudjihartati[1], T D Kurnia[1] and A W Setiawan[1]
Faculty of Agriculture, Satya Wacana Christian University, Jl. Diponegoro No. 52–60, Salatiga, Sidorejo, Kota Salatiga, Jawa Tengah 50711, Indonesia

[*] E-mail: marinakartika@gmail.com

Abstract. Artemisinin is a secondary metabolite contained in the potential genus Artemisia as an antimalarial. *Artemisia cina* Berg ex Poljakov is a species from Artemisia that grows as a weed in Indonesia and contains artemisinin. The problem is that the artemisinin content in *A. cina* is minimal. The efforts must be made to increase the content. Clone improvement through breeding through genetic manipulation and enhancement of technical culture are the strategies that can use to improve the artemisinin content of *A. cina.* The purpose of this study was to determine the agronomic performance of nine *A. cina* polyploid genotypes (genotypes A, B, C, D, E, F, G, H, I) obtained from polyploidy induction using colchicines. The results of this study showed that leaf area and chlorophyll content were not significantly different among genotypes.In contrast, shoot dry weight, and plant height showed that I genotype was significantly higher than other genotypes. Glandular trichomes density of genotype was more elevated than different genotypes. Genotypes F, H, I had high artemisinin content compared to other genotypes. The genotype I showed higher artemisinin weight than other genotypes.
Key Word: Genotype, colchicines, agronomic character, artemisinin.

1. Introduction

Artemisia is a plant that can use as medicine. Treatment exploration on Artemisia has primarily used artemisinin as a source of potential antimalarial. There was much research regarding this topic. However, one species of Artemisia, which grows a weed on highlands of Indonesia and contains artemisinin, is *Artemisia cina* Berg ex Polyakov. Artemisinin is a product from the secondary metabolism of *A. cina*. The artemisinin content on *A. cina* is minimal, ranging from 0.0075% to 0.066% [1–5]. The effort to increase artemisinin content on A. cina has been made through tissue culture or agronomic practices. However, that technique has not improved the artemisinin content of *A. cina* plants, because *A. cina* is a wild plant that has not cultivated.

The presence of artemisinin potential contained in *A. cina*, genetically engineered. The Improvements in technical culture are a great alternative to increase the content of artemisinin in the production of varieties of *A. cina*. Improvements to varieties through genetic engineering through artificial polyploidy is one method in improving biomass and secondary metabolites contents in plants. Colchicine is one of the chemicals commonly used to induce polyploidy in plants with a high success rate. Colchicine is an anti-mitotic agent is used for chromosome multiplication under in vitro conditions. Colchicine plays a role in inhibiting chromosome segregation during cell division in chromosome multiplication. The effect of colchicine for in vitro chromosome multiplication differs in terms of concentration, method, and duration of treatment, as well as plant genetic factors [6–7]. In addition to the different chromosomes, polyploid plants have different morphological characteristics compared to diploid plants [8]. Colchicine treatment in *A. annua* plants produces the highest tetraploid plants [9]. Polyploidy induction in *A. annua* plants to polyploid through colchicine treatment has 38% higher artemisinin content than diploid plants [10]. The artemisinin content in tetraploid *A. annua* is 1.5 times greater than that of diploid plants [9]. The artemisinin content in plants *A. annua* tetraploid average increased from 39% to 56% compared to diploid plants *A. annua* [11]. The content of artemisinin in *A. cina* tetraploid is higher (0.488%) versus diploid plant (0.04%) [12]

Agronomics performance is the growth and development of the plant-related field. Agronomic appearance is observed in *A. cina* polyploids and associated with artemisinin. They are broadleaves, shoot dry weight, root dry weight, density and size of trichomes glands, density and size of stomata, and leaf chlorophyll content. found that Artificial polyploidy induction using a 2.4-D plant growth regulator combined with BA was able to improve agronomic characters of *A. cina* polyploids compared that diploid i.e., on broadleaf, trichome glands density, the size of stomata, and chlorophyll content of leaves [13].

Artemisia spp. has been known as the plant that has a high economic value. The economic benefit is evident from the International Association of plant breeders' rights. The International Union for the Protection of New Varieties of Plants (UPOV) already enters the genus Artemisia spp. in a list of plants that need a protected creation results of new varieties in 2009. However, *A. cina* is not yet explicitly listed. Besides, there is no guideline what characters that must be observed for the development of new crop varieties/assemblers of Artemisia spp (The International Union for the Protection of New Varieties of Plants [14]. The purpose of this study was to determine the agronomic performance of nine *A. cina* polyploids (genotypes A, B, C, D, E, F, G, H, I) obtained from polyploidy induction using colchicines.

2. Material and method

A. cina used in this research was a putative polyploid. *A. cina* putative 1 (P1) is the result of the colchicine induction. *A cina* polyploid plants were planted in polybags with a 50 cm diameter filled with garden soil and manure ratio of 1: 1. After two weeks, he moved to the shading house with 50% shade. One week after moving to the shading house, the first fertilization was carried out, namely Urea 152 kg/ha, SP36 233 kg/ha, and KCl 250 kg/ha. At the first fertilization, given ½ dose for urea while the other one dose. The second urea fertilization is carried out when the plants are one month after the first fertilization. GA3 growth regulators were applied to all genotypes when the plants were 30 days after moving to the shading house.

2.1. Leaf area analysis

Measurement of leaf area using the scanner method with iDaun software. Broadleaf measurements were done when the plant was four weeks old after it moved into the shading house (aged eight weeks from start to grafting). The leaves were measured when it opened up correctly in the fourth position.

2.2. Trichome glandular density analysis

The method used to measure the glandular density of trichomes is the replication method [15]. The observation of trichome glands was done by spreading top coat nail polish on the surface leaf's epidermis. Then, using affixed duct tape to peel the green area and placed it above the glass objects of the microscope, and then calculated the density of the trichomes gland. The measurement of the density of trichomes glands was done using USB Optilab camera and raster image program.

2.3. Chlorophyll analysis

Analysis of chlorophyll content using a modified method [16]. The leaves that were used were the leaves that had opened up correctly. Weighing the leaves to 0.04 grams, then cut into small pieces, placed it on a ceramic bowl, added 5 mL of DMSO, and then incubated in the darkroom at room temperature for 48 hours. After 48 hours, the leaves filtered with a filter paper. The absorption value was measured using a spectrophotometer (UV-mini 1240, UV VIS Spectrophotometer, Shimadzu) at a wavelength of 649 nm (A665) and 665 nm (A649). Chlorophyll content expressed in mg.g^{-1} fresh weight.

$$\text{Chlorophyll a} = (12{,}19 \text{ x A665}) - (3{,}45 \text{ x A649}) \quad (1)$$

$$\text{Chlorophyll b} = (21{,}99 \text{ x A649}) - (5{,}32 \text{ x A665}) \quad (2)$$

$$\text{Total Chlorophyll} = (18{,}54 \text{ x A649}) + (6{,}87 \text{ x A665}) \quad (3)$$

2.4. Analysis of dry weight shoots
The shoot dry weight was measured using the oven drying method [17]. Observation of dry weight of shoots was done at the end of the study. The dry weight of shoots was measured by drying the leaves in the oven at a temperature of 40°C until constant weight. Then, it weighed with the units of grams (g).

2.5. Artemisinin content analysis by high performance liquid chromatography
Analysis of artemisinin content was performed 12 weeks after plantlets were transferred to the soil media. Shoots (the top of the plant) was cut then dried in an oven at a temperature of 40°C until dry. Furthermore, artemisinin content was analyzed using HPLC [18]. 100 mg dry weight of the sample was then extracted with 2 ml of Toluene, and filtered using filter paper and then inserted into the flakon. Analysis of artemisinin was quantitatively by the use of HPLC [18]. First, artemisinin was hydrolyzed in the alkaline solution. The hydrolysis product called Q260 was then measured at the wavelength of 260 nm. The procedure of quantitative analysis was as follows: 500 µL of the extraction with toluene was taken then evaporated / dried, the residue that remained was dissolved in 200 mL of methanol again. Next, 800 mL of NaOH solution (0.2% w / v) was added and the mixture was agitated with a vortex formed and heated in water bath for 30 minutes at the temperature of 50°C. After it got cold, 200 µL of methanol and 800 mL 0.2 M acetic acid were added, then artemisinin was measured by HPLC, at a wavelength of 260 nm, using a RP-18 column Licrospher length of 10 cm. The mobile phase used was methanol: 0:05 Kaliumdihidrogenphosphat mM (55: 45), a flow rate of 1.3 mL / min. Retention time was 40 minutes. Q260 product was stable for at least 4 days at 4°C storage temperature.

2.6. Data analysis
The experimental design used in this study was a random Groups Design with three replicates. It was analyzed using statistic analysis (ANOVA) and a variety of prints with advanced test DMRT on levels 5%. The data were analyzed using the program SAS 9.1.3.

3. Results and discussion
Based on the results of the analysis of the prints range (table 1) shows that the characteristics of agronomic crop for plant height on *A. cina* high-yield crops polyploids on genotype I differ markedly higher than those of the other genotype. The height of plants is the result of the Division and the unfolding of the cells. Application of Gibberellin (GA_3) to all observed genotypes also affected the increase in plant height. The sensitivity of each genotype to the administration of GA_3 is different. The application of growth regulators to plants depends on the plant's age, the dose of growth regulators given, and the part of the plant that is the target. Gibberellin has a role in supporting cell elongation, cambium activity, and supporting new RNA and protein synthesis. Plants respond to GA_3 with increased stem length. The GA_3 effect is mainly in the elongation of plant internodes caused by an increase in the expansion of cells in these segments. The role of Gibberellin in stem elongation is the result of 3 processes. The first process is cleavage at the end of the stem. The results of [19] show that cell division is caused by the gibberellin stimulus to cells in the G1 phase to enter the S phase and shorten the S phase immediately. The second process is that Gibberellin stimulates cell growth by

increasing the hydrolysis of starch, fructans, and sucrose into glucose and fructose to be used for energy-producing respiration. This energy will then be used to form cell walls and other cell components so that the cell formation process can take place quickly. The third process is increasing cell wall plasticity by Gibberellin. Gibberellins also reduce water potential to enter the cell more quickly, and cell expansion occurs. Agronomic perform for wide leaves, chlorophyll and the density of glandular trichome on Nine genotype *A. cina* observed do not differ markedly between genotype.

Agronomic appearance for the dry weight of shoots on genotype I showed real difference results is higher than the other genotype. The dry weight of a plant are affected biomass plant itself. On table 1. Biomass from Genotype I *A. cina* polyploids are indicated by high crop and broad leaves. Height of the plant determines the number of nodes owned by the plant and the number of leaves produced, whereas the broad-leaf plants capture light capability, so the effect on the dry weight of the plant. Koester et al. (2014) stated that broad leaf determines the amount of light that is captured and is an important parameter to determine the productivity of plants [20].

Table 1. Agronomic perform of *Artemisia cina* polyploid genotype

Genotype	Plant Height (cm)	Leaf area (cm)	Chlorophyll (mg/g)	Glandular Trichoma Density/μm	Shoot dry weigh (g)	Artemisinin content (%)
A	33.153 d	24.967 ab	2.960 a	60.000 a	6.690 cd	0.0145 f
B	52.053 bc	28.733 ab	2.897 a	58.750 a	5.163 b	0.0125 f
C	54.970 bc	31.930 ab	2.800 a	59.500 a	9.230 cb	0.0225 e
D	60.037 ab	37.083 a	2.647 a	61.250 a	10.010 cb	0.0355 b
E	51.127 bc	30300 ab	2.880 a	62.417 a	7.270 bcd	0.0320 c
F	56.247 abc	29.507 ab	2.643 a	65.140 a	7.310 bcd	0.0405 a
G	61.070 ab	28.990 ab	2.257 a	61.750 a	10.840 b	0.0270 d
H	43.987 cd	35.230 a	2.290 a	64.110 a	8.163 bcd	0.0400 a
I	70.833 a	35.990 a	2.120 a	70.250 a	16.347 a	0.0390 a

The mean value followed by different letters showed significant ($\alpha = 0.05$) in the DMRT test.

Table 1 shows that the levels of artemisinin on genotype F, H, and I differ markedly higher than those of the other genotypes. Artemisinin is a secondary metabolite produced by *A. cina*. The availability of precursor compound as starting material for the synthesis of artemisinin is to affect levels of artemisinin in *A. cina*. Acetyl Co-A is a precursor compound for the synthesis of artemisinin. Acetyl CoA synthesized from the sugars derived from the results of photosynthesis. It allegedly produced the dry ingredients used for the formation of artemisinin [12]. Figure 1 dry weight indicates that the highest-weight of artemisinin produced by I genotype. A dry weight of artemisinin is affected by the heading and levels of artemisinin. Besides, it is also affected by the number of glandular trichomes per unit area. From the observations, I genotype had a higher dry weight of shoots and a more elevated amount of glandular trichomes compared to other genotypes. Artemisinin biosynthesis and storage are present in glandular trichomes. The density of the glandular trichomes correlates with the artemisinin content. Artemisinin levels have a positive correlation; the correlation coefficient is 0.51 and significant with glandular trichome density. Thus, the higher the glandular density of trichomes, the artemisinin content will increase [12]. Correlation between trichome density and artemisinin content is in line with the results of research by [21], who stated that glandular trichome density correlates with artemisinin content.

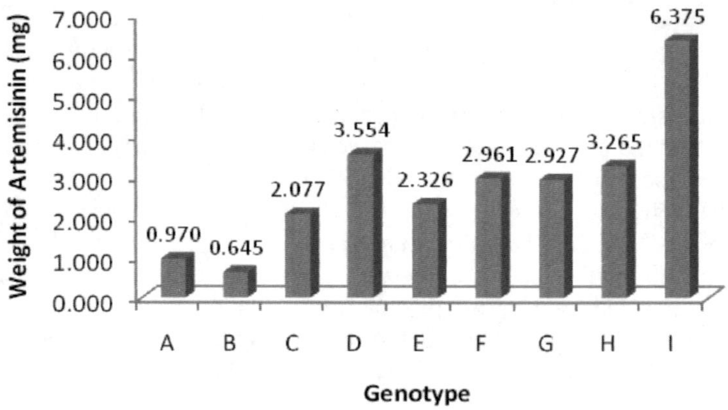

Figure 1. Artemisinin production on *A. cina* polyploidy genotype.

4. Conclusion

Agronomic appearance is the character of the growth and development of plants. The glandular trichomes, leaf area, and chlorophyll content are not significant for genotypes A, B, C, D, E, F, G, H, I. Agronomic appearances of plant height, dry weight shoot on genotype I significantly higher compared to the other genotype. Levels of artemisinin on genotype F, genotype H, and genotype I differ markedly higher than those of the different genotypes. The weight of artemisinin on genotype I is higher than most other genotypes.

Acknowledgments

This research is part of PTUPT Grants Research 2018 funded by the higher education DRPM. Thanks to Dr. Iwan Setyawan for the iDaun software to leaf area analysis.

Reference

[1] Aryanti 2001 *Variasi Kandungan Artemisinin dari Akar Rambut dan Regenerasi Artemisia cina Berg ex Poljakov Sebagai Anti Kanker* Master Thesis (Institut Pertanian Bogor)
[2] Herawati M M, Purwantoro A dan Soegihardjo C J 2003 *Produksi Artemisinin Melalui Kultur Pucuk Artemisia cina Berg ex Poljakov dengan Perlakuan Sukrosa dan Ekstrak Khamir* Master Thesis (Universitas Gadjah Mada Yogyakarta)
[3] Supardiyanti I and Herawati MM 2005 *Pengaruh Konsentrasi IBA terhadap Pertumbuhan dan Produksi Artemisinin dari Artemisia cina Berg ex Poljakovsecara Kultur Pucuk* Undegraduate Thesis (Universitas Kristen Satya Wacana)
[4] Manoppo R and Herawati M M 2006 *Pengaruh Pemberian Berbagai Konsentrasi Media MS terhadap Produksi Metabolit Sekunder Artemisinin pada Planlet Artemisia cina Berg ex Poljakov* Undergraduate Thesis (Universitas Kristen Satya Wacana Salatiga)
[5] Hidayat W and Herawati M M 2007 *Prosiding Seminar Nasional Tanaman Obat dan Obat Tradisional*
[6] Aina O, Quesenberry K and Gallo M 2012 *Crop Sci.* **52** 1076
[7] Dhooghe E, Van Laere K, Leus L and Van Huylenboeck J 2011 *Plant Cell, Tissue Organ Cult.* **104** 359
[8] Rantau D E, Hafiizh E A and Rahman W 2014 *Prosiding Seminar Nasional ke 53 Temu-Ilmiah Jaringan Kerjasama Kimia Indonesia*
[9] Banyai W, Sangthong R, Karaket N, Inthima P, Mii M and Supaibulwatana K 2010 *Plant Biothecnol.* **27** 427
[10] Wallaart T E, Pras N, Beekman A C and Quax W J 2000 *Planta Med.* **66** 57

[11] Lin X, Zhou Y, Zhang J, Zhang F, Shen Q, Wu S, Chen Y, Wang T and Tang K 2011 *Biotechnol. Appl. Biochem.* **58** 50
[12] Herawati M M 2016 *Peningkatan Hasil Artemisinin Melalui Poliploidisasi dan Kultur Teknik Artemisia cina Berg ex Poljakov* Doctor Thesis (Universitas Gadjah Mada)
[13] Herawati M M, Pudjihartati E, Pramono S, Sulistiyaningsih E and Purwantoro A 2015 *Agrivita* **37** 178
[14] The International Union for the Protection of New Varieties of Plants (UPOV) 2010 *Plant Variety Protection* Gazette and Newsletter No. 103
[15] Budiono R, Sugiarti D, Nurzaman M, Setiawati T., Supriatun T, Zaenal A and Mutaqin 2016 *Kerapatan stomata dan kadar klorofil tumbuhan Clausena excavate berdasarkan perbedaan intensitas cahaya* Seminar Nasional Pendidikan dan Saintek
[16] Hiscox J D and Israelstam G F 1979 *Can. J. Bot.* **57** 1332
[17] Miftahudin, Setyaningsih D S and Chikmawati T 2015 *Jurnal Sumberdaya HAYATI* **1** 1
[18] Pras N, Visser J F, Batterman S, Woerdenbag H J and Malingre T 1991 *Phytochem. Anal.* **2** 80
[19] Liu P B W and Loy B 1976 *Am. J. Bot.* **63** 700
[20] Koester R P, Skoneczka J A, Cary T R, Diers B W and Ainsworth E A 2014 *J. Exp. Bot.* **65** 3311
[21] Dangash A, Pandya N, Bharillya A and Jhala A 2014 *Notulae Scientia Biologicae* **6** 349

IOP Conf. Series: Materials Science and Engineering **959** (2020) 012033 doi:10.1088/1757-899X/959/1/012033

Synthesis of poly pyridine-2-ylmethyl 2-(eugenoxy) acetate (PMEOA) as a metal mixture carrier

M C Djunaidi[1,*], H Cahyono[1], Ismiyarto[1], D Siswanta[2] and Jumina[2]
[1] Chemistry Department, Faculty of Science and Mathematics, University of Diponegoro, Semarang, Indonesia
[2] Chemistry Department, Faculty of Science and Mathematics, Gadjah Mada University, Yogyakarta, Indonesia

[*] E-mail: choliddjunaidi@live.undip.ac.id

Abstract. The synthesis of poly pyridin-2-ylmethyl 2-(eugenoxy) acetate (poly (PMEOA)) as a metal mixture carrier by BLM (Bulk Liquid Membrane) method has been conducted. The synthesis began with the conversion of eugenol to acid followed by the synthesis of pyridin-2-ylmethyl 2-(eugenoxy) acetate (PMEOA) ester. The monomer synthesis results were characterized by FTIR and GC-MS along with NMR. Polymerization of the ester produced poly (PMEOA). The resulting polymer was characterized using FTIR and NMR. Both of the synthesis products (monomers and polymers) were tested for their selectivity towards metal mixtures with Bulk Liquid Membrane Method (BLM) extraction. The results obtained show the difference in selectivity and effectiveness of its extraction ability. Mainly in the improvement of the heavy metal ions transport effectiveness resulted by the use of poly (PMEOA).

1. Introduction

Indonesia is a country with vast tropical forest areas and a wide variety of natural resources. Thus, the development of biological resources is necessary, especially in the field of chemistry. One type of plant that can be utilized in regards to this study is cloves *(Eugenia caryophyllata)*. Cloves yield the most eugenol compared to cinnamon and nutmeg. Eugenol ($C_{10}H_{22}O_2$) (figure 1) has a higher value than clove oil but Indonesia still imports eugenol although being the largest clove oil exporter in the world [1].

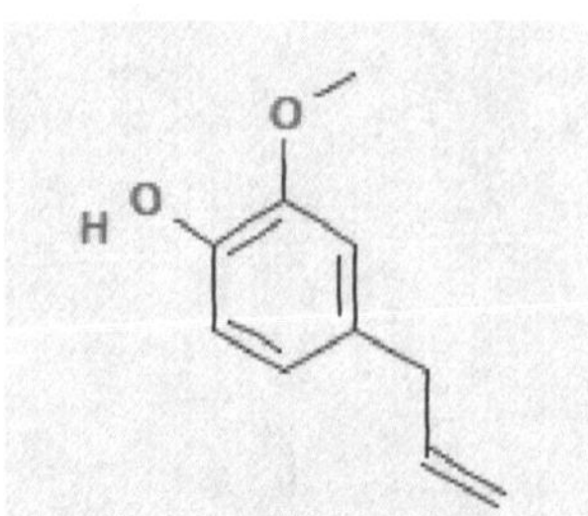

Figure 1. Eugenol structure (2).

Polymerization of natural compounds such as eugenol that produces polyeugenol is a relatively new thing to do. Earlier studies about the synthesis of poly methyl thiazole eugenoxy ethyl acetate) (PMTEEA) from eugenol showed the ability of eugenol derived membrane to transport metal ions thus proved its capability for the recovery of wastewater [3]. The development and utilization of such polymer now undergo further studies [3–4] one of which is the use of active N polyeugenol as a liquid membrane in heavy metal *recovery*.

The active N compound is hoped to be used as a carrier compound in heavy metal *recovery*, such as Cr^{3+}, Cd^{2+}, and Cu^{2+}, and be selective towards one of them. Each of those metals (Cr^{3+}, Cd^{2+}, and Cu^{2+}) represent groups of hard, soft, and borderline (between hard and soft) acids, respectively. The carrier compound works is based on the HSAB theory (theory of acid-base grouping based on hardness and softness) which states that hard metal ions (in this case Cr^{3+}) commonly form a stronger complexes with hard donor atoms and soft metal ions (in this case Cd^{2+}) will form a stronger complexes with soft donor atoms [3–6]. Thus, based on the theory of HSAB, it is hoped that the compounds with N donor atoms originating from pyridine carbinol will be selective towards borderline metal (Cu^{2+}).

Earlier studies done on heavy metal recovery using bulk liquid membrane (BLM) technique showed promising results as it was proven to be capable of removing heavy metal from wastewater with a satisfying result [7–10]. Another study used the polymer inclusion membrane to remove toxic heavy metal from wastewater [11–12]. So far, the bulk liquid membrane is the easiest to perform compared to other types of liquid membrane due to its simplicity and great membrane stability although it is known that BLM has inferior solute fluxes and high membrane resistance [7].

The selectivity of separation in a liquid membrane technique was determined by the complex stability of metal ions and carrier compounds. The donor atom type (active group) contained in the carrier compound (ligand structure), which matches the electron ion metal configuration [13–17] and pH of the solution [5–15] will determine the stability of the complex. The efficiency and selectivity of transport will be improved by the use of carrier compound [18]. In this study the synthesis of the carrier compound N-group and its polymerization will be carried out and the its effectiveness as a carrier compound will be studied.

2. Methodology

The active N polyugenol obtained from the synthesis would be used as a carrier compound in heavy metal *recovery* by a liquid membrane technique. The 30 ppm metal mixture containing Cr^{3+}, Cu^{2+}, and Cd^{2+} will act as the feed phase and HCl as the receiving phase. The pH of the receiving phase was made constant (pH 1). Quantitatively, the remaining heavy metals as well as those taken up were determined by AAS.

2.1. Materials

Eugenol (Sigma Aldrich), BF_3 diethyl ether (Sigma Aldrich), $SOCl_2$ (Sigma Aldrich), Pyridyl carbinol (Sigma Aldrich), NaOH (E-Merck), Chloroacetic acid (E-Merck), Chloroform (E-Merck), Methanol (E-Merck), Diethyl ether (E-Merck), Aquadest, $CrCl_3$. $6H_2O$ (E-Merck), $CuCl_2.2H_2O$ (E-Merck), $CdCl_2.2H_2O$ (E-Merck)

2.2. Instrumentation

Analytical balance (Mettler-200), pH meter (HACH E C20), Atomic Absorption Spectrophotometer (Perkin Elmer AA 400), FT-IR (Nicolet 360), GC-MS (Shimadzu prestige 21), ^1H NMR (Jeol PMX 60), reflux apparatus.

2.3. Synthesis of eugenoxy acetic acid

A total of 5 g (0.03 mol) of eugenol was put into a 100 mL boiling flask, then 33% NaOH solution (33 g NaOH in 100 mL) as much as 17.5 mL (0.144 mol) was added. The mixture was stirred for approximately 30 minutes, and 12.5 mL of 50% chloroacetic acid solution (50 g of 100 mL water)

(0.066 mol) was added gradually while being stirred constantly. The mixture was heated in a water bath with temperature 80-90°C. The heating took 2 hours, then the mixture was cooled and acidified with HCl 6 M to pH 1. Furthermore, it was extracted with diethyl ether 3 times, 50 mL each. The ether extract result was combined and extracted with 5% (w/v) of sodium bicarbonate 3 times, 30 mL each. Later the water layer was acidified with HCl 6 M until it reached pH 1. Filtering, drying, and weighing were done afterward. The results obtained were tested for melting point and identified with FTIR and GC-MS.

2.4. Synthesis of Pyridin-2-ylmetyl 2-eugenoxy acetate

A total of 3 g (0.0135 mol) of eugenoxy acetic acid (EOA) was introduced into a 100 mL three-neck flask with auxiliary equipment (funnel, reflux). The EOA was added with 1.6 mL (2,624 g, 0.022 mol) of thionyl chloride dropwise. Then the mixture was refluxed for 150 minutes in a warm water bath (40°C) and allowed to cool. Next, 2 mL (2,262 g, 0.0207 mol) of pyridyl carbinol was added dropwise to the mixture and refluxed again in a warm water bath (40°C) for 60 min. After it had cooled the results obtained were dissolved in chloroform and washed with water. The extraction results were dried with anhydrous sodium sulfate, filtered and then evaporated to remove the remaining solvent. The results were analyzed with FTIR and GC-MS.

2.5. Polymerization

A total of 2 g (0.00639 mol) of pyridin-2-ylmetyl 2-eugenoxyacetate was introduced in a three-necked flask (100 mL) and then added with 1 mL of BF_3 diethyl ether for every hour as much as 0.25 mL of BF_3 diethyl ether. After the reaction process had lasted for 4 hours, the polymerization was discontinued with the addition of 0.5 mL of methanol. The precipitate formed is dissolved in chloroform, then washed with distilled water until neutral. After that, it is dried using Na_2SO_4 anhydrous to attract the water. The solvent was then evaporated and weighed. The obtained polymer was analyzed by FTIR.

2.6. Measurement of molecular weight of the polymer

A total of 1 g of poly(pyridin-2-ylmetyl 2-eugenoxyacetate) was dissolved in methanol to a concentration of 0.04 g/mL. The solution was then diluted with methanol into three different concentration: 0.02 g/mL; 0.01 g/mL and 0.005 g/mL. The viscosities of the pure solvent (methanol) (t_o) and various concentrated polymer solutions were measured using Ubbelohde viscometer, thus t_0, t_1, t_2, t_3 and t_4 were obtained. Through calculations, relative viscosity (η_{rel}) and specific viscosity (η_{sp}) were obtained. Then, a reduced viscosity (η_{sp}/C) versus concentration (C) graph was created. The graph was then extrapolated to a zero concentration that an intrinsic viscosity [η_{sp}] was obtained. With the equation of Mark-Houwink-Sakurada $[\eta] = KM_v^a$ (23), the relative molecular mass of polyeugenol can be calculated with the price $K = 11x10^{-3}$ and $a = 0.725$ (19)

2.7. Preparation of the feed solution

Parent-Solution Preparation:

1. Cr (III) solution 500 ppm was prepared by dissolving $CrCl_3.6H_2O$ 0.6407 g in aquadest in a 250 mL volumetric flask.
2. Cu (II) solution 500 ppm was prepared by dissolving $CuCl_2$. $2H_2O$ 0.3352 g in aquadest in a 250 mL volumetric flask
3. Cd (II) solution 500 ppm was prepared by dissolving $CdCl_2$. H_2O 0.2240 g in aquadest in a 250 mL volumetric flask

2.8. Feed solution preparation

A metal mixture was prepared by taking Cr (III), Cu (II), and Cd (II) solutions from the parent solutions, each 6 mL, diluted in a 100 mL aquadest in a volumetric flask. The metal mixture solution was then measured using a pH meter.

2.9. Preparation of the receiving solution
A total of 0.8 mL of a concentrated HCl 12 M solution was diluted with aquadest in a 100 mL volumetric flask. The acidity of the diluted solution was tested using pH meter.

2.10. Preparation of the liquid membrane
In this study, BLM testing was conducted three times. Two of which were based on the two-separate mass of PMEOA used which were 0.3 g and 0.7 g in 30 ml of chloroform each, whilst the other used 0.7 g of Poly-PMEOA in 30 mL of chloroform.

2.11. Recovery process
A solution containing 30 mL of PMEOA or PolyPMEOA was inserted in a U-tube placed between the feed and receiving phases, 13 mL each, then stirred for 24 hours.

2.12. pH measurement
After the feed phase and the receiving phase underwent the stirring process, pH measurements were taken

2.13. AAS analysis
Analysis of the metal ion contents in the feed phase and the receiving phase after they had been separated was carried out with an atomic absorption spectrometer.

3. Results and discussions

3.1. Polyeugenol with active N group
The results of the synthesis of poly (pyridine methyl eugenoxy acetate) and its performance as carrier compounds in metal ion transport Cd^{2+}, Cu^{2+}, and Cr^{3+} will be discussed.

3.1.1. Synthesis of eugenoxy acetic acid (EOA)
Eugenoxy acetic acid was purified by extraction using diethyl ether to remove polar impurities and with $NaHCO_3$ to remove nonpolar impurities such as acidic compounds and forming water-soluble salts. Eugenoxy acetic acid formed from the synthesis was a brownish-white precipitate with a yield of 73% and a melting point of 88-93°C (figure 2). The solubility test with the organic compounds at temperature of 28°C indicates that the compound is soluble in ether, chloroform, methanol, and slightly soluble in benzene.

Figure 2. Eugenoxyacetic acid.

3.1.2. FTIR analysis of eugenoxy acetic acid

The results of the analysis with an infrared spectrometer can be seen in figure 3. The absorption bands at 3400 and 3500 cm^{-1} indicate the presence of OH groups. The spectrum range of 3100-3000 cm^{-1} is the C-H stretching band of aromatic compounds that indicates the presence of aromatic compounds, the absorption band 1600-1475 cm^{-1} is the C=C vibration of the aromatic ring and the range of 800-900 cm^{-1} indicates that the aromatics are substituted. The absorptions at 1635 cm^{-1}, 997 cm^{-1}, and 915 cm^{-1} are the absorption bands of the vinyl group ($RHC=CH_2$). While the carbonyl acid group is indicated by the absorption at 1727 cm^{-1}.

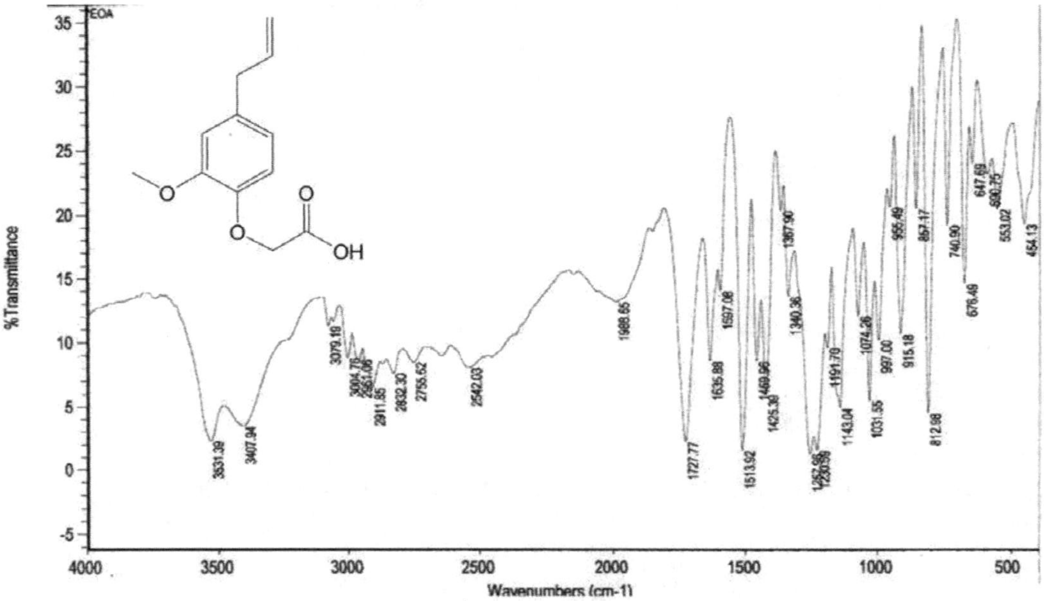

Figure 3. FTIR spectrum of EOA compound

Based on the absorption bands on the FTIR spectrum, it can be concluded that the compound being analyzed contains substituted aromatic nuclei, OH, methylene, vinyl and carbonyl groups of acids.

3.1.3. GC-MS Analysis of Eugenoxy acetic (EOA)

The gas chromatography-mass spectrometer (GC-MS) analysis of the EOA sample from the synthesis gave 4 peaks, each with a retention time of 3.491; 5,321; 7.941; and 9,388 minutes. By looking at the MS data it can be summarized that EOA appears at the fourth peak, with a retention time of 9.388 minutes and the area of 99.33% (figure 4).

The synthesized eugenoxy acetic acid resulted as a compound that has a mass spectrum with molecular ion M^+ of m/z = 222, which is the relative molecular mass of eugenoxy acetic acid. The fragmentation pattern formed from the molecular ion peak of eugenoxy acetic acid compound is shown as follows (figure 5).

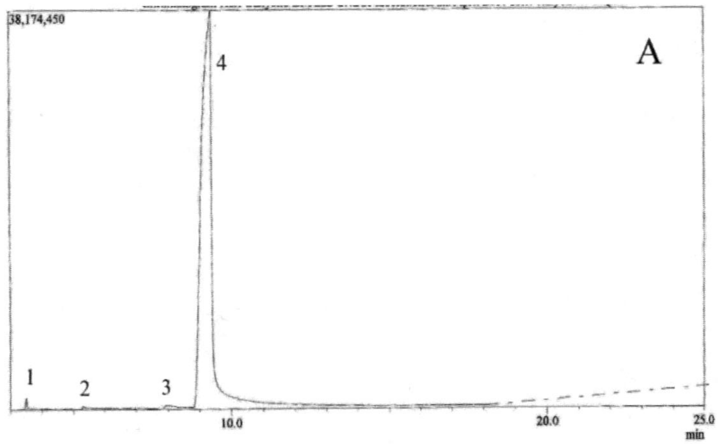

(a)

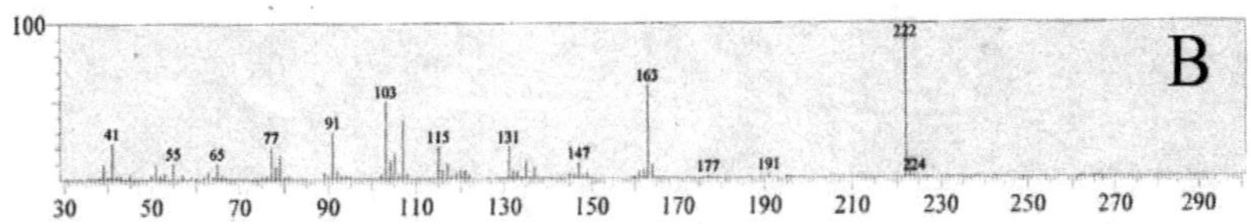

(b)

Figure 4. (a) Gas chromatography of the EOA compound (shown by peak number 4) and (b) mass spectrum of EOA compound at a retention time of 9.388 min.

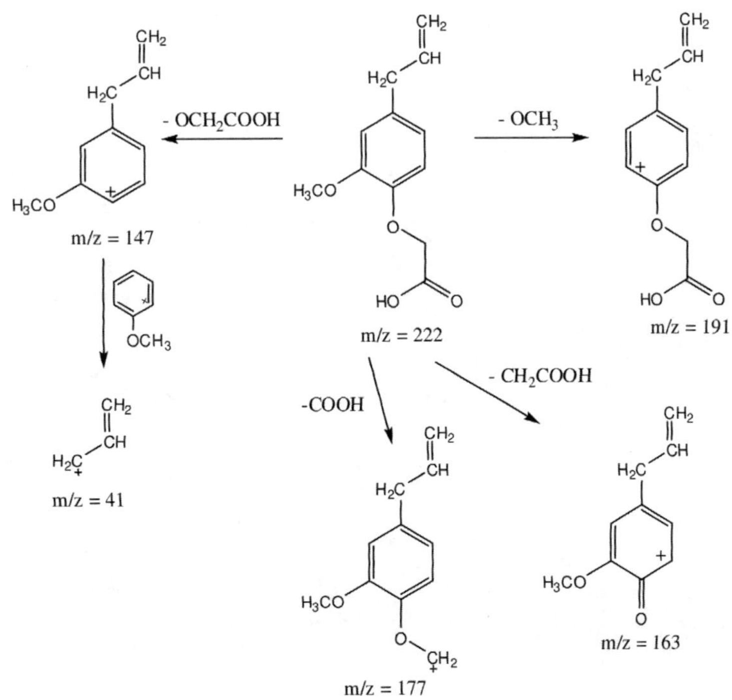

Figure 5. Fragmentation pattern of EOA compound.

3.2. Synthesis of pyridin-2-ylmethyl 2-(eugenoxy)acetate (PMEOA)

The pyridine methyl eugenoxy acetate (PMEOA) compound was synthesized from EOA. Eugenoxy acetic acid is a carboxylic acid that can react with alcohol to form esters. However, because the esterification reaction is reversible [20], the resulting yield was less than satisfactory. A good yield can be obtained using another esterification method that converts eugenoxy acetic acid to acid chloride using thionyl chloride first, then the newly formed acid chloride is reacted with alcohol. Chloride acid presence could be known due to its effect to the solution pH, making it acidic. This method can produce esters with a yield of above 80% [21].

Figure 6. Pyridine methyl eugenoxy acetate (PMEOA).

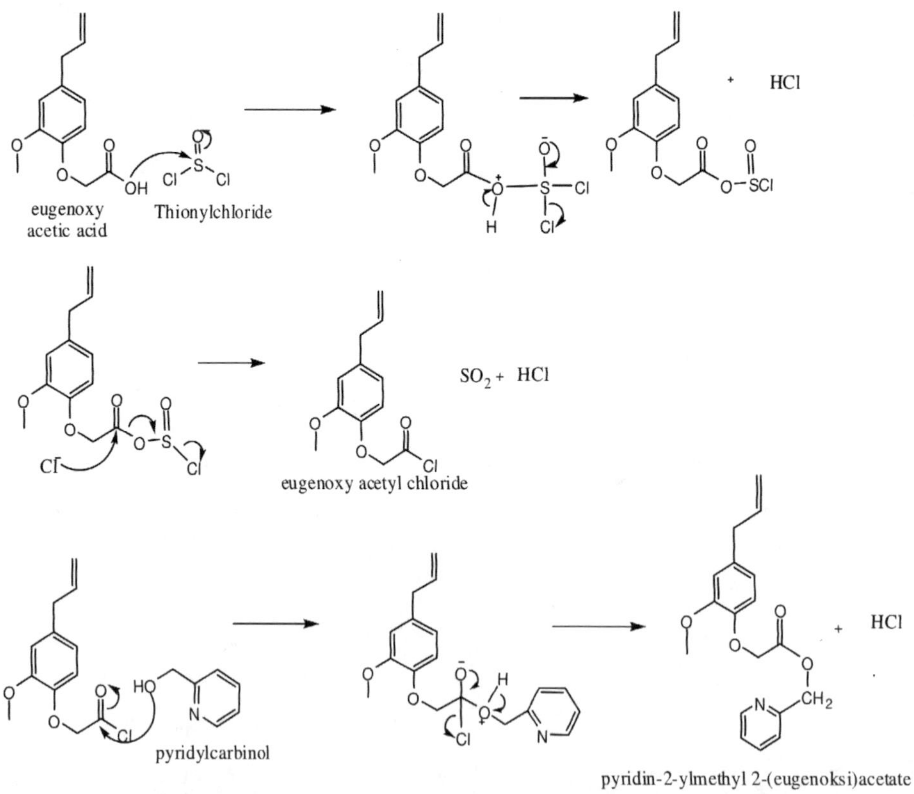

Figure 7. Formation reaction of pyridyl methyl eugenoxy acetate (PMEOA).

IOP Conf. Series: Materials Science and Engineering **959** (2020) 012033 doi:10.1088/1757-899X/959/1/012033

The PMEOA formed is a black fragrant-smelling thick liquid (figure 6) with a yield of 89%. The mass density of PMEOA is 1,246 g/mL. The solubility test using organic compounds at temperature 28°C showed that this compound dissolves in chloroform and benzene but is difficult to dissolve in n-hexane and methanol.

The PMEOA esterification mechanism can theoretically be explained by the following stages of the reaction (figure 7).

3.2.1. FTIR Analysis of pyridin-2-ylmethyl 2-(eugenoxy) acetate (PMEOA)
The analysis result of pyridin-2-ylmethyl 2-(eugenoxy) acetate using FTIR can be seen in figure 8. The formation of the ester can be known from the infrared spectrum at the wavenumber of 1763.64 cm^{-1}, which is the spectrum of the carbonyl ester group. As opposed to carbonyl acid, ester has a larger wave number. Another evidence of ester formation is the loss of the characteristic spectrum of the hydrogen bond of the OH group which forms a wide band at a wavenumber of about 3500 cm^{-1}.

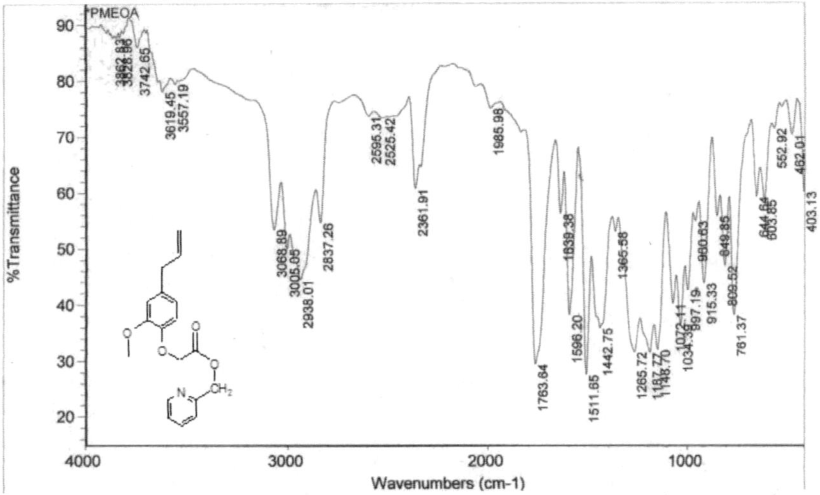

Figure 8. FTIR spectra of PMEOA compound.

3.2.2. GC-MS analysis of PMEAOA
The results of the synthesis PMEOA sample analysis by gas chromatography-mass spectrometer (GC-MS) (figure 9) gave 2 peaks, each with a retention time of 18.552 and 24.778 min. By looking at the MS data it can be seen that PMEOA appears at the second peak with retention time of 24.778 minutes and an area of 99.27%.

Based on the results of FTIR and GC-MS analysis, it can be concluded that pyridyl methyl eugenoxy acetate compound had formed with relative purity of 99.27% and a relative molecular weight of 313 g/mol.

IOP Conf. Series: Materials Science and Engineering **959** (2020) 012033 doi:10.1088/1757-899X/959/1/012033

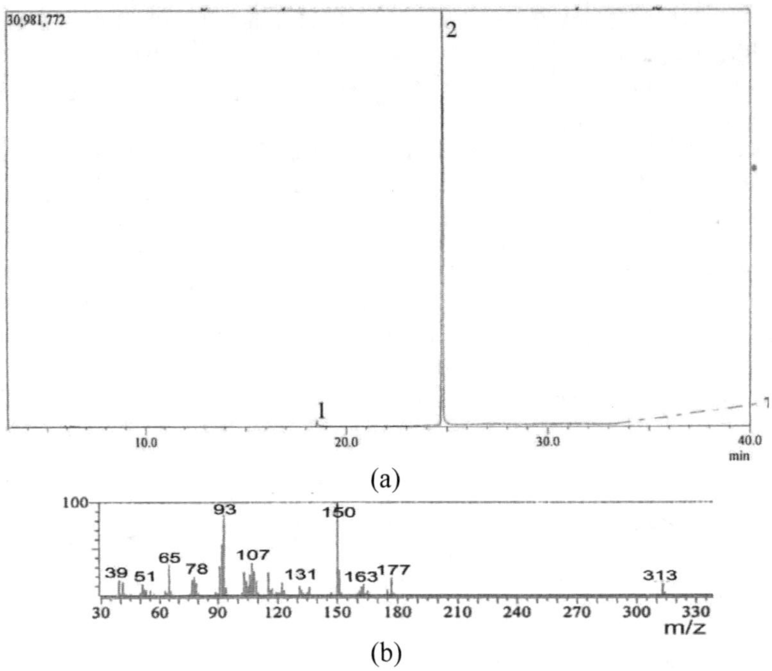

Figure 9. (a) Gas chromatography of PMEOA compound and (b) mass spectrum of PMEOA compound at a retention time of 24.778 min.

3.3. Polymerization of pyridin-2-ylmethyl 2-(eugenoxy)acetate (PMEOA)

The polymer of pyridin-2-ylmethyl 2-(eugenoxy) acetate (PMEOA) has successfully been synthesized using the BF_3-diethyl ether catalyst with monomer: catalyst = 2: 1. The polymerization of pyridyl methyl eugenoxy acetate resulted in the formation of a black gel-like solid with a 100% yield.

The determination of the relative molecular mass of poly(pyridyl methyl eugenoxy acetate) (poly (PMEOA)) was based on the viscosity measurement of each solution. Intrinsic viscosity [η] determination was based on the equation: $\eta_{sp}/C = [\eta]+k[\eta]^2 C$ [23]. Accordingly, the intercept of the intrinsic viscosity [η].

The relative molecular mass was calculated by the Mark-Houwink-Sakurada equation $[\eta] = KM_v^a$ [24], with the price $K = 11x10^{-3}$ and a = 0.725 [20]. From the calculation result, the relative molecular mass of poly (PMEOA) is 4924 with a degree of repetition n ≅ 16.

3.3.1. FTIR analysis and NMR of poly pyridin-2-ylmethyl 2-(eugenoxy)acetate poly (PMEOA)

FTIR analysis (figure 10) showed that the loss of vinyl spectra at wave number 915 cm^{-1}, 997 cm^{-1} and 1648-1638 cm^{-1} indicates that polymerization has occurred. Other evidence of polymerization can be seen from the ^1H NMR spectra of the poly (PMEOA) compound and compared it with its monomers. The ^1H NMR used the frequency of 60 MHz. Occurrence of polymerization is characterized by loss of shifting chemical δ = 5.2 ppm in a monomer which is a signal of a hydrogen attached to a vinyl and the emergence of δ = 1 ppm (multiplet) on the polymer spectra, which is a signal of hydrogen atoms bond to the polymer backbone ($-CH_2-CH_2-$) (figure 11). This reinforces the addition of the addition polymerization reaction.

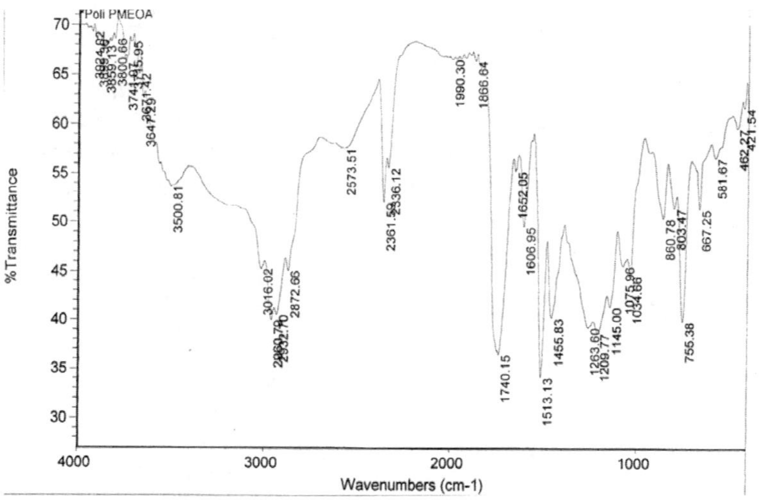

Figure 10. Spectra FTIR poly (PMEOA).

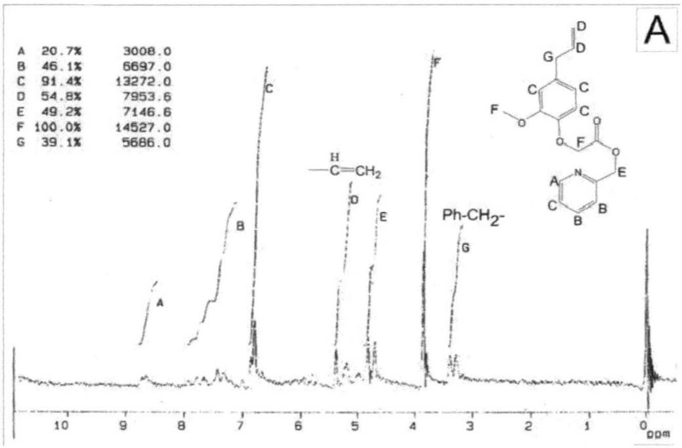

(a)

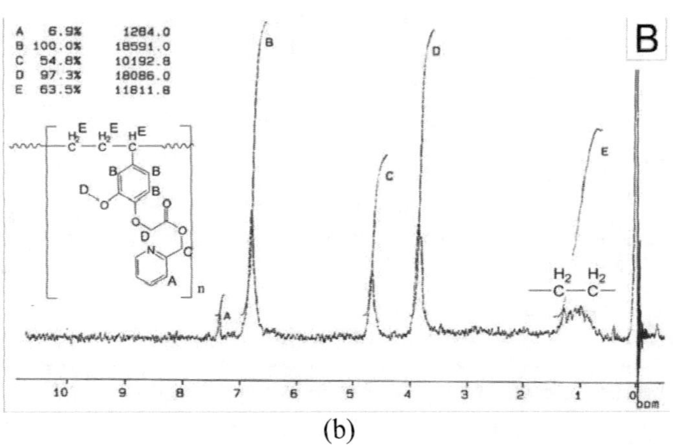

(b)

Figure 11. Spectra of ^{1}H NMR: (a) PMEOA and (b) polyPMEOA.

IOP Conf. Series: Materials Science and Engineering **959** (2020) 012033 doi:10.1088/1757-899X/959/1/012033

3.4. Metal mixture transport using the BLM technique

In this study, metal ion transport was studied by using the U pipe. The feed phase of 13 mL contained a mixture of metal ions, namely Cr (III), Cd (II), and Cu (II) each 30 ppm (pH 3.31). The liquid membranes used were PMEOA esters (0.0023 and 0.001 mol) and Poly PMEOA (0.0023 mol) in 30 mL of chloroform solvent. The receiving phase was controlled at pH 1 (13 ml of HCl 0.1 M solution). During the process, the system was stirred at a constant speed for 24 hours.

In the recovery process of heavy metals with the bulk liquid membrane (BLM) technique, there were changes in pH in the feed and receiver phases. These pH changes indicate an exchange mechanism of metals and H^+ ions between the two phases, as shown in the following reaction mechanisms (figure 12 and table 1).

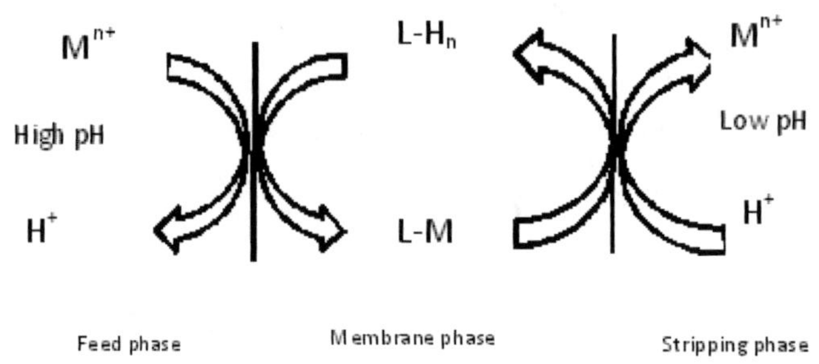

Figure 12. Mechanism of transport in liquid membrane.

Table 1. Changes of pH values between the initial and final condition in BLM.

	pH of feed phase		pH of receiving phase	
Membrane	Initial pH	Final pH	Initial pH	Final pH
PMEOA-1	3.31	1.79	1.00	1.50
PMEOA-2	3.31	2.54	1.00	1.41
PolyPMEOA	3.31	1.37	1.00	1.37

Description:

PMEOA-1 : PMEOA with a mass of 0.7 g
PMEOA-2 : PMEOA with a mass of 0.3 g
Poly (PMEOA) : Polymer of PMEOA with a mass of 0.7 g

3.4.1. Metal transport by monomeric ligands

In this study, PMEOA monomers were used in two different levels of concentrations to test the selectivity and effectiveness of metal transport by using the BLM technique. The metal mixture transport results are shown in table 2.

Table 2. Influence of carrier membrane concentrations in metal transport.

Metal ions	% Transport in Feed Phase		% left in Feed phase		% left in Membrane	
	PMEOA-1	PMEOA-2	PMEOA-1	PMEOA-2	PMEOA-1	PMEOA-2
Cd (II)	43,53	25,70	59,00	84,50	~ 0	~ 0
Cu (II)	87,54	73,03	16,39	17,42	28,85	9,55
Cr (III)	0,60	2,01	92,51	97,99	6,89	~ 0

Based on the above table, it can be seen that Cu (II) metal ions were the most metal to be transported to the receiving phase, followed by Cd (II) and Cr (III) metal ions, respectively. This conforms with the theory of HSAB which states that hard acid metals bind more strongly with hard bases, while soft acid binds more strongly with a soft base, and borderline acid binds more strongly with a borderline base. Cu (II) is a borderline acid metal ion, thus it can bind to an active N ligand which is a stronger borderline base than any other metal ion.

In this study, we studied the influence of the carrier concentration on metal transport by comparing the number of metal ions transported based on the different membrane concentrations which are 0.023 g/mL and 0.01 g/mL. The experimental results showed that, in general, the performance of metal transport would increase with the increase of the carrier compound concentrations. This behavior may be related to the greater chance of metal ions to bind to the carrier at greater concentration.

The advantage of BLM compared to SLM or ELM techniques are that this method sufficiently enables the study of the transport mechanism and the influence of the carrier structure in its efficiency and selectivity [13] Looking at the data of the percentage of metal ions left in the membrane, Cu(II) appeared to be having the most ions left in the membrane (28.85 in PMEOA-1 and 9.55 in PMEOA-2) which shows that the ligand creates a stronger bond with Cu (II) in comparison to the other metal ions. This suggests the selectivity of the PMEOA ligand towards Cu (II) and that the transport mechanism occurred is based on the HSAB theory. HSAB theory was also applied in the studies of carriers made from the derivatives of eugenol with eugenol monomer of active N and S groups for the separation of Cu (II) and Cd (II) [4].

3.4.2. Metal transport by polymeric ligands

The first polymer used in the metal ion transport test in this study was soluble in chloroform. The second polymer could not be tested as it did not find a suitable solvent for the recovery process. A suitable solvent should have a greater density than water and is insoluble or has very small solubility in water. The recovery process of metal ions is shown in table 3.

Table 3. Percentage of metal ions in feed phase, receiving phase and membrane.

Metal ions	% transport in the feed phase	% left in the feed phase	% left in the membrane
Cd (II)	54,98	63,32	~ 0
Cu (II)	47,17	53,17	~ 0
Cr (III)	45,58	51,48	2,94

Based on the data above, it can be seen that the amounts of transported metal ions were relatively the same. Thus, there is no indication that shows transport selectivity. However, the overall effectiveness of each of the metal ions were increased compared to their monomer. This suggests that the transport mechanism that occurs is no longer HSAB. Rather, it's the capture of a metal ion in the bulky polymer structure. The results are different when compared with the eugenol derivative polymers with the active groups N and S that still follow the HSAB theory [3].

4. Conclusions

Based on this research, the following can be concluded:

1. Pyridine methyl eugenoxy acetate carrier compound is selective towards Cu (II) metal ions.
2. Poly (pyridine methyl eugenoxy acetate) can improve the transport effectiveness but not its selectivity towards heavy metal ions.

Acknowledgement
We would like to express sincere gratitude to the Ministry of Research and Technology of Indonesia for the Competitive Grant.

References

[1] Towaha J 2011 *Perspektif* **11** 79
[2] National Center for Biotechnology Information. PubChem Compound Database; CID=3314, https://pubchem.ncbi.nlm.nih.gov/compound/3314 (accessed Feb. 16, 2019).
[3] Djunaidi M C, Khabibi and Ulumudin I 2017 *IOP Conf. Ser. Mat. Sci. Eng.* **172** 012032.
[4] Djunaidi M C, Wibawa P J and Murti R H 2016 *Indones. J. Chem.* **1** 121
[5] Djunaidi M C, Lusiana R A, Wibawa P J, Siswanta D and Jumina 2010 *Reaktor* **13** 16
[6] Djunaidi M C, Khabibi and Trisna D 2010 *Jurnal Kimia Sains Aplikasi* **1391** 12
[7] Chang S H 2016 *Desalination Water Treat.* **57** 19785
[8] Spreti N, Brinchi L, Germani R, Mancini M and Savelli G 2004 *Eur. J. Org. Chem.* **2004** 3865
[9] Bhatluri K K, Chakraborty S, Manna M S, Ghoshal A K and Saha P 2015 *RSC Adv.* **5** 88331
[10] Nezhadali A, Es'haghi Z, Bahar S, Banaei A and Shiran J A 2016 *J. Braz. Chem. Soc.* **27** 99
[11] Sgarlata C, Arena G, Longo E, Zhang D, Yang Y and Bartsch R A 2008 *J. Membr. Sci.* **323** 444
[12] Nghiem L, Mornane P, Potter I, Perera J, Cattrall R W and Kolev S D 2006 *J. Membr. Sci.* **281** 7
[13] Peterson R T and Lamb J D 1996 *Design of macrocyclic carriers for liquid membrane in chemical separation with liquid membrane* ed R A Bartsch and J D Way (Washington: American Chemical Society) p 130
[14] Nezhadali A and Pour F M 2013 *Alexandria Eng. J.* **52** 123
[15] Kazemi S V and Shamsipur M 2005 *Bull. Korean Chem. Soc.* **26** 930
[16] Kazemi S V and Hamidi A S 2011 *J. Chem. Eng. Data.* **56** 222
[17] Eljaddi T, Hor M, Benjjar A, Riri M, Mouadli H, Mountassir Y and Hlaibi M 2015 *BAOJ Chem.* **1** 003
[18] Bartsch R A and Way J D 1996 *An overview in chemical separation with liquid membrane* ed R A Bartsch and J D Way (Washington: American Chemical Society) p 1
[19] Brandrup J and Immergut 1975 *Polymer handbook* (second edition) (New York: John Willey & Sons)
[20] Fessenden R H and Fessenden J S 1997 *Kimia Organik* (Jakarta: Erlangga)
[21] Vogel A I 1978 *Vogel's textbook of practical organic chemistry* (London: Longman Scientific & Technical)
[22] Rosenthal L C 1990 *J. Chem. Educ.* **67** 70
[23] Hartomo 1993 *Dasar-dasar profesi politeknik pemrosesan polimer praktis* (Yogyakarta: Andi)

AUTHOR INDEX

Abdullah, I 104
Abdulloh, S H 226
Amir, H 155
Amiruddin, E 142
Andayani, I G A S 56
Andriani, Y 155
Anggraini, R 142
Anh, P N Q 183
Asnani, A 231
Awaluddin, A 142
Azahra, S N A R 110
Bach, L G 183, 192, 201
Cahyono, H 244
Cang, M H 183, 192
Danh, P H 172, 192
Delsy, E V Y 6
Dewi, N 142
Djunaidi, M C 163, 244
Edoway, V 129
Fajar, M 91
Fauziyah, R H 226
Febriantini, D 42, 68
Giyanto 33
Gunawan 148
Gustianthy, A P 42
Haris, A 148
Hartini, S 211
Heliawaty, L 33
Helmiyati 110
Herawati, M M 238
Hermanto, D 56
Hidayat, W 1
Hien, T T 172, 183, 192, 201
Hong-Tham, N T 120
Honiar, R 56
Ismillayli, N 56
Ismiyarto 244
Istiqomaturohmah 226
Iswanto, P 6, 231
Jan'ah, A M 135
Jumina 244
Karwur, F F 10
Kassim, MNI 155
Khikmah, A M 104
Khoirunnisa, N 68
Krisnandi, Y K 104
Kurnia, I 17
Kurnia, T D 238
Kurniawan, Y S 98
Laksono, A 231
Lestari, G A D 129
Lusiana, R A 163
Mardhani, R 129
Mariana, B 56
Martono, Y 219
Marzuki, A 1
Muninggar, J 10, 219
Muzayanha, S U 42
Nawfa, R 47
Ngan, T T K 172, 183, 192
Nhan, L T H 172, 183, 192, 201
Nhan, N P T 201
Nhat, D D 183
Nhut, P T 201
Noverdi, R 42
Nugraha, A S 68
Nugrahaningtyas, K D 74
Nurhayati 142
Pardoyo 163
Pebriani, S 68
Phuc, N D 183, 192
Pranoto 74
Prasetoko, D 91
Prasetyo, E 1
Priyangga, K T A 98
Pudjihartati, E 238
Purnomo, A S 47
Purnomo, B 42, 68
Putra, S R 47
Putri, R N O 74
Quyen, N T C 183
Rahayu, D M 47
Redhana, I W 25
Rianto, A K 68
Riniati 226
Rondonuwu, F S 10
Rubini, B 33
Sababathy, M 155
Sanjaya, R K 56
Saputri, D D 135
Saraswati, T E 135
Sarjono, PR 155
Septina, W 148
Septiyarini, C P 211
Setiawan, A W 238
Setiawan, A 10

Setyarini, W 142
Setyawan, A 1
Siahaan, P 17
Siregar, S S 142
Siswanta, D 244
Soetjipto, H 211
Suardana, I N 25
Sudria, I B N 25
Sugiwati, S 155
Suja, I W 25
Suprihatin, I E 129
Suseno, A 17
Sy, D T 62
Syamsumir, DF 155
Tan, L V 62, 120
Tham, N T H 62
Thinh, P V 120
Tien, L X 172
Ulfa, M 91
Usman 68
Wahyuni, T 163
Widodo, D S 163
Yanuar, H P 1
Yuliati, L 98